AN INTRODUCTION TO MOLECULAR BIOTECHNOLOGY

Edited by

Natassia Correa

Toronto Academic Press

AN INTRODUCTION TO MOLECULAR BIOTECHNOLOGY

Editor: **Natassia Correa**

Toronto Academic Press
4164 Lakeshore Road
Burlington ON L7L 1A4
Canada
www.tap-books.com

Email: orders@arclereducation.com

ISBN: 978-1-77956-326-2

ABOUT THE EDITOR

Natássia Corrêa is a biologist with a background in Molecular Biology, Cellular Biology, and Biotechnology. She is currently a public servant at the Federal University of Uberlândia, Minas Gerais, Brazil. Her academic journey includes earning her Master's and Doctorate degrees in Molecular Biology from the Federal University of Minas Gerais. In addition to working in the biotechnology industry, she completed her postdoctoral studies at FIOCRUZ and served as a faculty member at the Federal University of Paraná. Dr. Natássia Corrêa has always valued the diversity of her learning experiences, driven by a passion for new knowledge, as evidenced by her diverse and in-depth curriculum in Molecular and Cellular Biology.

Table of Contents

List of Figures *ix*

List of Tables *xxi*

Preface *xxiii*

1 INTRODUCTION TO MOLECULAR BIOTECHNOLOGY 1

Introduction **2**

1.1. The Emergence of Molecular Biotechnology **3**

1.2. Recombinant DNA Technology **6**

1.3. Applications of Molecular Biotechnology **10**

1.3.1. Applications in Medicine 10

1.3.2. Impact on Agriculture 12

1.3.3. Environmental Applications 15

1.3.4. Industrial Applications 17

1.4. Ethical, Legal, and Social Implications **20**

1.4.1. Ethical Implications 20

1.4.2. Social Implications 21

1.4.3. Economic Implications 21

Summary **25**

Review Questions **25**

References **26**

2 MOLECULAR BIOLOGY 27

Introduction **28**

2.1. DNA Structure and Function **29**

2.1.1. Properties of DNA 29

2.1.2. Types of DNA 30

2.1.3. DNA Structure 31

2.1.4. Functions of DNA 35

2.1.5. Chemical Modification 37

2.1.6. DNA Translation 38

2.2. RNA Structure and Function **41**

2.2.1. RNA Structure 41

2.2.2. Types of RNA 42

2.2.3. Functions of RNA 45

2.3. The Genetic Code and the Concept of A Gene **46**

2.3.1. The Genetic Code 47

2.3.2. Evidence for the Triplet Code 47

2.3.3. Cracking the Code 48

2.3.4. Start and Stop Codons 49

2.3.5. Expanding the Genetic Code 49

2.3.6. Current Concept of the Gene 49

2.3.7. Non-coding RNAs 50

2.4. Translation of RNA into Proteins **52**

2.4.1. Transfer RNA 52

2.4.2. Structure of the Ribosome 53

2.4.3. Protein Synthesis 54

Summary **61**

Review Questions **62**

References **62**

3 STRUCTURE AND FUNCTIONS OF CELL 65

Introduction **66**

3.1. Overview of the Cell **67**

3.1.1. The Evolution of Cells 68

3.1.2. The History of Cell Theory 70

3.2. Types of Cells **76**

3.2.1. Prokaryotic Cell 76

3.2.2. Eukaryotic Cells 79

3.2.3. Differences Between Eukaryotic and Prokaryotic Cells 83

3.3. The Molecular Composition of Cells **85**

3.3.1. Carbohydrates 86

3.3.2. Lipids 88

3.3.3. Nucleic Acids 90

3.3.4. Proteins 91

3.4. Cell Cycle **95**

3.4.1. The Cell-Cycle Control System Is Similar in All Eukaryotes 97

3.4.2. The Cell-Cycle Control System Can Be Dissected Genetically in Yeasts 97

3.4.3. The Cell-Cycle Control System Can Be Analyzed Biochemically in Animal Embryos 99

3.4.4. The Cell-Cycle Control System of Mammals Can Be Studied in Culture 100

3.4.5. Cell-Cycle Progression Can Be Studied in Various Ways 101

Summary **105**

Review Questions **105**

References **106**

4 GENOMICS AND PROTEOMICS 109

Introduction **110**

4.1. Functional Genomics **111**

4.1.1. DNA Microarray Technology 111

4.1.2. Serial Analysis of Gene Expression 117

4.2. Proteomics **121**

4.2.1. Proteomics Origins 122

4.2.2. Why Proteomics? 122

4.2.3. Types of Proteomics 124

4.2.4. Technology of Proteomics 131

4.2.5. Acquisition of Protein Structure Information 134

Summary **141**

Review Questions **141**

References **142**

5 RECOMBINANT DNA AND GENETIC ANALYSIS 145

Introduction **146**

5.1. Constructing Gene Libraries **147**

5.1.1. Digesting Genomic DNA Molecules 147

5.1.2. Ligating DNA Molecules 148

5.1.3. Aspects of Gene Libraries 149

5.1.4. Genomic DNA Libraries 151

5.1.5. cDNA Libraries 152

5.1.6. Treatment of Blunt cDNA Ends 154

5.1.7. Enrichment Methods for RNA 155

5.1.8. Subtractive Hybridization 155

5.1.9. Cloning PCR Products 156

5.2. Cloning Vectors **157**

5.2.1. Plasmids 158

5.2.2. Virus-based Vectors 162

5.2.3. M13 and Phagemid-based Vectors 166

5.2.4. Cosmid-Based Vectors 168

5.2.5. Large Insert Capacity Vectors 168

5.2.6. Yeast Artificial Chromosome (YAC) Vectors 168

Summary **172**

Review Questions **173**

References **173**

6 CHEMICAL SYNTHESIS, AMPLIFICATION, AND SEQUENCING OF DNA 175

Introduction **176**

6.1. Chemical Synthesis of DNA **177**

6.1.1. The Phosphoramidite Method 177

6.1.2. Uses of Synthesized Oligonucleotides 182

6.2. Polymerase Chain Reaction **189**

6.2.1. Principles of PCR 189
6.2.2. The Components of PCR 190
6.2.3. Procedure 193
6.2.4. Validating PCR 195
6.2.5. Types of PCR 195
6.3. DNA-Sequencing Techniques 197
6.3.1. Sanger Sequencing (Chain Termination Method) 197
6.3.2. Next-Generation Sequencing (NGS) 198
6.3.3. Third-Generation Sequencing (Single-Molecule Sequencing) 199
6.3.4. Shotgun Sequencing 201
6.3.5. Metagenomic Sequencing 203
6.3.6. Cyclic Array Sequencing 205
Summary 214
Review Questions 215
References 215

7 MOLECULAR DIAGNOSTICS 217

Introduction 218
7.1. Immunological Diagnostic Procedures 219
7.1.1. Enzyme-Linked Immunosorbent Assay (ELISA) 219
7.2. Monoclonal Antibodies 222
7.2.1. Formation and Selection of Hybrid Cells 223
7.2.2. Identification of Specific Antibody-Producing Hybrid Cell Lines 224
7.3. Biofluorescent and Bioluminescent Systems 227
7.3.1. Colored Fluorescent Proteins 227
7.3.2. Luciferase 229
7.3.3. Microbial Biosensors 230
7.4. Molecular Diagnosis of Genetic Disease 233
7.4.1. Screening for Cystic Fibrosis 233
7.4.2. Sickle-Cell Anemia 235
7.4.3. Padlock Probes 236
7.4.4. Genotyping with Fluorescence-Labeled PCR Primers 238
7.4.5. TaqMan Assay 239
7.5. Diagnostic Techniques in Molecular Biology 240
7.5.1. Nucleic Acid Amplification Techniques 243
7.5.2. Nucleic Acid Hybridization Techniques 245
7.5.3. Microarrays 245
7.5.4. Genotyping 246
7.5.5. Mass Spectrometry 247
7.5.6. Key Factors Affecting Test Performance and Reliability 247
7.6. Applications of Molecular Diagnostic Techniques 249
7.6.1. Prenatal Tests 249
7.6.2. Infectious Diseases 249
7.6.3. Disease Risk Management 249
Summary 254
Review Questions 255
References 255

8 PROTEIN SYNTHESIS 257

Introduction 258
8.1. Translation of mRNA 259
8.1.1. Transfer RNAs 259
8.1.2. The Ribosome 262
8.1.3. The Organization of mRNAs and the Initiation of Translation 263
8.1.4. The Process of Translation 265
8.2. Protein Structure 271
8.2.1. Primary Structure 271
8.2.2. Secondary Structure 272
8.2.3. Tertiary Structure 273
8.2.4. Quaternary Structure 274
8.3. Protein Folding and Processing 276
8.3.1. Chaperones and Protein Folding 276
8.3.2. Enzymes and Protein Folding 279
8.3.3. Protein Cleavage 281
8.3.4. Glycosylation 283
8.3.5. Attachment of Lipids 285
8.4. Cell-Free Protein Synthesis 288
8.4.1. Early History of Cell-Free Protein Synthesis 288
8.4.2. Landmark Advances in the Development of Cell-Free Protein Synthesis Systems 288

8.4.3. Use of Commercially Available Cell-Free Protein Synthesis Systems 291
8.4.4. Application of Cell-Free Proteins 291
8.4.5. Comparison of Protein Synthesis in Prokaryotes and Eukaryotes 293
Summary **298**
Review Questions **298**
References **299**

9 HETEROLOGOUS PROTEIN PRODUCTION IN EUKARYOTIC CELLS 303

Introduction **304**
9.1. Posttranslational Modification of Eukaryotic Proteins **305**
9.2. General Features of Eukaryotic Expression Systems **308**
9.3. Fungus-Based Expression Systems **310**
9.3.1. Saccharomyces cerevisiae Expression Systems 310
9.3.2. Pichia Pastoris Expression Systems 319
9.4. Baculovirus–Insect Cell Expression Systems **323**
9.4.1. Baculovirus Expression Vectors 325
9.4.2. Increasing the Yield of Recombinant Baculovirus 326
9.4.3. Integration of Target Genes into Baculovirus by Site-Specific Recombination 327
Summary **333**
Review Questions **334**
References **334**

10 BIOCHEMISTRY AND MOLECULAR BIOLOGY TECHNIQUES 337

Introduction **338**
10.1. Mass Spectroscopic Techniques **339**
10.1.1. Electrospray Ionization (ESI) 340
10.1.2. Matrix-Assisted Laser Desorption/Ionization (MALDI) 341
10.1.3. Tandem Mass Spectrometry 342
10.1.4. High-Resolution Mass Spectrometry (HRMS) 342
10.1.5. Quantitative Mass Spectrometry 343
10.1.6. Imaging Mass Spectrometry 344
10.2. Electrophoretic Techniques **346**
10.2.1. Agarose Gel Electrophoresis 347
10.2.2. Polyacrylamide Gel Electrophoresis (PAGE) 348
10.2.3. Capillary Electrophoresis (CE) 348
10.2.4. Isoelectric Focusing (IEF) 350
10.2.5. Protein Blotting Techniques (Western Blotting) 350
10.2.6. DNA Gel Electrophoresis 351
10.2.7. RNA Gel Electrophoresis 352
10.3. Chromatographic Techniques **354**
10.3.1. Reversed-Phase Chromatography 355
10.3.2. Size Exclusion Chromatography 356
10.3.3. Ion Exchange Chromatography 357
10.3.4. Hydrophobic Interaction Chromatography 358
10.3.5. Gas Chromatography 359
10.3.6. High-Performance Liquid Chromatography (HPLC) 361
10.4. Spectroscopic Techniques **363**
10.4.1. Ultraviolet-Visible Spectroscopy (UV-Vis) 364
10.4.2. Infrared Spectroscopy (IR) 365
10.4.3. Nuclear Magnetic Resonance (NMR) Spectroscopy 366
10.4.4. Fluorescence Spectroscopy 367
10.4.5. Raman Spectroscopy 367
Summary **371**
Review Questions **373**
References **373**

INDEX 375

List of Figures

Figure 1.1. Principal steps of a bioengineered biotechnology process. Parenthetically, Karl Ereky's scheme entailed using inexpensive sugar beets (raw material) to feed pigs (biotransformation) for the production of pork (downstream processing)

Figure 1.2. Many scientific disciplines contribute to molecular biotechnology, which generates a wide range of commercial products

Figure 2.1. DNA. Deoxyribonucleic acid is a polymer chain of nucleotides connected by 5' to 3' phosphodiester bonds. DNA normally exists as a two antiparallel complementary strands held together by hydrogen bonds between adenines (A) and thymines (T) and between guanines (G) and cytosines (C)

Figure 2.2. Three conformations of DNA. B-DNA is most common, A-DNA is likely an artifact of crystallization *in vitro*, and Z-DNA may form transiently in parts of the chromosome

Figure 2.3. DNA packaging

Figure 2.4. The nucleosome is composed of slightly over two turns of DNA around a histone core containing two copies each of H2A, H2B, H3, and H4 histones. The H1 histone is not part of the core unit and functions in coordinating interaction between nucleosomes

Figure 2.5. (a) Ribonucleotides contain the pentose sugar ribose instead of the deoxyribose found in deoxyribonucleotides. (b) RNA contains the pyrimidine uracil in place of thymine found in DNA

Figure 2.6. (a) DNA is typically double-stranded, whereas RNA is typically single-stranded, (b) Although it is single-stranded, RNA can fold upon itself, with the folds stabilized by short areas of complementary base pairing within the molecule, forming a three-dimensional structure

Figure 2.7. mRNA (messenger RNA)

Figure 2.8. rRNA (ribosomal RNA)

Figure 2.9. tRNA (transfer RNA)

Figure 2.10. The flow of genetic information

Figure 2.11. DNA sequence showing the amino acid translation underneath

Figure 2.12. The structure of a protein-coding eukaryotic gene

Figure 2.13. Transfer RNA: (A) Tertiary structure of the phenylalanine tRNA from yeast showing the anticodon (grey), the acceptor stem (violet) with the nucleotides CAA at the 3 OH end (yellow). (B) Clover-leaf representation of the secondary structure of tRNA

Figure 2.14. The structure of the ribosome of Thermus thermophilus showing the small subunit (green), large subunit (blue), mRNA (red) and three tRNAs in the acceptor, peptidyl and exit sites (yellow). In (A) the new tRNA is delivered to the ribosome by elongation factor EF-Tu (purple). In (B) the amino acid on the incoming tRNA is brought close to the amino acid on the tRNA in the peptidly site to facilitate peptide bond formation (bright green)

Figure 2.15. Protein synthesis

Figure 2.16. Amino acids and peptide bonds

Figure 3.1. Cells are the fundamental unit of life. They range in size from 0.0001 mm to nearly 150 mm across

Figure 3.2. Lipid structure

Figure 3.3. Rudolf Virchow

Figure 3.4. French chemist and microbiologist Louis Pasteur performing a scientific experiment. Pasteur's studies helped establish the principle of biogenesis, the development of new organisms from the reproduction of other organisms

Figure 3.5. Prokaryotic cell

Figure 3.6. Relative sizes of molecules, cells and organisms on a logarithmic scale

Figure 3.7. Cell surface size

Figure 3.8. Since prokaryotes do not have nuclei, their DNA is housed in a nucleoid

Figure 3.9. Eukaryotic cell

Figure 3.10. Characteristics of water

Figure 3.11. Structure of simple sugars

Figure 3.12. Formation of a glycosidic bond

Figure 3.13. Structure of polysaccharides

Figure 3.14. Structure of fatty acids

Figure 3.15. Structure of triacylglycerols

Figure 3.16. Components of nucleic acids

Figure 3.17. Polymerization of nucleotides

Figure 3.18. Complementary pairing between nucleic acid bases

Figure 3.19. Structure of amino acids

Figure 3.20. Formation of a peptide bond

Figure 3.21. Tertiary structure of ribonuclease

Figure 3.22. The events of eucaryotic cell division as seen under a microscope

Figure 3.23. The phases of the cell cycle. The cell grows continuously in interphase, which consists of three phases: DNA replication is confined to the S phase; G1 is the gap between the M phase and the S phase, while G2 is the gap between the S phase and the M phase. In the M phase, the nucleus and then the cytoplasm divide

Figure 3.24. A comparison of the cell cycles of fission yeasts and budding yeasts

Figure 3.25. The behavior of a temperature-sensitive CDC mutant

Figure 3.26. The morphology of budding yeast cells arrested by a CDC mutation

Figure 3.27. A mature Xenopus egg, ready for fertilization

Figure 3.28. Studying the cell cycle in a cell-free system

Figure 3.29. Mammalian cells in culture

Figure 3.30. Labeling S-phase cells

Figure 3.31. Analysis of DNA content with a flow cytometer

Figure 4.1. Gene expression profiling with a DNA microarray. mRNA is extracted from two samples (sample 1 and sample 2), and during reverse transcription, the first cDNA strands are labeled with the fluorescent dyes Cy3 and Cy5, respectively. The cDNA samples are mixed and hybridized to an ordered array of either gene sequences or gene-specific oligonucleotides. After the hybridization reaction, each probe cell is scanned for both fluorescent dyes and the separate emissions are recorded. Probe cells that produce only a green or red emission represent genes that are transcribed only in samples 1 and 2, respectively; yellow emissions denote genes that are active in both samples; and no emissions (black) represent genes that are not transcribed in either sample

Figure 4.2. Fluorescence image of a DNA microarray hybridized with Cy3- and Cy5-labeled cDNA

Figure 4.3. Gene expression profiling with an oligonucleotide microarray. mRNA is purified with a poly(dT) sequence that has a T7 RNA polymerase primer sequence extension. After two-stranded cDNA synthesis, the second cDNA strand acts as a template for the synthesis of cRNA by T7 RNA polymerase in the presence of ATP, cytidine triphosphate (CTP), guanosine triphosphate (GTP), uridine triphosphate (UTP), biotinylated CTP, and biotinylated UTP. The gray circles represent incorporated biotinylated nucleotides. The biotinylated cRNA is purified, fragmented into pieces from 50 to 100 nucleotides in length, and hybridized to an oligonucleotide microarray. The microarray is treated with streptavidin–phycoerythrin, and the probe cells (black squares) are scanned for emission (yellow) from the biotin-bound streptavidin–phycoerythrin

Figure 4.4. Gene expression profile of cirrhotic liver tissue. Columns 1 to 7 and 8 to 15 are liver samples from patients with ethanol- and hepatitis C virus-induced cirrhosis of the liver, respectively. Each patient's sample was compared to normal liver tissue. The cluster consists of 2,965 genes. The asterisks denote patients with severe cirrhosis of the liver.

Figure 4.5. Gene expression profile of lymphocyte-specific genes from cirrhotic liver tissue. Columns 1 to 7 and 8 to 15 are liver samples from the patients shown in Figure 4.4 with ethanol- and hepatitis C virus-induced cirrhosis of the liver, respectively. Each patient's sample was compared to normal liver tissue. The cluster consists of about 70 genes. The asterisks denote patients with severe cirrhosis of the liver

Figure 4.6. SAGE. Poly(A)+ mRNA transcripts are isolated by poly(dT) hybridization. The poly(dT) sequence is biotinylated (B) and bound to streptavidin (SA)- coated magnetic beads (yellow circles), or the poly(dT) is attached directly to magnetic beads. Double-stranded cDNA is synthesized from the captured mRNAs and then cut with the restriction endonuclease NlaIII. The fragments of cDNA that are not bound to the magnetic beads are eluted. The NlaIII-cut cDNA sample that is attached to the magnetic beads is divided in two, and one sample is ligated with adaptor A and the other with adaptor B. Each adaptor contains a

4-base extension that is complementary to the 4-base extension produced by cleavage with NlaIII, a recognition site for the restriction endonuclease BsmFI, and its specific primer sequence (primer A and primer B). The ligated "adaptor–NlaIII-cut" molecules are treated with BsmFI, which cleaves the DNA 10 and 14 nucleotides downstream from its recognition site (open arrows). The extensions of the BsmFI fragments are filled in by DNA synthesis, and the mixture is blunt-end ligated. Some of the ligated molecules consist of two joined segments from different cDNAs (ditags) that are flanked by sequences for primers A and B

Figure 4.7. Types of proteomics and their applications to biology

Figure 4.8. Mechanisms by which a single gene can give rise to multiple gene products. Multiple protein isoforms can be generated by RNA processing when RNA is alternatively spliced or edited to form mature mRNA. mRNA, in turn, can be regulated by stability and efficiency of translation. Proteins can be regulated by additional mechanisms, including posttranslational modification, proteolysis, or compartmentalization

Figure 4.9. Protein expression profiling by 2-DE. Whole-cell lysates from nontransformed and Abelson murine leukemia virus (AMuLV)-transformed mouse fibroblasts were resolved by 2-DE, and proteins were visualized by silver staining. Differentially expressed proteins were excised from the gel and identified by MS

Figure 4.10. Strategies for protein identification. The identification of proteins from a polyacrylamide gel by mixed-peptide sequencing or MS is depicted. For mixed-peptide sequencing, proteins are transferred to a membrane and cleaved with CNB or skatole, and the resulting peptides are sequenced simultaneously by Edman degradation. For MS, proteins are in-gel digested with proteases and the resulting peptides are mass fingerprinted or sequenced. Information from all these methods is used to search nucleotide and protein databases for protein identification

Figure 4.11. The FASTF and FASTS search programs. (A) Example of a FASTF search where the amino acid sequence is obtained by Edman sequencing of a mixture of peptides. The information is then deconvoluted by a computer algorithm, and the results are given an expectation score (e). (B) With the FASTS program, a similar type of search is conducted except that peptide sequences obtained from MS are used

Figure 5.1. Ligation molecules with cohesive ends

Figure 5.2. Outline of gene cloning

Figure 5.3. Comparison of the general steps involved in the construction of genomic and complementary DNA (cDNA) libraries

Figure 5.4. Comparison of (a) partial and (b) complete digestion of DNA molecules at restriction enzyme sites (E)

Figure 5.5. Strategies for producing first-strand cDNA from mRNA

Figure 5.6. Second-strand cDNA synthesis using the RNase H method

Figure 5.7. Use of linkers. In this example, blunt-ended DNA is inserted into a specific restriction site on a plasmid, after ligation to a linker containing the same restriction site

Figure 5.8. Scheme of analyzing specific mRNA molecules by subtractive hybridization

Figure 5.9. Cloning of PCR products using dA: dT cloning

Figure 5.10. Production of multiple copies of a single clone from a stable gene bank or library

Figure 5.11. Map and important features of pBR322

Figure 5.12. Replica plating to detect recombinant plasmids. A sterile velvet pad is pressed onto the surface of an agar plate, picking up some cells from each colony growing on that plate

Figure 5.13. Map and important features of pUC18

Figure 5.14. Principle of blue/white selection for the detection of recombinant vectors

Figure 5.15. The lysogenic and lytic cycles of bacteriophage λ.

Figure 5.16. Two strategies for producing *in vitro* packaging extracts for bacteriophage λ

Figure 5.17. General schemes used for cloning in λ insertion and λ replacement vectors. Cl857 is a temperature-sensitive mutation that promotes lysis at 42°C after incubation at 37°C

Figure 5.18. General map of λZap cloning vector, indicating important areas of the vector. The multiple cloning site is based on the lacZ gene, providing blue/white selection based on the β-galactosidase gene. In between the initiator (I) site and terminator (T) site lie sequences encoding the phagemid bluescript

Figure 5.19. Single-stranded DNA rescue of phagemid from λZap. The single-stranded phagemid pBluescript SK may be excised from λZap by addition of helper phage. This provides the necessary proteins and factors for transcription between the I and T sites in the parent phage to produce the phagemid with the DNA cloned into the parent vector

Figure 6.1. Flowchart for the chemical synthesis of DNA oligonucleotides. After n coupling reactions (cycles), a single-stranded piece of DNA with n + 1 nucleotides is produced

Figure 6.2. Starting complex for the chemical synthesis of a DNA strand. The initial nucleoside has a protective DMT group attached to the 5′ hydroxyl group of the deoxyribose moiety and a spacer molecule attached to the hydroxyl group of the 3′ carbon of the deoxyribose. The spacer unit is attached to a solid support, which is usually a CPG bead

Figure 6.3. Structure of a phosphoramidite. Phosphoramidites are available for each of the four bases (A, C, G, and T) that are used for the chemical synthesis of a DNA strand. A diisopropylamine group is attached to the 3′ phosphite group of the nucleoside. A β-cyanoethyl group protects the 3′ phosphite group, and a DMT group is bound to the 5′ hydroxyl group of the deoxyribose sugar

Figure 6.4. Detritylation. The 5′ DMT group is removed by treatment with TCA. In this example, the detritylation of the first nucleoside is depicted

Figure 6.5. Activation and coupling. The activation of a phosphoramidite enables its 3′ phosphite group to attach to the 5′ hydroxyl group of the bound detritylated nucleoside

Figure 6.6. Capping. The available 5′ hydroxyl groups of unreacted detritylated nucleosides are acetylated to prevent them from participating in the coupling reaction of the next cycle

Figure 6.7. Oxidation. The phosphite triester internucleotide linkage is oxidized to the pentavalent phosphate triester. This reaction stabilizes the phosphodiester bond and makes it less susceptible to cleavage under either acidic or basic conditions

Figure 6.8. (A) All possible DNA sequences deduced from a protein sequence. For simplicity, only 3 amino acids, 2 of which have twofold degeneracy and 1 that has a single codon, are shown. (B) Formation of sets of mixed probes. A question mark denotes a site where equal concentrations of A, C, G, and T are supplied during chemical synthesis. The remaining sites have the same deoxynucleotide (C). Four and 16 different oligonucleotides are produced when there is a 25% probability that any 1 of the 4 nucleotides will be incorporated at one or two sites, respectively

Figure 6.9. Typical linker and adaptor sequences

Figure 6.10. Cloning with a linker. A palindromic oligonucleotide (red) with a BamHI recognition site base pairs to itself to form a BamHI linker. During the ligation reaction, linkers are successively attached to the source DNA and to each other. The ligation mixture is treated with BamHI. The large DNA molecules are purified and cloned into the BamHI site of a cloning vector. DNA molecules without linkers are not cloned

Figure 6.11. Creating a restriction endonuclease site in a vector with an adaptor. After self-hybridization, an adaptor molecule (red) with two BamHI 5′ extensions and a SmaI site (BamHI–SmaI adaptor) is formed. The BamHI–SmaI adaptor is inserted into a unique BamHI site of a vector to create a unique SmaI site. DNA (blue) is cloned into the SmaI site by blunt-end ligation. Although the SmaI site is destroyed by the insertion of a DNA molecule, the insert can be retrieved by cutting the vector with BamHI

Figure 6.12. Assembly of a synthetic gene from short oligonucleotides. Individual oligonucleotides (20- to 60-mers) are synthesized chemically. Their sequences are designed so that they form a double-stranded array after annealing. T4 DNA ligase is used to seal the nicks and produce an intact version of the gene

Figure 6.13. Assembly and *in vitro* enzymatic DNA synthesis of a gene. Individual oligonucleotides are synthesized chemically. Their sequences are designed to enable them to form a stable molecule, with base-paired regions separated by open regions (gaps). The gaps are filled in by *in vitro* enzymatic DNA synthesis. The nicks are sealed with T4 DNA ligase

Figure 6.14. The component of PCR

Figure 6.15. A basic PCR protocol – DNA synthesis cycle

Figure 6.16. PCR – a simple thermocycling protocol

Figure 6.17. Features of adaptors A and B that are used for template preparation, PCR amplification, and sequencing using a cyclic array-sequencing strategy with the 454 sequencing platform

Figure 6.18. Preparation of single-stranded genomic DNA fragments for the 454 sequencing platform

Figure 6.19. DNA capture bead. Oligomers that are complementary to the PCR amplification sequence of adaptor B are attached at their 5′ ends to a bead. Each DNA capture bead hybridizes with only one adaptor A–genomic DNA–adaptor B strand. The inset highlights the available 3′ end of the immobilized oligomer

Figure 6.20. Emulsion PCR

Figure 6.21. Wells of a picotiter plate used for 454 sequencing. Reproduced with the kind permission of Roche Diagnostics North America, Indianapolis, IN

Figure 7.1. Generalized ELISA protocol for detecting a target antigen. The primary antibody is often obtained from rabbits that have been immunized with the target antigen, while the secondary antibody is from goats immunized with rabbit antibodies. The enzyme (E) is conjugated to the secondary antibody. (A) Indirect ELISA; (B) direct ELISA

Figure 7.2. Schematic representation of a target antigen. The surface of the antigen depicted has seven (numbered 1 to 7) different antigenic determinants (epitopes). When this antigen is used to immunize an animal, each antigenic determinant elicits the synthesis of a different antibody. Together, the different antibodies that interact with an antigen constitute a polyclonal antibody directed against that antigen

Figure 7.3. The HAT procedure for selecting hybrid spleen–myeloma (hybridoma) cells

Figure 7.4. Screening for the production of a monoclonal antibody. Spleen cells from a mouse that was immunized with a specific antigen are isolated and fused in culture with myeloma cells that do not produce antibodies. Fused cells are selected for the ability to grow on HAT medium, which contains hypoxanthine, aminopterin, and thymidine. Cells that produce a specific antibody to the immunizing antigen (hybridomas) are identified by an immunoassay and individually subcultured. A hybridoma, which grows in culture and secretes a single type of antibody molecules, is the source of a monoclonal antibody

Figure 7.5. Schematic representation of a bacterial cell, engineered to respond to a particular environmental compound, producing a detectable fluorescent or bioluminescent signal

Figure 7.6. Construction of a modified monomeric red fluorescent protein (modified mRFP). The regions of the gene encoding the N and C termini of the green fluorescent protein (GFP) were spliced onto the gene for mRFP following the removal of the portion of the mRFP gene encoding the N terminus and the addition of an oligonucleotide encoding a short linker peptide (L)

Figure 7.7. Various colored monomeric fluorescent proteins derived from monomeric red fluorescent protein showing their emission wavelengths and color maxima following excitation. The colors have been called, from derivative 1 through 7, honeydew, banana, orange, tomato, tangerine, strawberry, and cherry, respectively

Figure 7.8. Assaying for the presence of pollutants with genetically engineered bioluminescent P. fluorescens

Figure 7.9. Schematic representation of a strain of the yeast Saccharomyces cerevisiae that has been engineered to detect low levels of estrogenic substances in the environment. Each yeast cell contains the human estrogen receptor (hER) gene integrated into its chromosomal DNA, one plasmid (pUTK404) that contains constitutively expressed bacterial luxCDE genes and the flavin oxidoreductase (frp) gene from the bacterium Vibrio harveyi, and one plasmid (pUTK407) that contains luxA and luxB genes under the regulatory control of estrogen response elements (ERE) and constitutive bacterial promoters (not shown). Following the binding of the hER protein to an estrogenic compound, the complex binds to an ERE and activates luxAB transcription. Expression of the luxAB, luxCDE, and frp genes leads to the production of a measurable bioluminescent signal

Figure 7.10. Labeled oligonucleotide probe hybridizes to a completely complementary genomic DNA but, under stringent conditions (A), not to a DNA sequence in which one of the bases in the middle has been altered (B). In this case, the probe was designed to hybridize to wild-type DNA. Other probes, designed to hybridize to the genomic DNA of known mutations, would not bind to the wild-type DNA

Figure 7.11. Allele-specific oligonucleotide dot blot to diagnose individuals who are either carriers of a mutant CFTR gene (heterozygotes) or affected by the disease (homozygotes). The dark dot blot indicates that the labeled oligonucleotide has bound to the individual's DNA

Figure 7.12. Detection of the sickle-cell anemia gene at the DNA level

Figure 7.13. Schematic representation of the functioning of a padlock probe

Figure 7.14. Detection of a single-base mutation with fluorescence-labeled PCR primers

Figure 7.15. TaqMan assay

Figure 7.16. Principle of polymerase chain reaction (PCR)

Figure 7.17. Principle of nucleic acid sequence-based amplification

Figure 7.18. This example is a modification of a microarray for the detection of human papilloma virus. In this example, biotin-labeled PCR products are hybridized with immobilized oligonucleotides. The chip is incubated in an anti-biotin antibody-conjugated gold nanoparticle solution. The silver enhancement by the gold nanoparticles bound to the biotin precipitates silver metal particles at the chip surfaces. This blocks light irradiated from above, which can be measured, allowing detection and quantitative analysis of the target DNA

Figure 8.1. The structure of tRNA is illustrated in open "cloverleaf" form (A) to show complementary base pairing. Modified bases are indicated as mG, methylguanosine; mC, methylcytosine; DHU, dihydrouridine; T, ribothymidine; Y, a modified purine (usually adenosine); and ψ, pseudouridine. The folded form of the molecule is shown in (B) and a space-filling model

Figure 8.2. Attachment of amino acids to tRNAs

Figure 8.3. Base pairing at the third codon position is relaxed, allowing G to pair with U, and inosine (I) in the anticodon to pair with U, C, or A. Two examples of abnormal base pairing, allowing phenylalanyl (Phe) tRNA to recognize either UUC or UUU codons and alanyl (Ala) tRNA to recognize GCU, GCC, or GCA, are illustrated

Figure 8.4. Prokaryotic and eukaryotic mRNAs

Figure 8.5. Initiation sites in prokaryotic mRNAs are characterized by a Shine-Delgarno sequence that precedes the AUG initiation codon. Base pairing between the Shine-Delgarno sequence and a complementary sequence near the 3′ terminus of 16S rRNA aligns the mRNA on the ribosome. In contrast, eukaryotic mRNAs are bound to the 40S ribosomal subunit by their 5′ 7-methylguanosine caps. The ribosome then scans along the mRNA until it encounters an AUG initiation codon

Figure 8.6. Overview of translation

Figure 8.7. Three initiation factors (IF-1, IF-2, and IF-3) first bind to the 30S ribosomal subunit. This step is followed by binding of the mRNA and the initiator N-formylmethionyl (fMet) tRNA, which is recognized by IF-2 bound to GTP. IF-3 is then released, and a 50S subunit binds to the complex, triggering the hydrolysis of bound GTP, followed by the release of IF-1 and IF-2 bound to GDP

Figure 8.8. Initiation of translation in eukaryotic cells

Figure 8.9. Elongation stage of translation

Figure 8.10. EF-Tu complexed to GTP escorts the aminoacyl tRNA to the ribosome. The bound GTP is hydrolyzed as the correct tRNA is inserted, so EF-Tu complexed to GDP is released. The EF-Tu/GDP complex is inactive and unable to bind another tRNA. In order for translation to continue, the active EF-Tu/GTP complex must be regenerated by another factor, EF-Ts, which stimulates the exchange of the bound GDP for free GTP

Figure 8.11. A termination codon (e.g., UAA) at the A site is recognized by a release factor rather than by a tRNA. The result is the release of the completed polypeptide chain, followed by the dissociation of tRNA and mRNA from the ribosome

Figure 8.12. Primary structure: The A chain of insulin is 21 amino acids long and the B chain is 30 amino acids long, and each sequence is unique to the insulin protein

Figure 8.13. Sickle cell disease: Sickle cells are crescent-shaped, while normal cells are disc-shaped

Figure 8.14. Secondary structure: The α-helix and β-pleated sheet form because of hydrogen bonding between carbonyl and amino groups in the peptide backbone. Certain amino acids have a propensity to form an α-helix, while others have a propensity to form a β-pleated sheet

Figure 8.15. Tertiary structure: The tertiary structure of proteins is determined by hydrophobic interactions, ionic bonding, hydrogen bonding, and disulfide linkages

Figure 8.16. Four levels of protein structure: The four levels of protein structure can be observed in these illustrations

Figure 8.17. Chaperones bind to the amino (N) terminus of the growing polypeptide chain, stabilizing it in an unfolded configuration until synthesis of the polypeptide is completed. The completed protein is then released from the ribosome and is able to fold into its correct three-dimensional conformation

Figure 8.18. A partially unfolded polypeptide is transported from the cytosol to a mitochondrion. Cytosolic chaperones stabilize the unfolded configuration. Mitochondrial chaperones facilitate transport and subsequent folding of the polypeptide chain within the organelle

Figure 8.19. Structure of a chaperonin. GroEL, a member of the Hsp60 family, is a porous cylinder composed of two stacked rings. Each ring consists of seven subunits

Figure 8.20. Chaperones of the Hsp70 family bind to and stabilize unfolded polypeptide chains during translation. The unfolded polypeptide is then transferred to chaperones of the Hsp60 family, within which protein folding takes place. ATP hydrolysis is required for release of the unfolded polypeptide from Hsp70 as well as for folding within Hsp60

Figure 8.21. The Protein disulfide isomerase (PDI) catalyzes the breakage and rejoining of disulfide bonds, resulting in exchanges between paired disulfides in a polypeptide chain. The enzyme forms a disulfide bond with a cysteine residue of the polypeptide and then exchanges its paired disulfide with another cysteine residue. In this example, PDI catalyzes the conversion of two incorrect disulfide bonds (1–2 and 3–4) to the correct pairing (1–3 and 2–4)

Figure 8.22. The action of peptidyl prolyl isomerase. Peptidyl prolyl isomerase catalyzes the isomerization of peptide bonds that involve proline between the cis and trans conformations

Figure 8.23. The role of signal sequences in membrane translocation

Figure 8.24. Proteolytic processing of insulin

Figure 8.25. The carbohydrate chains of N-linked glycoproteins are attached to asparagine; those of O-linked glycoproteins are attached to either serine (shown) or threonine. The sugars joined to the amino acids are usually either N-acetylglucosamine (N-linked) or N-acetylgalactosamine (O-linked)

Figure 8.26. Synthesis of N-linked glycoproteins

Figure 8.27. Examples of N-linked oligosaccharides

Figure 8.28. Examples of O-linked oligosaccharides. O-linked oligosaccharides usually consist of only a few carbohydrate residues, which are added one sugar at a time

Figure 8.29. Addition of a fatty acid by N-myristoylation. The initiating methionine is removed, leaving glycine at the N terminus of the polypeptide chain. Myristic acid (a 14-carbon fatty acid) is then added

Figure 8.30. Prenylation of a C-terminal cysteine residue

Figure 8.31. Palmitoylation. Palmitate (a 16-carbon fatty acid) is added to the side chain of an internal cysteine residue

Figure 8.32. Structure of a GPI anchor

Figure 8.33. Cell-free protein synthesis (CFPS0 for producing proteins)

Figure 9.1. Cleavage of inactive preproinsulin to yield active mature insulin. Proteases remove the leader peptide (L) and an internal peptide (C), yielding a peptide that consists of chains A and B

Figure 9.2. Examples of some O-linked oligosaccharides in yeasts (A), insects (B), and mammals (C). O-linked oligosaccharides have a number of arrangements with different combinations of sugars. Some of the more prevalent forms are shown here. S, serine; T, threonine; red circles, mannose; darkblue squares, N-acetylglucosamine; light-blue squares, N-acetylgalactosamine; green squares, galactose; orange squares, sialic acid

Figure 9.3. Examples of some N-linked oligosaccharides in yeasts (A), insects (B), and mammals (C). All N-linked glycosylations in eukaryotes start with the same initial group, which is subsequently trimmed and then elaborated in diverse ways within and among species. Some yeast sites have 15 or fewer mannose units (core series), and others have more (outer-chain family). In *S. cerevisiae,* the chains frequently have 50 or more mannose units. An asparagine (N) residue next to any amino acid (X) followed by either threonine (T) or serine (S) can be targeted for glycosylation. Red circles, mannose; dark blue squares, N-acetylglucosamine; yellow triangles, glucose; green squares, galactose; orange squares, sialic acid; maroon triangle, fucose

Figure 9.4. Generalized eukaryotic expression vector. The major features of a eukaryotic expression vector are a eukaryotic transcription unit with a promoter (p), a multiple cloning site (MCS) for a gene of interest, and a DNA segment with termination and polyadenylation signals (t); a eukaryotic selectable marker (ESM) gene system; an origin of replication that functions in the eukaryotic cell (orieuk); an origin of replication that functions in *E. coli* (oriE); and an *E. coli* selectable marker (Ampr) gene

Figure 9.5. Schematic representation of integration of DNA with a YIp vector. A selectable marker gene (LEU2) and a gene of interest (GOI) with transcription and translation control elements (not shown) are inserted into a YIp vector between two segments from the ends of a nonessential yeast gene (A1 and A2). The ampicillin resistance (Ampr) gene and the origin of replication (oriE) function in *E. coli*. A leucine-requiring (leu2) yeast strain is transformed with restriction endonucleasedigested (RE) vector DNA because chromosomal DNA is more likely to recombine with linearized DNA than with circular DNA. The restriction endonuclease sites flank the segments from the nonessential gene. The DNA sequences at the ends of nonessential gene A undergo recombination (×) that leads to the incorporation of both the gene of interest and the LEU2 gene into the corresponding chromosome site. Transformants grow on medium that is not supplemented with leucine. Nonrecombined DNA is degraded

Figure 9.6. YAC cloning system. The YAC plasmid (pYAC) has an *E. coli* selectable marker (Ampr) gene; an origin of replication that functions in *E. coli* (oriE); and yeast DNA sequences, including URA3, CEN, TRP1, and ARS. CEN provides centromere function, ARS is a yeast autonomous replicating sequence that is equivalent to a yeast origin of replication, URA3 is a functional gene of the uracil biosynthesis pathway, and TRP1 is a functional gene of the tryptophan biosynthesis pathway. The T regions are yeast chromosome telomeric sequences. The SmaI site is the cloning insertion site. pYAC is first treated with SmaI, BamHI, and alkaline phosphatase and then ligated with size-fractionated (100-kb) input DNA. The final construct carries cloned DNA and can be stably maintained in double-mutant ura3 and trp1 cells

Figure 9.7. *S. cerevisiae* expression vector. The cDNA for human Cu/Zn-SOD was cloned between the promoter (GAPDp) and termination–polyadenylation sequence (GAPDt) of the *S. cerevisiae* glyceraldehyde phosphate dehydrogenase gene. The LEU2 gene that was cloned between segments of the yeast 2µm plasmid DNA encodes a functional enzyme of the leucine biosynthesis pathway. The yeast origin of replication is included in the 2µm plasmid DNA. The ampicillin resistance (Ampr) gene and the *E. coli* origin of replication (oriE) are derived from plasmid pBR322

Figure 9.8. Summary of protein folding in the endoplasmic reticulum of yeast cells. During synthesis on ribosomes associated with the endoplasmic reticulum (ER), nascent proteins are bound by the chaperones BiP and calnexin, which aid in the correct folding of the protein. Protein disulfide isomerases (PDI) catalyze the formation of disulfide bonds between cysteine amino acids that are nearby in the folded protein. Quality control systems ensure that only correctly folded proteins are released from the ER. Proteins released from the ER are transported to the Golgi apparatus for further processing. Prolonged binding of BiP to misfolded proteins leads to activation of the *S. cerevisiae* transcription factor Hac1, which controls the expression of several proteins that mediate the unfolded-protein response (UPR)

Figure 9.9. Differential processing of glycoproteins in *P. pastoris*, humans, and "humanized" *P. pastoris*. Initial additions of sugar residues to glycoproteins in the endoplasmic reticulum are similar in human and *P. pastoris* cells

Figure 9.10. *P. pastoris* integrating expression vector. The gene of interest (GOI) is cloned between the promoter (AOX1p) and termination-polyadenylation sequence (AOX1t) of the *P. pastoris* alcohol oxidase 1 gene. The HIS4 gene encodes a functional histidinol dehydrogenase of the histidine biosynthesis pathway. The ampicillin resistance (Ampr) gene and an origin of replication (oriE) function in *E. coli*. The segment marked 3′ AOX1 is a piece of DNA from the 3′ end of the alcohol oxidase 1 gene of *P. pastoris*. A double recombination event between the AOX1p and 3′ AOX1 regions of the vector and the homologous segments of chromosome DNA results in the insertion of the DNA carrying the gene of interest and the HIS4 gene

Figure 9.11. Integration of DNA into a specific *P. pastoris* chromosome site by single (A) or double (B) recombination

Figure 9.12. Budded (A) and occluded (B) forms of AcMNPV. During budding, a nucleocapsid becomes enveloped by the membrane of an infected cell. A polyhedron consists of clusters of nucleocapsids (occluded virions) embedded in various orientations in a polyhedrin matrix

Figure 9.13. Organization of the expression unit of a baculovirus (AcMNPV) transfer vector. The gene of interest is inserted into the multiple cloning site (MCS) that lies between the polyhedrin gene promoter (Pp) and polyhedrin gene transcription termination (Pt) sequences. The AcMNPV DNA upstream from the polyhedrin promoter (5′ AcMNPV DNA) and downstream from the polyhedrin transcription termination sequence (3′ AcMNPV DNA) provides sequences for integration of the expression unit by homologous recombination into an AcMNPV genome

Figure 9.14. Replacement of the AcMNPV polyhedrin gene with an expression unit from a transfer vector. A double crossover event (×) between homologous DNA segments of the transfer vector and the AcMNPV genome results in the integration of the expression unit into the AcMNPV genome. GOI, gene of interest; Pp, polyhedrin promoter; Pt, polyhedrin polyadenylation and termination sequence

Figure 9.15. Production of recombinant baculovirus. Single Bsu36I sites are engineered into gene 603 and a gene (1629) that is essential for AcMNPV replication. These genes flank the polyhedrin gene in the AcMNPV genome. After a baculovirus with two engineered Bsu36I sites is treated with Bsu36I, the segment between the Bsu36I sites is deleted. Insect cells are cotransfected with a Bsu36I-treated baculovirus DNA and a transfer vector with a gene of interest under the control of the promoter (p) and terminator (t) elements of the polyhedrin gene and the complete sequences of both genes 603 and 1629. A double crossover event (dashed lines) generates a recombinant baculovirus with a functional gene 1629. With this system, almost all of the progeny baculoviruses are recombinant

Figure 9.16. Construction of a recombinant bacmid

Figure 9.17. Construction of a recombinant baculovirus expression vector *in vitro*

Figure 10.1. Illustration of the separation process by size exclusion chromatography

Figure 10.2. A simplified diagram of a gas chromatograph showing: (1) carrier gas, (2) autosampler, (3) inlet, (4) analytical column, (5) detector and (6) PC. Credit: anthias consulting

Figure 10.3. Overview of HPLC

List of Tables

Table 1.1. Selected developments in the history of molecular biotechnology

Table 2.1. The genetic code

Table 2.2. Properties of ribosomes

Table 3.1. The structure and function of eukaryotic organelles

Table 4.1. Converting Cy3 and Cy5 intensities to log2 ratios

Table 4.2. Different SAGE systems

Table 5.1. Numbers of clones required for representation of DNA in a genome library

Table 5.2. Comparison of vectors generally available for cloning DNA fragments

Table 6.1. Overall yields of chemically synthesized oligonucleotides with different coupling efficiencies

Table 6.2. Concentrations of PCR buffers

Table 7.1. Targets for diagnostic monoclonal antibodies

Table 7.2. Comparison of laboratory methods for direct detection of infections

Table 7.3. Advantages and disadvantages of different molecular diagnostic techniques

Table 8.1. Translation factors

Table 9.1. Recombinant proteins produced by *S. cerevisiae* expression systems

Table 9.2. Promoters for *S. cerevisiae* expression vectors

Table 9.3. Some of the recombinant proteins that have been produced by the baculovirus expression vector system

PREFACE

Molecular biology is a crucial part of modern biology. It helps us understand the structures, functions, and internal workings of individual cells. This knowledge is vital for developing new drugs, diagnosing diseases, and gaining deeper insights into how cells function.

Molecular biologists study genetics in plants, animals, and humans, exploring their variations and relationships. They apply their understanding of biology, chemistry, physics, physiology, and genetics to study different types of cells, DNA, and their behaviors.

This field of study is still relatively new and offers abundant opportunities for research in medicine, agriculture, animal husbandry, and related areas. Various advanced molecular techniques based on omics (such as transcriptomics, proteomics, and genomics) and phylogenetics have revolutionized biological sciences. These methods have evolved into sophisticated quantitative tools that researchers across disciplines can utilize to design diverse studies. Leveraging these molecular-based approaches, scientists can comprehensively and systematically identify genetic variations, perform molecular docking, and construct predictive models for ecological niches and taxonomic classifications.

Investigations in genomics, proteomics, and phylogenetics often involve novel DNA elements like microsatellites from mitochondria and chloroplasts, as well as retrotransposons, which contribute to genetic diversity using molecular data. Moreover, the advantages and limitations of these molecular approaches have been extensively studied and acknowledged. Integrating molecular phylogenetic and omics techniques with expression and pathway analyses enhances our understanding and ability to develop new molecular mechanisms and stress responses in biological systems. These techniques also present significant opportunities for developing targeted therapies and disease diagnostics based on molecular data.

It is essential to evaluate how data from diverse molecular techniques can be applied across various biological studies. The study and application of molecular approaches hold tremendous potential for advancing genomics, proteomics, and phylogenetic techniques in biological sciences.

The molecular realm is a captivating field where scientists continually uncover intricate systems governing life within the dynamic landscape of biological sciences. This book explores the latest advancements in this rapidly evolving field while identifying knowledge gaps and charting future research directions. Over the past decade, remarkable progress has been made in understanding the molecular intricacies of living organisms. Breakthroughs in imaging technologies, genomics, and proteomics have revealed previously unknown details about cellular functions, signaling pathways, and the molecular underpinnings of diseases. Further advancements have been propelled by the integration of artificial intelligence and big data analytics, deepening our comprehension of complex biological phenomena.

Yet, behind these successes lies a complex reality with many unanswered questions. Each layer of complexity unveils new inquiries, compelling scientists to dig deeper and continue their quest for knowledge.

Content/Structure of the Book

This comprehensive book is organized into 10 chapters, each exploring fundamental aspects of Molecular Biotechnology, catering to both novices and experts in the field.

Chapter 1, Introduction to Molecular Biotechnology, serves as a gateway to the discipline, providing an overview of its historical evolution, foundational concepts, and contemporary applications. It sets the stage by elucidating the significance of molecular techniques in various scientific and industrial domains.

Chapter 2, Molecular Biology, examines the intricacies of molecular mechanisms within biological systems. It covers essential topics such as DNA replication, transcription, and translation, laying the groundwork for understanding advanced biotechnological methodologies.

Structure and Functions of Cells are dissected in Chapter 3, offering insights into cellular architecture and the roles played by organelles in cellular processes. This foundational knowledge is crucial for comprehending subsequent chapters on genetic manipulation and protein synthesis.

Chapter 4, Genomics and Proteomics, explores the vast landscapes of genomes and proteomes, emphasizing their roles in health, agriculture, and environmental sciences. It highlights cutting-edge technologies driving advancements in molecular biotechnology.

Recombinant DNA and Genetic Analysis take center stage in Chapter 5, elucidating techniques for DNA manipulation and genetic engineering. This chapter bridges theoretical knowledge with practical applications, essential for biotechnological innovations.

Chapter 6, Chemical Synthesis, Amplification, and Sequencing of DNA, provides a detailed exploration of techniques like PCR (Polymerase Chain Reaction) and DNA sequencing methods, pivotal for molecular diagnostics and genetic research.

Molecular Diagnostics is extensively covered in Chapter 7, focusing on the application of molecular techniques in medical diagnostics, disease surveillance, and personalized medicine. It underscores the transformative impact of biotechnology on healthcare.

Chapter 8, Protein Synthesis, unravels the mechanisms of protein production and regulation within cells, offering insights into biopharmaceutical development and industrial biotechnology.

Heterologous Protein Production in Eukaryotic Cells is the focus of Chapter 9, detailing strategies for producing recombinant proteins in eukaryotic hosts, essential for therapeutic proteins and biotechnological applications.

Finally, Chapter 10, Biochemistry and Molecular Biology Techniques, consolidates practical methodologies essential for experimental design, data analysis, and interpretation in molecular biotechnology research.

Together, these chapters form a cohesive narrative, equipping readers with a deep understanding of Molecular Biotechnology's theoretical foundations and practical applications across diverse scientific disciplines and industrial sectors.

Audience

This book is designed for a diverse readership that spans students, researchers, and professionals within the fields of biotechnology, molecular biology, and related life sciences. Undergraduate and graduate students will find it an invaluable resource for understanding the fundamental principles and advanced techniques that underpin molecular biotechnology. Researchers and academics can utilize the book as a reference guide to stay updated with the latest developments and methodologies in the field. Additionally, professionals working in the biotech industry, pharmaceutical companies, and healthcare sectors will benefit from the practical insights and applications discussed throughout the chapters. The book's comprehensive coverage of topics such as genomics, proteomics, recombinant DNA technology, and molecular diagnostics makes it an essential tool for those involved in biotechnological research and development, as well as those seeking to apply molecular techniques in clinical and industrial settings. By bridging theoretical knowledge with practical applications, this book caters to a broad audience, fostering a deeper understanding of the molecular mechanisms driving biotechnological innovations.

Learning Features

- Start-of-chapter learning objectives
- Key terms in each chapter
- Summary Points at the end of each chapter
- Review Questions at the end of each chapter
- Case Studies and Activities in each chapter for solving real problems
- References at the end of each chapter

CHAPTER

1 Introduction to Molecular Biotechnology

LEARNING OBJECTIVES

After studying this chapter, you will be able to:

- Define the emergence of molecular biotechnology;
- Understand the recombinant DNA technology;
- Explain the applications of molecular biotechnology;
- Discuss ethical, legal, and social implications.

KEY TERMS FROM THIS CHAPTER

Bacterial genetics
Cancer biomarkers
Deoxyribonucleic acid (DNA)
Metabolic functions
Microbiology
Molecular biology
Biochemistry
Crossbreeding
DNA technology
Microarray analysis
Microorganisms
Molecular level

INTRODUCTION

Molecular biotechnology is a dynamic and interdisciplinary field that explores the manipulation and understanding of biological molecules at the molecular level to develop innovative solutions for various scientific and practical challenges. At its core, molecular biotechnology harnesses the principles of molecular biology, genetics, biochemistry, and other related disciplines to engineer biological systems for applications in healthcare, agriculture, industry, and environmental sustainability.

Molecular biotechnology plays a crucial role in the emerging fields of personalized medicine and precision agriculture, where treatments and interventions are tailored to individual genetic profiles or specific environmental conditions. By leveraging molecular insights, researchers and practitioners can develop targeted therapies, diagnostic tools, and crop varieties with enhanced traits, thereby improving outcomes for patients, farmers, and the environment.

Overall, molecular biotechnology continues to drive innovation and advancements in various sectors, offering promising solutions to some of the most pressing issues facing society today. As our understanding of molecular processes deepens and technological capabilities expand, the potential for molecular biotechnology to shape the future of science, medicine, and industry remains immense.

1.1. THE EMERGENCE OF MOLECULAR BIOTECHNOLOGY

To find ways to produce more food, preserve food, and heal the sick, humans looked to the natural world long before we understood that microorganisms existed or that genes were the units of inheritance. The discovery made by our ancestors included the ability to preserve grains by fermenting them into beer, the ability to heal infected saddle sores by storing horse saddles in a warm, damp corner of the stable, and the possibility that purposeful contact with a "contagion" would shield a subsequent exposure victim from an infectious disease.

Microorganisms have played a significant role in the development of many practical procedures and goods since the 17th century when the microscopic world was discovered. We have a lot of these in our backyards and homes. Probiotics and yogurt are made with lactic acid bacteria, many of the plants from which the vegetables in our refrigerator are harvested are sprayed with bacteria that produce insecticides, legume cultivation requires the addition of nitrogen-fixing bacteria to the soil, laundry detergent's enzymatic stain removers are derived from common soil microbes, and infectious diseases are treated with antibiotics derived from these microbes. These are only a handful of the conventional biotechnologies that have made our lives better. However, traditional biotechnology was not a well-established scientific field until the early 1970s and the majority of the research in this field was conducted in chemical engineering departments and occasionally in specialized microbiology programs.

Broadly speaking, biotechnology is the process of producing goods for use in commerce that are produced by microorganisms during their metabolic processes. "The application of scientific and engineering principles to the processing of material by biological agents to provide goods and services" is the formal definition of biotechnology. Hungarian engineer Karl Ereky coined the term "biotechnology" in 1917 to refer to an integrated method for producing pigs on a large-scale using sugar beets as their food source. "All lines of work by which products are produced from raw materials with the aid of living things" is what Ereky defined as biotechnology. This fairly exact definition was largely disregarded. The terms "biotechnology" and "two very different engineering disciplines" were used interchangeably for several years. It alludes, on the one hand, to industrial fermentation. Conversely, it was employed in the study of workplace efficiency, or ergonomics as it is currently known. This confusion was resolved in 1961 when the Swedish microbiologist Carl Göran Hedén suggested changing the name of the Journal of Microbiological and Biochemical Engineering and Technology to Biotechnology and Bioengineering, a scientific journal devoted to publishing research in the fields of industrial fermentation and applied microbiology. Since then, the field of biotechnology—which has its roots in the fields of microbiology, biochemistry, and chemical engineering—has become inextricably linked to the study of "the industrial production of goods and services by processes using biological organisms, systems, and processes."

An industrial biotechnology process that uses microorganisms for producing a commercial product typically has three key stages (Figure 1.1):

- The preparation of the microorganism and the raw materials needed for it to grow and produce the intended product is known as upstream processing.
- Fermentation and transformation: the target microorganism is grown (fermented) in a sizable bioreactor (often larger than 100 liters), leading to the biotransformation of the desired compound—which could be an antibiotic, an amino acid, or a protein, for example.
- Purification of the target compound from the cell mass or the cell medium is known as downstream processing.

The goal of biotechnology research is to find microorganisms that can produce products that can be used to prepare drugs, food supplements, and foods while also optimizing the overall efficiency of each of these steps. This research concentrated on downstream processing, bioreactor design, and upstream processing in the 1960s and 1970s. The results of these studies led to the development of effective large-scale growth facilities that raised the yields of numerous products, as well as improved bioinstrumentation for tracking and managing the fermentation process.

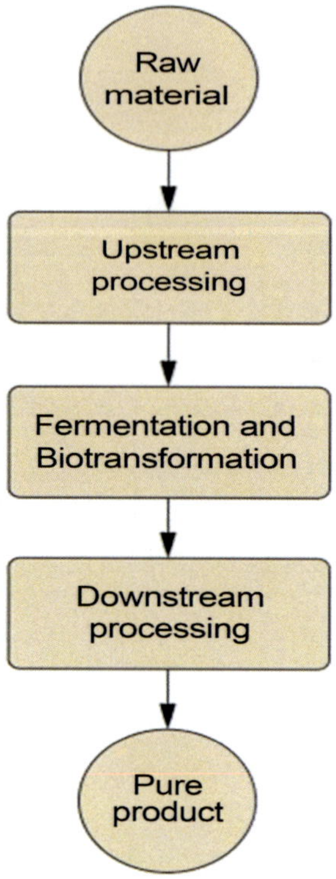

Figure 1.1. *Principal steps of a bioengineered biotechnology process. Parenthetically, Karl Ereky's scheme entailed using inexpensive sugar beets (raw material) to feed pigs (biotransformation) for the production of pork (downstream processing).*

Source: Bernard R. Glick, Jack J. Pasternak, and Cheryl L. Patten.; Molecular biotechnology: principles and applications of recombinant DNA; ISBN 978-1-55581-498-4.

The hardest part of the entire process to manipulate was the biotransformation part. Large-scale production of commodities by naturally occurring microbial strains was frequently far from ideal. Initially, efforts to increase product yields concentrated on producing variants, or mutants, by inducing changes in the genetic makeup of strains that were already in existence using chemical mutagens or UV radiation. The degree of improvement that could be attained in this manner was typically constrained by biological factors. For instance, if a mutated strain produced an excessive amount of a compound, other metabolic processes were frequently compromised, which resulted in less than ideal strain growth during large-scale fermentation. The conventional "induced mutagenesis and selection" methods of strain improvement were remarkably effective for many processes, including the synthesis of antibiotics, despite this limitation.

Due to the enormous number of colonies that needed to be chosen, screened, and tested, the conventional genetic improvement regimens were laborious, expensive, and time-consuming. Furthermore, rather than increasing a strain's genetic potential, the greatest outcome that could be anticipated with this strategy was the enhancement of an already-existing inherited trait. Notwithstanding these drawbacks, efficient methods for the large-scale manufacturing of an extensive variety of commercial goods had been refined by the late 1970s.

The biochemistry, genetics, and molecular biology of microorganisms are now sufficiently understood to enable the creation of new products that would not otherwise exist, as well as to speed up the development of beneficial and enhanced biological processes and products. Molecular biotechnology is the field that deals with manipulating genes to produce useful goods and services using living organisms. Molecular biotechnology is different from traditional biotechnology in that modern methods require knowledge of and manipulation of genes, the functional units of inheritance. The creation of methods for separating genes and transferring them from one organism to another was a crucial advancement that made this technology possible. Recombinant deoxyribonucleic acid (DNA) technology was born out of a lunchtime discussion between two scientists from different fields who met in 1973 during a scientific conference. Stanley Cohen had been creating techniques in his lab at Stanford University in California to introduce plasmids, which are tiny circular DNA molecules, into bacterial cells. The University of California, San Francisco's Herbert Boyer was experimenting with enzymes that cleave DNA at particular nucleotide sequences in the meantime. They reasoned during a scientific meeting over lunch that Boyer's enzyme could be used to splice a particular DNA segment into a plasmid and that Cohen's method could then be used to introduce the recombinant plasmid into a host bacterium (Russel & Sambrook, 2001; Campbell & Reece, 2002).

1.2. RECOMBINANT DNA TECHNOLOGY

Cohen, Boyer, and others saw the wide-ranging potential of recombinant DNA technology. At the time, Cohen observed, "It might be able to introduce in *E. coli*, genes specifying metabolic or synthetic functions such as photosynthesis or antibiotic production indigenous to other biological classes. Recombinant DNA technology was first used to produce human insulin, a medication used to treat diabetes, for commercial use. The synthesis of the DNA sequence encoding human insulin was accomplished, which was an amazing achievement at the time, and the resulting plasmid was introduced into a common bacterium, *Escherichia coli*, to sustain it. The two peptide chains of human insulin were produced by the bacterial host cells, which functioned as biological factories. These chains were then combined, purified, and administered to diabetics who were allergic to the pig insulin that was sold commercially. This accomplishment would have seemed unattainable 10 years ago. By today's standards, however, this type of genetic engineering is considered commonplace.

The discovery of recombinant DNA technology fundamentally altered the field of biotechnology. These methods provided a more direct means of optimizing the biotransformation stage of a biotechnology process. The ability to generate extremely productive strains rather than just isolate them was made possible by genetic engineering. Soon after the first commercial preparation of recombinant human insulin was produced, a range of therapeutic proteins, including interferon, growth hormone, insulin, and viral antigens, were produced using bacteria and subsequently eukaryotic cells. Recombinant DNA technology may also be utilized to support the large-scale biological synthesis of valuable macromolecules and low-molecular-weight compounds that are found in minute amounts in nature. It became possible to use plants and animals as natural bioreactors to produce novel or modified gene products that would not have been possible to produce through crossbreeding or mutagenesis and selection. The industry standard for creating living systems with unique properties and functions for the production of significant commercial goods is now molecular biotechnology.

The majority of newly developed scientific fields do not emerge fully on their own. They are frequently created by combining information from several study fields. Industrial microbiologists and chemical engineers developed the biotechnology aspect of molecular biotechnology, while advances in molecular biology, bacterial genetics, and nucleic acid enzymology have greatly contributed to the development of recombinant DNA technology (Table 1.1). Broadly speaking, molecular biotechnology uses information from several basic scientific fields to produce goods that are beneficial for a variety of uses in the marketplace (Figure 1.2).

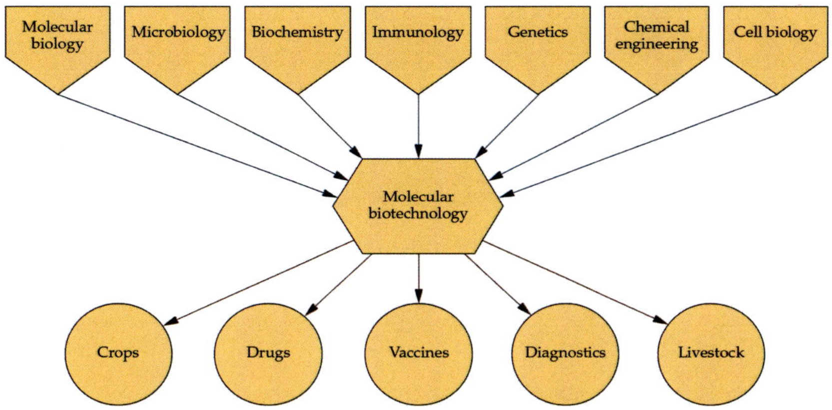

Figure 1.2. *Many scientific disciplines contribute to molecular biotechnology, which generates a wide range of commercial products.*

Source: Bernard R. Glick, Jack J. Pasternak, and Cheryl L. Patten.; Molecular biotechnology: principles and applications of recombinant DNA; ISBN 978-1-55581-498-4.

Table 1.1. *Selected Developments in the History of Molecular Biotechnology*

1917 Karl Ereky coins the term "biotechnology"
1940 A. Jost coins the term "genetic engineering"
1943 Penicillin is produced on an industrial scale
1944 Avery, MacLeod, and McCarty demonstrate that DNA is the genetic material
1953 Watson and Crick determine the structure of DNA
1961 The journal Biotechnology and Bioengineering is established
1961–1966 Entire genetic code is deciphered
1970 First restriction endonuclease is isolated
1972 Khorana and coworkers synthesize an entire tRNA gene
1973 Boyer and Cohen establish recombinant DNA technology
1975 Kohler and Milstein describe the production of monoclonal antibodies
1976 First guidelines for the conduct of recombinant DNA research are issued
1976 Techniques are developed to determine the sequence of DNA
1978 Genentech produces human insulin in *E. coli*

1980 U.S. Supreme Court rules in the case of Diamond v. Chakrabarty that genetically manipulated microorganisms can be patented

1981 First commercial, automated DNA synthesizers are sold

1981 First monoclonal antibody-based diagnostic kit is approved for use in the United States

1982 First animal vaccine produced by recombinant DNA methodologies is approved for use in Europe

1983 Engineered Ti plasmids are used to transform plants

1988 U.S. patent is granted for a genetically engineered mouse susceptible to cancer

1988 PCR method is published

1990 Approval is granted in the United States for a trial of human somatic cell gene therapy

1990 Human Genome Project is officially initiated

1990 Recombinant chymosin is used for cheese making in the United States

1994–1995 Detailed genetic and physical maps of human chromosomes are published

1994 FDA announces that genetically engineered tomatoes are as safe as conventionally bred tomatoes

1995 First genome sequence of a cellular organism, the bacterium Haemophilus influenzae, is completed

1996 First recombinant protein, erythropoietin, exceeds $1 billion in annual sales

1996 Complete DNA sequence of all the chromosomes of a eukaryotic organism, the yeast Saccharomyces cerevisiae,

is determined

1996 Commercial planting of genetically modified crops begins

1997 Nuclear cloning of a mammal (a sheep) with a differentiated cell nucleus is accomplished

1998 FDA approves first antisense drug

1999 FDA approves recombinant fusion protein (diphtheria toxin–interleukin-2) for cutaneous T-cell lymphoma

2000 Arabidopsis genome is sequenced

2000 Monoclonal antibodies exceed $2 billion in annual sales

2000 Development of "golden rice" (provitamin-A-producing rice) is announced

2000 Over $33 billion is invested in U.S. biotechnology companies

2001 Human genome is sequenced

2002 Complete human gene microarrays (gene chips) become commercially available

2002 FDA approves first nucleic acid test system to screen whole blood from donors for HIV and HCV

2004	Large-scale sequencing of the Sargasso Sea metagenome begins
2005	NCBI announces that there are 100 gigabases of nucleotides in the GenBank sequence database
2006	Recombinant cancer vaccine becomes available to protect against cervical cancer
2008	Two-billionth acre of genetically engineered crops is planted
2009	FDA approves first drug produced in a genetically engineered animal (a goat)

Source: Bernard R. Glick, Jack J. Pasternak, and Cheryl L. Patten.; Molecular Biotechnology: Principles and Applications of Recombinant DNA; ISBN 978-1-55581-498-4.

The gene cloning method developed by Cohen and Boyer was a "heard round the world" experiment. Upon the public release of their concept, numerous other researchers recognized the potential benefits of gene cloning. As a result, researchers developed a wide range of experimental procedures that improved the efficiency and relative ease of identifying, isolating, characterizing, and employing genes. New discoveries in almost every field of biology, including animal behavior, developmental biology, molecular evolution, cell biology, and human genetics, have been greatly influenced by these technological advancements. In fact, the ability to clone sizable DNA fragments into plasmids to prepare them for sequence determination was essential to the emergence of the field of genomics (Russel & Sambrook, 2001; Campbell & Reece, 2002).

1.3. APPLICATIONS OF MOLECULAR BIOTECHNOLOGY

Molecular biotechnology has diverse applications across various fields, including medicine, agriculture, environmental science, and industry.

1.3.1. Applications in Medicine

One of the most significant areas of application for molecular biotechnology is medicine, where it has revolutionized diagnostics, therapeutics, and personalized medicine.

1.3.1.1. Diagnostics

- **Genetic Testing**: Molecular biotechnology facilitates the identification of genetic mutations associated with inherited diseases, cancer predisposition, and pharmacogenomic variations. Techniques such as polymerase chain reaction (PCR), DNA sequencing, and microarray analysis are used to detect genetic variants and assess disease risk.
- **Infectious Disease Diagnosis**: Molecular biotechnology plays a crucial role in the detection and characterization of infectious pathogens, including viruses, bacteria, fungi, and parasites. Molecular diagnostic tests such as nucleic acid amplification assays (e.g., PCR), next-generation sequencing (NGS), and serological assays enable rapid and accurate diagnosis of infectious diseases, guiding treatment decisions and outbreak management.
- **Cancer Biomarkers**: Molecular biotechnology facilitates the identification of cancer biomarkers, including genetic mutations, gene expression patterns, and epigenetic alterations associated with different cancer types. Biomarker-based assays aid in cancer diagnosis, prognosis, and treatment selection, enabling personalized oncology care.

1.3.1.2. Therapeutics

- **Gene Therapy**: Molecular biotechnology enables the delivery of therapeutic genes into target cells to treat genetic disorders, cancer, and infectious diseases. Viral vectors, non-viral vectors, and genome editing tools such as CRISPR-Cas9 are used to introduce corrective or therapeutic genes into patient cells, restoring normal cellular function and treating genetic diseases at the molecular level.

- **RNA-based Therapeutics**: Molecular biotechnology has led to the development of RNA-based therapeutics, including messenger RNA (mRNA) vaccines, small interfering RNA (siRNA) therapeutics, and antisense oligonucleotides (ASOs). RNA-based therapies target specific genes or gene products involved in disease pathogenesis, offering potential treatments for cancer, genetic disorders, and viral infections.
- **Monoclonal Antibodies**: Molecular biotechnology plays a central role in the production of monoclonal antibodies (mAbs) for therapeutic use. Recombinant DNA technology enables the generation of mAbs targeting specific antigens implicated in autoimmune diseases, cancer, inflammatory disorders, and infectious diseases, providing targeted therapies with high specificity and efficacy.

1.3.1.3. Drug Discovery and Development

- **Target Identification and Validation**: Molecular biotechnology facilitates the identification and validation of drug targets involved in disease pathogenesis. Genomic, transcriptomic, and proteomic technologies enable the characterization of disease mechanisms, signaling pathways, and molecular targets for drug development.
- **High-throughput Screening**: Molecular biotechnology accelerates the drug discovery process through high-throughput screening (HTS) of compound libraries against molecular targets or disease models. Automated assays, robotic platforms, and bioinformatics tools enable the screening of large chemical libraries for lead identification and optimization.
- **Precision Medicine**: Molecular biotechnology enables the development of precision medicine approaches tailored to individual patients' genetic profiles, biomarker signatures, and disease characteristics. Pharmacogenomics, companion diagnostics, and biomarker-guided therapies optimize treatment selection, dosing regimens, and therapeutic responses, improving patient outcomes and minimizing adverse effects.

1.3.1.4. Personalized Medicine

- **Pharmacogenomics**: Molecular biotechnology plays a critical role in pharmacogenomics; the study of how genetic variations influence drug response and metabolism. Genetic testing and molecular profiling identify genetic variants associated with drug efficacy, toxicity, and adverse reactions, guiding personalized drug selection, dosing, and monitoring.
- **Liquid Biopsies**: Molecular biotechnology enables the non-invasive detection and analysis of circulating tumor DNA (ctDNA), circulating tumor cells (CTCs), and other biomarkers in blood, urine, or other bodily fluids. Liquid biopsies offer a minimally invasive approach to cancer diagnosis, monitoring, and

treatment response assessment, providing real-time information on tumor dynamics and evolution.

- **Cellular Therapies**: Molecular biotechnology drives advances in cellular therapies, including adoptive cell transfer (ACT), chimeric antigen receptor (CAR) T-cell therapy, and stem cell transplantation. Genetically engineered cells, such as CAR T cells targeting specific antigens on cancer cells, offer personalized treatments for hematological malignancies and solid tumors, with promising results in clinical trials.

1.3.1.5. Disease Modeling and Drug Testing

- **Organ-on-a-Chip**: Molecular biotechnology enables the development of organ-on-a-chip and human-on-a-chip systems that replicate the structure and function of human organs and tissues *in vitro*. These microfluidic platforms serve as models for drug testing, disease modeling, and toxicity screening, providing insights into drug efficacy, pharmacokinetics, and adverse effects in a physiologically relevant context.
- **Patient-Derived Models**: Molecular biotechnology facilitates the generation of patient-derived cell models, including patient-derived xenografts (PDX), organoids, and induced pluripotent stem cells (iPSCs), for disease modeling and drug discovery. These models recapitulate patient-specific genetic and phenotypic characteristics, enabling personalized medicine approaches and preclinical testing of novel therapeutics.
- **Drug Repurposing**: Molecular biotechnology contributes to drug repurposing efforts by identifying new indications for existing drugs based on molecular mechanisms, target validation, and pathway analysis. Computational biology, data mining, and high-throughput screening platforms accelerate the discovery of repurposable drugs with therapeutic potential for different diseases, including cancer, neurodegenerative disorders, and infectious diseases.

1.3.2. Impact on Agriculture

In agriculture, molecular biotechnology has led to the development of genetically modified organisms (GMOs) with enhanced traits such as improved yield, resistance to pests and diseases, and nutritional quality.

1.3.2.1. Crop Improvement

- **Genetically Modified Organisms (GMOs)**: Molecular biotechnology facilitates the development of genetically modified crops with desirable traits, including herbicide tolerance, insect resistance, drought tolerance, disease resistance, and improved nutritional content. Genetic engineering techniques, such as gene insertion, gene silencing, and genome editing, enable the precise manipulation of crop genomes to introduce or enhance beneficial traits.
- **Trait Stacking**: Molecular biotechnology allows for the stacking of multiple traits in crop plants through genetic engineering, enabling the development of multifunctional crops with enhanced resilience to biotic and

abiotic stresses. Trait stacking combines traits such as insect resistance, herbicide tolerance, and drought tolerance to improve crop performance and yield stability under changing environmental conditions.

- **Nutritional Enhancement**: Molecular biotechnology is used to enhance the nutritional quality of crops by increasing the levels of essential vitamins, minerals, and micronutrients. Biofortified crops, such as Golden Rice (rich in vitamin A), high-iron beans, and zinc-enriched maize, address nutrient deficiencies and improve human health, particularly in regions where staple crops lack essential nutrients.

1.3.2.2. Pest and Disease Management

- **Resistance Breeding**: Molecular biotechnology accelerates the breeding of crop varieties with innate resistance to pests, diseases, and pathogens. Genetic markers linked to resistance genes enable marker-assisted selection (MAS) of resistant genotypes, facilitating the development of disease-resistant and pest-resistant crop varieties with improved yield stability and reduced reliance on chemical pesticides.
- **RNA Interference (RNAi)**: Molecular biotechnology harnesses RNAi technology to control pest populations by targeting specific genes essential for pest survival and reproduction. RNAi-based insecticides and biopesticides deliver double-stranded RNA molecules that silence target genes in pests, leading to mortality or reduced fitness without harming beneficial organisms or non-target species.
- **Pathogen Detection and Surveillance**: Molecular biotechnology enables the rapid and sensitive detection of plant pathogens, including viruses, bacteria, fungi, and nematodes. Molecular diagnostic techniques, such as PCR, loop-mediated isothermal amplification (LAMP), and high-throughput sequencing, are used for pathogen detection, surveillance, and monitoring in agricultural systems, facilitating early disease detection and management strategies.

1.3.2.3. Sustainable Agriculture

- **Precision Agriculture**: Molecular biotechnology contributes to precision agriculture approaches by providing molecular tools and technologies for site-specific management of crops, soils, and resources. Molecular markers, remote sensing, and data analytics optimize crop inputs, irrigation, fertilization, and pest control practices, reducing resource use, minimizing environmental impact, and increasing agricultural productivity.
- **Biological Control**: Molecular biotechnology supports biological control strategies for pest management by enhancing the efficacy and specificity of biological control agents, such as entomopathogenic fungi, parasitoid wasps, and microbial biopesticides. Genetic engineering and microbial biotechnology improve the performance and persistence

of biological control agents, enhancing their suitability for integrated pest management (IPM) programs.

- **Soil Health Management**: Molecular biotechnology contributes to soil health management practices by enhancing nutrient cycling, soil fertility, and microbial activity in agricultural soils. Molecular tools, such as DNA sequencing and metagenomic analysis, elucidate the composition and function of soil microbiomes, guiding soil conservation practices, organic amendments, and microbial inoculants to improve soil health and productivity.

1.3.2.4. Environmental Sustainability

- **Reduced Chemical Inputs**: Molecular biotechnology reduces reliance on chemical inputs in agriculture by developing crop varieties with built-in resistance to pests and diseases. Pest-resistant and disease-resistant crops require fewer insecticides, fungicides, and herbicides, leading to reduced chemical usage, lower environmental pollution, and improved safety for farmers and consumers.
- **Carbon Sequestration**: Molecular biotechnology contributes to carbon sequestration and climate change mitigation through the development of carbon-efficient crops and agronomic practices. Crop breeding and genetic engineering strategies optimize photosynthetic efficiency, carbon fixation, and biomass production, enhancing carbon sequestration in agricultural ecosystems and mitigating greenhouse gas emissions from agricultural activities.
- **Biodiversity Conservation**: Molecular biotechnology supports biodiversity conservation efforts by promoting sustainable farming practices, habitat restoration, and genetic diversity conservation in agricultural landscapes. Crop diversity, genetic resources conservation, and ecosystem-based approaches contribute to biodiversity conservation, ecosystem resilience, and ecological balance in agroecosystems.

1.3.2.5. Food Security and Resilience

- **Climate-Resilient Crops**: Molecular biotechnology enhances crop resilience to climate change by developing climate-smart crop varieties adapted to changing environmental conditions, including heat stress, drought, salinity, and waterlogging. Crop breeding, gene editing, and genetic engineering techniques improve crop tolerance to abiotic stresses, ensuring food security and livelihood resilience for farmers in climate-vulnerable regions.
- **Drought-Tolerant Crops**: Molecular biotechnology addresses water scarcity and drought stress in agriculture by developing drought-tolerant crop varieties with improved water-use efficiency and stress tolerance mechanisms. Genomic selection, marker-assisted breeding, and transgenic approaches enhance drought resilience in crops, enabling sustainable crop production in arid and semi-arid regions.
- **Crop Wild Relatives**: Molecular biotechnology harnesses genetic resources from crop wild relatives to develop climate-resilient and stress-tolerant

crop varieties with novel traits and genetic diversity. Genomic characterization, gene introgression, and pre-breeding efforts enhance the adaptive capacity of crops to environmental challenges, safeguarding food security and agricultural sustainability in a changing climate.

1.3.3. Environmental Applications

Molecular biotechnology offers a wide array of applications in environmental science and conservation, contributing to solutions for pollution remediation, renewable energy production, biodiversity conservation, and ecosystem restoration. By harnessing the power of biological systems and molecular tools, molecular biotechnology addresses environmental challenges while promoting sustainability and ecological resilience. Below are some key applications of molecular biotechnology in environmental science:

1.3.3.1. Bioremediation

- **Microbial Biotechnology**: Molecular biotechnology leverages the metabolic capabilities of microorganisms for bioremediation of contaminated environments. Engineered bacteria, fungi, and algae are used to degrade organic pollutants, detoxify hazardous chemicals, and remediate contaminated soils, sediments, and water bodies. Genetic engineering and synthetic biology approaches enhance microbial degradation pathways and tolerance to environmental stressors, improving the efficiency and effectiveness of bioremediation strategies.
- **Phytoremediation**: Molecular biotechnology enhances phytoremediation, a green technology that utilizes plants to remove, degrade, or immobilize contaminants from soil, water, and air. Transgenic plants with enhanced pollutant uptake, metabolism, and sequestration capabilities are engineered to remediate heavy metals, organic pollutants, and radioactive contaminants in polluted sites. Molecular markers and genetic engineering tools enable the selection and optimization of phytoremediation plant species for specific contaminants and environmental conditions.

1.3.3.2. Bioenergy Production

- **Biofuel Production**: Molecular biotechnology contributes to the development of biofuels derived from renewable biomass, including lignocellulosic feedstocks, algae, and microbial fermentation. Engineered microorganisms, such as bacteria, yeast, and algae, are optimized for efficient conversion of biomass into biofuels, including ethanol, biodiesel, methane, and hydrogen. Metabolic engineering, synthetic biology, and enzyme biotechnology enable the design and optimization of biofuel production pathways, improving yields, conversion efficiencies, and sustainability of bioenergy systems.
- **Biogas Generation**: Molecular biotechnology enhances anaerobic digestion and biogas production from organic wastes, agricultural residues, and wastewater streams. Microbial communities involved in biogas production are engineered

for enhanced methane production, substrate utilization, and process stability. Metagenomic analysis, microbial consortia engineering, and bioprocess optimization improve biogas yields, quality, and environmental performance, promoting the transition to a circular bioeconomy and reducing greenhouse gas emissions from organic waste streams.

1.3.3.3. Environmental Monitoring

- **Biosensors and Bioindicators**: Molecular biotechnology develops biosensors and bioindicators for environmental monitoring of pollutants, toxins, and contaminants in air, water, soil, and food. Genetically engineered organisms, such as bacteria, yeast, and plants, are designed to detect and respond to specific environmental cues, including chemical pollutants, heavy metals, and environmental stressors. Biosensor technologies enable real-time, on-site monitoring of environmental quality and health, facilitating early detection of pollution incidents and timely intervention measures.
- **DNA Barcoding and Metagenomics**: Molecular biotechnology employs DNA barcoding and metagenomics to characterize and monitor biodiversity, ecosystem health, and microbial communities in natural and anthropogenic environments. DNA barcodes enable species identification and tracking of invasive species, endangered species, and ecosystem indicators, informing conservation and management strategies. Metagenomic analysis reveals the genetic diversity, functional potential, and ecological roles of microbial communities in diverse habitats, providing insights into ecosystem dynamics and resilience.

1.3.3.4. Conservation Biology

- **Genetic Monitoring and Management**: Molecular biotechnology supports genetic monitoring and management of endangered species, populations at risk, and biodiversity hotspots. Genetic markers, such as microsatellites, single nucleotide polymorphisms (SNPs), and mitochondrial DNA sequences, are used to assess population structure, genetic diversity, and evolutionary history of species. Molecular tools inform conservation prioritization, breeding programs, and habitat restoration efforts, enhancing the resilience and adaptive capacity of vulnerable populations to environmental threats.
- **Assisted Reproduction Technologies**: Molecular biotechnology develops assisted reproduction technologies (ARTs) for conservation breeding and genetic rescue of endangered species. *In vitro* fertilization (IVF), embryo cryopreservation, and sperm banking preserve the genetic diversity and reproductive potential of threatened species in captivity. Genome editing and cloning techniques offer potential interventions for genetic rescue, disease management, and population augmentation of endangered species, addressing challenges of small population size and genetic bottleneck effects.

1.3.3.5. Ecosystem Restoration

- **Microbial Ecology**: Molecular biotechnology investigates microbial ecology and ecosystem processes in natural and restored habitats, including soil, water, wetlands, and mangroves. Microbial communities play critical roles in nutrient cycling, carbon sequestration, and ecosystem functioning, influencing soil fertility, plant growth, and ecosystem resilience. Molecular tools, such as DNA sequencing, stable isotope analysis, and metatranscriptomics, elucidate microbial interactions, community dynamics, and biogeochemical cycles, guiding ecosystem restoration and conservation strategies.
- **Ecological Engineering**: Molecular biotechnology integrates ecological engineering principles with biotechnological approaches for ecosystem restoration and habitat enhancement. Engineered plants, microbes, and microbial consortia promote ecological functions, such as nitrogen fixation, phytostabilization, and soil aggregation, in degraded or contaminated environments. Molecular tools enable the design and implementation of bio-based solutions for erosion control, revegetation, and ecological restoration, fostering ecosystem resilience and biodiversity conservation.

1.3.4. Industrial Applications

Molecular biotechnology has revolutionized industrial processes by providing innovative solutions for biomanufacturing, bioenergy production, bioremediation, and the synthesis of bio-based materials and chemicals. By harnessing the power of biological systems and molecular tools, molecular biotechnology enables sustainable and environmentally friendly approaches to industrial production. Below are some key applications of molecular biotechnology in the industrial sector:

1.3.4.1. Biomanufacturing

- **Industrial Enzymes**: Molecular biotechnology produces industrial enzymes for diverse applications in food processing, textile manufacturing, pulp and paper production, detergent formulation, and biofuel production. Engineered enzymes, such as amylases, proteases, cellulases, and lipases, catalyze biochemical reactions with high specificity and efficiency, enabling cost-effective and eco-friendly production processes.
- **Biopharmaceuticals**: Molecular biotechnology plays a critical role in the production of biopharmaceuticals, including therapeutic proteins, monoclonal antibodies, vaccines, and gene therapies. Recombinant DNA technology and mammalian cell culture systems produce complex biologics with precise post-translational modifications and therapeutic efficacy, addressing unmet medical needs and improving patient outcomes.
- **Bio-based Materials**: Molecular biotechnology develops bio-based materials, including biodegradable polymers, bioplastics, and biomaterials, for applications in packaging, biomedical devices, 3D printing, and construction.

Microbial fermentation, metabolic engineering, and biomimetic synthesis produce bio-based materials with tailored properties, such as biocompatibility, mechanical strength, and biodegradability, reducing reliance on fossil fuels and mitigating environmental impact.

1.3.4.2. Bioenergy Production

- **Biofuels**: Molecular biotechnology contributes to biofuel production from renewable biomass, including lignocellulosic feedstocks, algae, and agricultural residues. Engineered microorganisms, such as bacteria, yeast, and algae, metabolize biomass into biofuels, including ethanol, biodiesel, methane, and hydrogen, through fermentation, enzymatic hydrolysis, and metabolic engineering approaches. Biofuels offer sustainable alternatives to fossil fuels, reducing greenhouse gas emissions and dependence on finite resources.
- **Biogas Generation**: Molecular biotechnology enhances biogas production from organic wastes, agricultural residues, and wastewater streams through anaerobic digestion processes. Microbial consortia, genetically engineered bacteria, and pretreatment technologies optimize biogas yields, composition, and process stability, producing renewable biogas for electricity generation, heat production, and vehicle fuel, while mitigating organic waste disposal and environmental pollution.
- **Biohydrogen Production**: Molecular biotechnology develops biohydrogen production technologies using photosynthetic microorganisms, such as cyanobacteria and green algae, for renewable hydrogen generation. Genetic engineering, synthetic biology, and bioprocess optimization improve light harvesting, carbon fixation, and hydrogen production efficiency in microalgal and cyanobacterial systems, offering clean and sustainable energy sources for transportation, industry, and power generation.

1.3.4.3. Bioremediation

- **Microbial Bioremediation**: Molecular biotechnology employs microbial bioremediation strategies for environmental cleanup of contaminated sites, including soils, sediments, groundwater, and wastewater. Engineered bacteria, fungi, and algae degrade pollutants, such as hydrocarbons, heavy metals, pesticides, and industrial chemicals, through metabolic pathways and enzymatic reactions, restoring ecosystem health and remediation efficiency.
- **Phytoremediation**: Molecular biotechnology enhances phytoremediation, a green technology that uses plants to remove, detoxify, or sequester contaminants from soil, water, and air. Transgenic plants with enhanced pollutant uptake, metabolism, and tolerance traits facilitate phytoremediation of heavy metals, organic pollutants, and radioactive contaminants in polluted sites, offering cost-effective and sustainable solutions for environmental restoration.

1.3.4.4. Industrial Biocatalysis

- **Enzyme Biocatalysis**: Molecular biotechnology develops enzyme biocatalysis for sustainable production of fine chemicals, pharmaceuticals, flavors, fragrances, and specialty compounds. Enzymatic synthesis, chemoenzymatic cascades, and enzyme immobilization technologies enable selective and efficient conversion of substrates into value-added products, replacing traditional chemical processes with environmentally friendly and biocompatible alternatives.
- **Biocatalytic Cascades**: Molecular biotechnology designs biocatalytic cascades and metabolic pathways for multi-step synthesis of complex molecules from renewable feedstocks. Enzyme cascades, co-factor recycling systems, and compartmentalization strategies orchestrate sequential reactions and intermediate channeling, mimicking natural metabolic pathways and enabling efficient biotransformation of substrates into high-value chemicals and pharmaceuticals.

1.3.4.5. Sustainable Manufacturing

- **Green Chemistry**: Molecular biotechnology promotes green chemistry principles in industrial manufacturing processes, emphasizing the design, synthesis, and utilization of environmentally benign and sustainable chemicals. Biocatalysts, bio-based solvents, and bio-based intermediates replace hazardous reagents and toxic chemicals, minimizing waste generation, energy consumption, and environmental pollution in industrial production.
- **Circular Economy**: Molecular biotechnology contributes to the transition to a circular economy by valorizing waste streams and by-products from industrial processes into renewable resources and value-added products. Microbial fermentation, enzymatic hydrolysis, and metabolic engineering convert organic wastes into bio-based materials, biofuels, and bioproducts, closing the loop on resource utilization and promoting resource efficiency and waste reduction. (Russel & Sambrook, 2001; Campbell & Reece, 2002)

1.4. ETHICAL, LEGAL, AND SOCIAL IMPLICATIONS

The advancement of molecular biotechnology brings with it a host of ethical, social, and economic implications that warrant careful consideration and management. These implications cut across various domains, including healthcare, agriculture, environment, industry, and society at large. Below, we explore some of the key ethical, social, and economic implications associated with molecular biotechnology:

1.4.1. Ethical Implications

1.4.1.1. Genetic Privacy

Molecular biotechnology raises concerns about genetic privacy and the confidentiality of personal genetic information. The widespread use of genetic testing and sequencing technologies raises questions about who should have access to individuals' genetic data, how it should be stored and protected, and the potential implications for discrimination, stigmatization, and privacy breaches.

1.4.1.2. Informed Consent

Molecular biotechnology interventions, such as gene editing, stem cell therapy, and genetic testing, raise complex issues related to informed consent and autonomy. Ensuring that individuals fully understand the risks, benefits, and implications of molecular biotechnological procedures is essential for respecting their autonomy and right to make informed decisions about their health and genetic information.

1.4.1.3. Equitable Access

Molecular biotechnology has the potential to exacerbate existing disparities in access to healthcare, genetic testing, and innovative therapies. Addressing issues of equitable access requires consideration of socio-economic factors, cultural beliefs, geographical barriers, and healthcare infrastructure to ensure that the benefits of molecular biotechnology are accessible to all individuals, regardless of their background or circumstances.

1.4.1.4. Dual-Use Dilemma

Molecular biotechnology poses a dual-use dilemma, where biotechnological innovations intended for beneficial purposes could be misused for harmful or malicious ends. Concerns about biosecurity, bioterrorism, and the proliferation of biological weapons underscore the need for responsible oversight, regulation, and international cooperation to prevent the misuse of molecular biotechnology for destructive purposes.

1.4.2. Social Implications

1.4.2.1. Healthcare Disparities

Molecular biotechnology has the potential to exacerbate disparities in healthcare access and outcomes, particularly in low-income and marginalized communities. Addressing healthcare disparities requires equitable distribution of resources, access to affordable healthcare services, and culturally competent approaches to healthcare delivery to ensure that all individuals benefit from advances in molecular biotechnology.

1.4.2.2. Cultural and Ethical Values

Molecular biotechnology intersects with diverse cultural, religious, and ethical values, raising questions about the acceptability and appropriateness of biotechnological interventions in different cultural contexts. Recognizing and respecting cultural diversity, religious beliefs, and ethical perspectives is essential for fostering inclusive dialogue, consensus-building, and ethical decision-making in the development and application of molecular biotechnology.

1.4.2.3. Social Perception and Acceptance

Molecular biotechnology innovations may face public skepticism, fear, or opposition due to perceived risks, ethical concerns, and uncertainty about their long-term implications. Building public trust, fostering science literacy, and engaging stakeholders in transparent and inclusive dialogue are critical for promoting informed decision-making, public acceptance, and responsible governance of molecular biotechnology.

1.4.2.4. Human Enhancement and Equity

Molecular biotechnology raises ethical questions about human enhancement interventions, such as cognitive enhancement, genetic enhancement, and lifespan extension, and their potential impact on equity, social justice, and human dignity. Ensuring equitable access to enhancement technologies, addressing concerns about fairness and distributive justice, and safeguarding human rights and dignity are essential considerations in the ethical governance of human enhancement.

1.4.3. Economic Implications

1.4.3.1. Intellectual Property Rights

Molecular biotechnology innovations generate intellectual property rights (IPRs) in the form of patents, copyrights, and trade secrets, which incentivize innovation and investment in biotechnological research and development. Balancing the need for innovation incentives with concerns about accessibility, affordability, and fair use of biotechnological inventions requires careful consideration of patent policies, licensing agreements, and technology transfer mechanisms.

1.4.3.2. Market Monopolies

Molecular biotechnology can lead to market monopolies and concentration of economic power in the hands of a few companies or institutions, limiting competition, innovation, and consumer choice. Antitrust regulations, competition policies, and measures to promote open access to biotechnological tools and resources are essential for fostering a competitive and dynamic biotechnology industry that benefits society as a whole.

1.4.3.3. Healthcare Costs

Molecular biotechnology interventions, such as gene therapies, personalized medicine, and biopharmaceuticals, may contribute to rising healthcare costs due to the high development, manufacturing, and distribution expenses associated with biotechnological innovations. Balancing the need for innovation with concerns about affordability, accessibility, and sustainability of healthcare requires policy interventions, pricing strategies, and reimbursement mechanisms to ensure equitable access to life-saving biotechnological therapies.

1.4.3.4. Global Health Disparities

Molecular biotechnology has implications for global health disparities, where access to innovative biotechnological therapies and technologies is unevenly distributed between developed and developing countries. Addressing global health inequities requires international collaboration, technology transfer, capacity building, and access to essential medicines and vaccines to ensure that all individuals, regardless of their geographic location or socio-economic status, benefit from advances in molecular biotechnology. (Russel & Sambrook, 2001; Campbell & Reece, 2002).

ROLE MODEL

JAMES WATSON: AMERICAN GENETICIST

James Watson (born April 6, 1928, Chicago, Illinois, U.S.) is an American geneticist and biophysicist who played a crucial role in the discovery of the molecular structure of deoxyribonucleic acid (DNA), the substance that is the basis of heredity. For this accomplishment, he was awarded the 1962 Nobel Prize for Physiology or Medicine with Francis Crick and Maurice Wilkins.

Watson enrolled at the University of Chicago when only 15 and graduated in 1947. From his virus research at Indiana University (PhD in 1950), and from the experiments of Canadian-born American bacteriologist Oswald Avery, which proved that DNA affects hereditary traits, Watson became convinced that the gene could be understood only after something was known about nucleic acid molecules. He learned that scientists working in the Cavendish Laboratory at the University of Cambridge were using photographic patterns made by X-rays that had been shot through protein crystals to study the structure of protein molecules.

After working at the University of Copenhagen, where he first determined to investigate DNA, he did research at the Cavendish Laboratory (1951–53). There Watson learned X-ray diffraction techniques and worked with Crick on the problem of DNA structure. In 1952 he determined the structure of the protein coat surrounding the tobacco mosaic virus but made no dramatic progress with DNA. Suddenly, in the spring of 1953, Watson saw that the essential DNA components—four organic bases—must be linked in definite pairs. This discovery was the key factor that enabled Watson and Crick to formulate a molecular model for DNA—a double helix, which can be likened to a spiraling staircase or a twisting ladder. The DNA double helix consists of two intertwined sugar-phosphate chains, with the flat base pairs forming the steps between them. Watson and Crick's model also shows how the DNA molecule could duplicate itself. Thus, it became known how genes, and eventually chromosomes, duplicate themselves. Watson and Crick published their epochal discovery in two papers in the British journal Nature in April–May 1953. Their research answered one of the fundamental questions in genetics.

Watson subsequently taught at Harvard University (1955–76), where he served as a professor of biology (1961–76). He conducted research on the role of nucleic acids in the synthesis of proteins. In 1965 he published Molecular Biology of the Gene, one of the most extensively used modern biology texts. He later wrote The Double Helix (1968), an informal personal account of the DNA discovery and the roles of the people involved in it, which aroused some controversy.

In 1968 Watson assumed the leadership of the Laboratory of Quantitative Biology at Cold Spring Harbor, Long Island, New York, and made it a world center for research in molecular biology. He concentrated his efforts on cancer research. In 1981 his The DNA Story (written with John Tooze) was published. From 1988 to 1992 at the National Institutes of Health, Watson helped direct the Human Genome Project, a project to map and decipher all the genes in the human chromosomes, but he eventually resigned because of alleged conflicts of interest involving his investments in private biotechnology companies.

In early 2007 Watson's own genome was sequenced and made publicly available on the Internet. He was the second person in history to have a personal genome sequenced in its entirety. In October of the same year, he sparked controversy by making a public statement referring to the idea that the intelligence of Africans might not be the same as that of other peoples and that intellectual differences between geographically separated peoples might arise over time as a result of genetic divergence. Watson's remarks were immediately denounced as racist. Though he denied this charge, he resigned from his position at Cold Spring Harbor and formally announced his retirement less than two weeks later. In January 2019 the Cold Spring Harbor administration divested him of the honorary titles the laboratory had bestowed upon him. The decision came after the airing of the PBS documentary American Masters: Decoding Watson, in which Watson stated that his views on race and intelligence had not changed.

SUMMARY

- Molecular biotechnology features the use of laboratory techniques to study and modify nucleic acids and proteins. The practice of molecular biotechnology results from the convergence of many areas of research, such as molecular biology, microbiology, biochemistry, immunology, genetics and cell biology.
- Biotechnology is concerned with the production of commercial products generated by the metabolic action of microorganisms. More formally, biotechnology may be defined as "the application of scientific and engineering principles to the processing of material by biological agents to provide goods and services." The term "biotechnology" was first used in 1917 by a Hungarian engineer, Karl Ereky, to describe an integrated process for the large-scale production of pigs by using sugar beets as the source of food.
- Biotechnology research is dedicated to maximizing the overall efficiency of each of these steps and to finding microorganisms that make products that are useful in the preparation of foods, food supplements, and drugs. During the 1960s and 1970s, this research focused on upstream processing, bioreactor design, and downstream processing. These studies led to enhanced bioinstrumentation for monitoring and controlling the fermentation process and to efficient large-scale growth facilities that increased the yields of various products.
- Molecular biotechnology captured the imagination of the public. Many small companies dedicated to gene cloning (recombinant DNA technology) were established with funding from private investors. Although these biotechnology companies took somewhat longer than expected to bring their products to the marketplace, a large number of recombinant DNA-based products are currently available, and many more are expected soon.

REVIEW QUESTIONS

1. What is biotechnology?
2. What is recombinant DNA technology?
3. Describe the basic steps of a bioengineered biotechnology process.
4. What are some key applications of molecular biotechnology in medicine, agriculture, and industry?
5. How do ethical considerations influence the development and implementation of molecular biotechnology applications in these fields?

REFERENCES

1. Brondyk, W. H. (2009). Selecting an appropriate method for expressing a recombinant protein. In *Guide to Protein Purification* (2nd ed., Vol. 463, pp. 131–147). Academic Press.
2. Brooks, B. A., Foster, J., Sandwell, D., Wolfe, C. J., Okubo, P., Poland, M., & Myer, D. (2008). Magmatically triggered slow slip at Kilauea Volcano, Hawaii. *Science, 321*(5893), 1177. https://doi.org/10.1126/science.1161957.
3. Campbell, N. A., & Reece, J. B. (2002). *Biology* (6th ed.). Addison Wesley.
4. Hannig, G., & Makrides, S. (1998). Strategies for optimizing heterologous protein expression in *Escherichia coli. Trends in Biotechnology, 16*(2), 54–60. https://doi.org/10.1016/S0167-7799(97)01124-7.
5. Hughes, S. (2001). Making dollars out of DNA: The first major patent in biotechnology and the commercialization of molecular biology, 1974–1980. *Isis: An International Review Devoted to the History of Science and Its Cultural Influences, 92*(3), 541–575. https://doi.org/10.1086/385281.
6. Mendelsohn, M., Kough, J., Vaituzis, Z., & Matthews, K. (2003). Are Bt crops safe? *Nature Biotechnology, 21*(9), 1003–1009. https://doi.org/10.1038/nbt0903-1003.
7. Paine, J. A., Shipton, C. A., Chaggar, S., Howells, R. M., Kennedy, M. J., Vernon, G., Wright, S. Y., Hinchliffe, E., Adams, J. L., Silverstone, A. L., & Drake, R. (2005). Improving the nutritional value of Golden Rice through increased pro-vitamin A content. *Nature Biotechnology, 23*(4), 482–487. https://doi.org/10.1038/nbt1082.
8. Russell, D. W., & Sambrook, J. (2001). *Molecular Cloning: A Laboratory Manual*. Cold Spring Harbor Laboratory Press.
9. Vecsey, G. (2010). A manager who exceeded his own expectations. *The New York Times*, p. D2.
10. Von Fange, T., McDiarmid, T., MacKler, L., & Zolotor, A. (2008). Clinical inquiries: Can recombinant growth hormone effectively treat idiopathic short stature? *The Journal of Family Practice, 57*(9), 611, 612.
11. Yang, Y. C., & Clark, S. C. (1989). Interleukin-3: Molecular biology and biologic activities. *Hematology/Oncology Clinics of North America, 3*(3), 441–452. https://doi.org/10.1016/S0889-8588(18)30540-9.

CHAPTER

2 Molecular Biology

LEARNING OBJECTIVES

After studying this chapter, you will be able to:

- Understand the DNA structure and function;
- Comprehend the RNA structure and function;
- Familiarize yourself with the genetic code and the concept of a gene;
- Learn the process of translation of RNA into proteins.

KEY TERMS FROM THIS CHAPTER

Allele	Chromosome	Codon
CRISPR	DNA	Gene expression
Genome	Mutation	Plasmid
Protein	Replication	Restriction enzyme
RNA	Transcription	Translation

INTRODUCTION

Molecular biology is a field of biology that explores the intricate mechanisms underlying the structure, function, and interactions of biological molecules within cells. Molecular biology delves into the molecular processes that drive life, examining the fundamental building blocks of living organisms at the smallest scales. One of its central focuses is the study of nucleic acids, particularly DNA and RNA, which carry the genetic information responsible for an organism's development, function, and inheritance.

The discovery of the DNA double helix by James Watson and Francis Crick in 1953 marked a watershed moment in molecular biology, providing a foundational framework for understanding genetic information storage and transfer. DNA, deoxyribonucleic acid, is a long, double-stranded molecule composed of nucleotide units. These nucleotides consist of a sugar-phosphate backbone and nitrogenous bases—adenine (A), thymine (T), cytosine (C), and guanine (G)—that form complementary base pairs, maintaining the structural integrity of the DNA helix.

RNA, or ribonucleic acid, plays a crucial role in molecular biology by serving as a messenger that translates the genetic instructions encoded in DNA into functional proteins. Transcription, the process by which RNA is synthesized from a DNA template, is a key step in this information transfer. The resulting messenger RNA (mRNA) carries the genetic code from the nucleus to the ribosomes, where translation occurs. During translation, transfer RNA (tRNA) molecules read the mRNA code and facilitate the assembly of amino acids into proteins.

Molecular biology extends beyond nucleic acids to encompass the study of proteins and other macromolecules. Proteins, composed of amino acids, play diverse roles in cellular structure, function, and signaling. Understanding the intricate interactions between DNA, RNA, and proteins is essential for unraveling the complexities of cellular processes, including cell division, metabolism, and response to environmental stimuli.

2.1. DNA STRUCTURE AND FUNCTION

Deoxyribonucleic acid, or DNA, is an organic compound that contains information needed for protein synthesis as well as genetic instructions. The majority of cells in all living things contain DNA. Genetic heredity takes place there, and it is an essential aspect of reproduction. Transferring the inheritance from the older generation to the future generations is, therefore, the primary goal.

Additionally, viruses experience the same issue. The majority of these organisms have genetic material in the form of DNA or RNA. For instance, the genetic material of many viruses is DNA, but RNA may be present in the genetic material of other viruses. The Human Immunodeficiency Virus (HIV) is RNA-based, but once it binds to a host cell, it changes into DNA.

Furthermore, DNA is essential for the synthesis of proteins. The DNA found in the nucleus of every cell is known as nuclear DNA. Eukaryotic organisms are the name of the organisms that have these cells. The majority of the genomes of the organisms can be coded by it. The DNA of the plastid and mitochondria handles the remainder in the interim. In addition, the DNA found in the cell's mitochondria is known as mitochondrial DNA, and it is passed down from mothers to their offspring. If we take human subjects, we can find about 16,000 base points in their mitochondrial DNA. Similarly, plastids have the ability to contain DNA. They are also essential to photosynthesis. (Alberts et al., 2004)

2.1.1. Properties of DNA

DNA (deoxyribonucleic acid) possesses several key properties that are fundamental to its role as the carrier of genetic information. Here are some important properties of DNA:

- **Double Helix Structure:** DNA has a double-stranded helical structure, resembling a twisted ladder or spiral staircase. The two strands are held together by hydrogen bonds between complementary base pairs: adenine (A) pairs with thymine (T), and cytosine (C) pairs with guanine (G).
- Antiparallel Orientation:
  - The two strands of DNA run in opposite directions, meaning they are antiparallel. One strand runs from 5' to 3,' while the complementary strand runs from 3' to 5.' This antiparallel arrangement is crucial for DNA replication and other cellular processes.
- Base Pairing Specificity:
  - Adenine pairs specifically with thymine, forming two hydrogen bonds, while cytosine pairs specifically with guanine, forming three hydrogen bonds. This base-pairing specificity ensures the accurate transmission of genetic

information during DNA replication and transcription.

- Complementary Base Pairing:
  - The sequence of bases on one strand of DNA determines the sequence on the complementary strand. The principle of complementary base pairing allows for the faithful replication and transcription of genetic information.
- Genetic Code:
  - The sequence of nucleotide bases in DNA encodes genetic information. The genetic code specifies the order of amino acids in proteins, and variations in the DNA sequence contribute to genetic diversity.
- Semiconservative Replication:
  - During DNA replication, each parental DNA strand serves as a template for the synthesis of a new complementary strand. The resulting DNA molecule consists of one original (parental) strand and one newly synthesized (daughter) strand. This semiconservative replication ensures the preservation of genetic information.
- High Stability:
  - The hydrogen bonds between complementary base pairs, along with the helical structure and interactions with proteins, contribute to the high stability of the DNA molecule. This stability is essential for the long-term storage of genetic information.
- Compact Packaging:
  - DNA is organized into structures called chromosomes. DNA molecules are tightly wound around histone proteins, forming nucleosomes. This compact packaging helps fit the extensive DNA strands into the cell's nucleus.
- Mutability:
  - While DNA is generally stable, it can undergo mutations—changes in the sequence of nucleotide bases. Mutations can occur due to errors during replication, exposure to mutagenic agents, or other factors. Mutations contribute to genetic diversity and evolution.
- Replication and Repair Mechanisms:
  - DNA has sophisticated mechanisms for replication and repair. DNA replication ensures accurate transmission of genetic information during cell division, while repair mechanisms fix errors and damage to maintain DNA integrity.

2.1.2. Types of DNA

There are several types of DNA based on their structural and functional characteristics. Here are some of the main types of DNA:

- **Genomic DNA:** This is the most well-known type of DNA, and it is found in the cell nucleus of eukaryotic organisms. Genomic DNA contains the complete set of genetic instructions for an organism.

- **Mitochondrial DNA (mtDNA):** Found in the mitochondria of eukaryotic cells, mtDNA is separate from genomic DNA and is inherited maternally. It carries genetic information essential for mitochondrial function, which is crucial for energy production within the cell.
- **Plasmid DNA:** Plasmids are small, circular DNA molecules often found in bacteria. They can carry extra genes that may provide advantages, such as antibiotic resistance. Plasmids are also widely used in genetic engineering and biotechnology.
- **Chloroplast DNA (cpDNA):** Found in the chloroplasts of plant cells, cpDNA carries genetic information essential for photosynthesis. Like mitochondrial DNA, chloroplast DNA is inherited maternally.
- **Viral DNA:** Viruses can contain DNA as their genetic material. Depending on the virus, the DNA may be single-stranded or double-stranded and can be linear or circular.
- **cDNA (Complementary DNA):** cDNA is synthesized from messenger RNA (mRNA) through a process called reverse transcription. It represents the coding sequences of genes and is often used in molecular biology research and genetic engineering.
- **Satellite DNA:** This type of DNA consists of short repetitive sequences repeated in tandem. Satellite DNA is often found in the centromeric and telomeric regions of chromosomes and plays a role in chromosome structure and segregation.
- **Single-Copy DNA:** This refers to unique, non-repetitive DNA sequences that code for specific genes. The majority of genomic DNA is single-copy DNA.

2.1.3. DNA Structure

These bases are adenine, guanine, cytosine, and thymine in the case of DNA. Take note of how adenine and guanine, as well as cytosine and thymine, have similar shapes. C and T are categorized as pyrimidines, whereas A and G are purines. While we're naming things, take note of the terms "ribose" and "deoxyribose." Deoxyribose is simply a ribose without an oxygen atom, as the name suggests. More precisely, the 2-carbon of deoxyribose only has a hydrogen atom attached to it, whereas the 2-carbon of ribose has a hydroxyl group attached to it. There isn't any other distinction between the two sugars.

There are no specific guidelines for the arrangement of nucleotides in relation to their bases when creating a single strand of nucleic acid *in vitro*. Because phosphodiester bonds are formed between the phosphate group and the pentose, the identities of their nitrogenous bases are inconsequential. For this reason, it's frequently called the sugar-phosphate backbone. When we dissect the term "phosphodiester," we find that it accurately characterizes the relationship: two ester bonds (—O—) with phosphorous positioned between the sugars hold them together. The directionality of this bond and, by extension, of nucleic acids in general, is one of the concepts that frequently stumps students. For instance, we say that DNA polymerase functions in a 5-prime (5') to 3-prime (3') direction when discussing the enzyme that catalyzes the addition of nucleotides in living cells. Though it may sound like obscure molecular biology jargon, this is actually very straightforward. Examine two nucleotides that are connected by a phosphodiester bond once more (Figure 2.1, bottom left). A cytosine and an adenine nucleotide are joined. The 5-carbon of one deoxyribose (or ribose in RNA) will always be joined to the 3-carbon of the subsequent sugar by a phosphodiester bond. This also implies that a chain of linked nucleotides will have a free 3' hydroxyl (-OH) group at one end and a free 5' phosphate (-PO_4) group at the other. These specify an RNA or DNA strand's directionality.

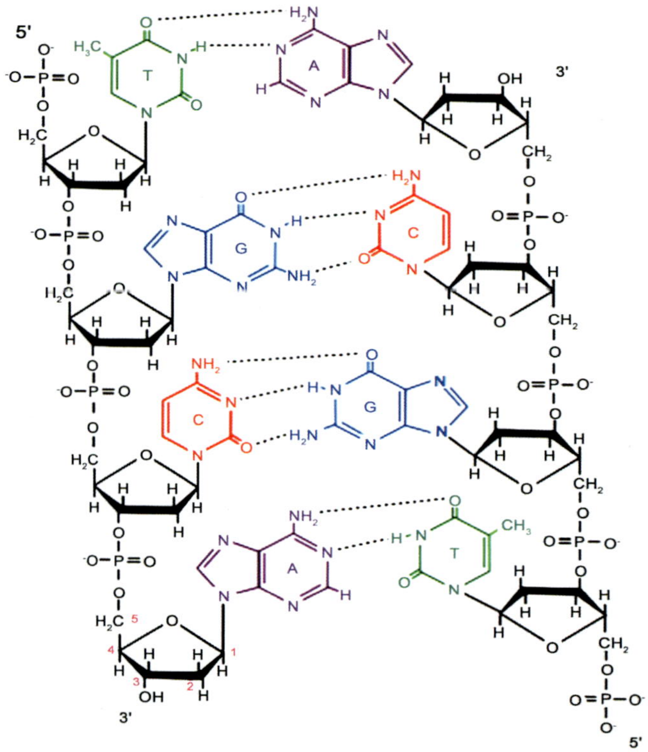

Figure 2.1. *DNA. Deoxyribonucleic acid is a polymer chain of nucleotides connected by 5' to 3' phosphodiester bonds. DNA normally exists as a two antiparallel complementary strands held together by hydrogen bonds between adenines (A) and thymines (T) and between guanines (G) and cytosines (C).*

Source: E. V. Wong, DNA Structure, bio.libretexts.org.

While RNA is primarily single-stranded, DNA is typically found in cells as double-stranded molecules. It is crucial to realize, however, that RNA and DNA can both become single-stranded or double-stranded under the right circumstances. Indeed, the molecules are so similar that one DNA strand and one RNA strand can be combined to form double-stranded hybrid molecules. Remarkably, compared to the more common DNA-DNA double helix, RNA-RNA and RNA-DNA double helices are marginally more stable. Base pairing is the fundamental mechanism underlying the double-stranded structure of DNA and, more broadly, the function of nucleic acids as a genetic information storage and transfer medium. Base-pairing is the process by which guanines, cytosines, adenines and thymines form hydrogen bonds. Compared to any association formed with the other potential bases, these pairs are noticeably more stable. Moreover, these base-pair associations exhibit uniform and highly stable spacing when they arise within the framework of two nucleic acid strands. As you may remember, hydrogen bonds are not very strong. But when it comes to DNA, hydrogen bonds are what give DNA its remarkable stability and suitability as a long-term genetic information storage medium. DNA double helices are at least thousands of nucleotides long, even in simple prokaryotes, which indicates that several thousand hydrogen bonds are holding the two strands together. Even though a tiny temperature increase or change in the ionic strength of the solution could easily temporarily disrupt any individual nucleotide-to-nucleotide hydrogen bonding interaction, a complete double-helix of DNA requires very high temperatures (usually over 90°C) to completely denature the double helix into individual strands.

It turns out that the two strands are essentially backup copies of each other because the nucleotides are paired exactly one-to-one, providing a safety net in case nucleotides are lost from one strand. As long as the other strand remains intact in the damaged area, the essential information can still be written into the complementary sequence of the other strand, even if parts of both strands are damaged. However, bear in mind that although one DNA strand can serve as a "backup" for the other, the two strands are complementary rather than identical. The two strands in this system of complementary and antiparallel strands can each carry unique information, which is an intriguing result. A DNA double helix has antiparallel strands. This indicates that one strand of a DNA double helix would be built in the 5' to 3' direction, while the complementary strand would be built in the 3' to 5' direction, if we were to view the molecule from left to right. We will talk about the role of these enzymes in creating and repairing DNA shortly. The left strand in Figure 2.1 is 5' to 3' from top to bottom, while the other strand is 5' to 3' from bottom to top.

DNA molecules are negatively charged (all those phosphates), and normally a double-helix with a right-handed twist. In this normal (also called the "B" conformation) state, one full twist of the molecule encompasses 11 base pairs, with 0.34 nm between each nucleotide base. Each of the nitrogenous bases are planar, and when paired with the complementary base, forms a at planar "rung" on the "ladder" of DNA. These are perpendicular to the longitudinal axis of the DNA. Most of the free-floating DNA in a cell, and most DNA in an aqueous solution of near-physiological osmolarity and pH, is found in this B conformation. However, other conformations have been found, usually under very specific environmental circumstances. A compressed conformation, A-DNA, was observed as an artifact of *in vitro* crystallization, with slightly more bases per turn, shorter turn length, and base-pairs that are not perpendicular to the longitudinal axis. Another, Z-DNA, appears to form transiently in GC-rich stretches of DNA in which, interestingly, the DNA twists in the opposite direction.

It has been proposed that these A and Z forms are both physiologically significant. The A form may be present in RNA-DNA hybrid double helices and when DNA is complexed with specific enzymes, according to available data. DNA methylation may cause the Z conformation to happen. Moreover, the "normal" conformation of B-DNA is essentially an idealized structure predicated on being completely hydrated, which is most likely the case inside a cell.

However, because of the minute fluctuations in that hydration state, the DNA conformation frequently deviates slightly from the B-conformation parameters shown in Figure 2.2.

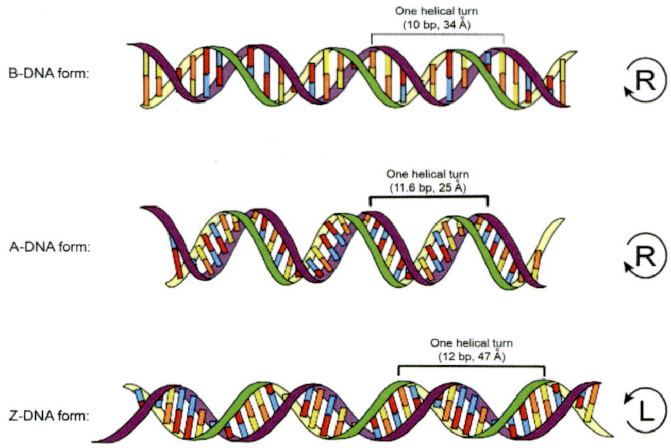

Figure 2.2. *Three conformations of DNA. B-DNA is most common, A-DNA is likely an artifact of crystallization in vitro, and Z-DNA may form transiently in parts of the chromosome.*

Source: E. V. Wong, DNA Structure, bio.libretexts.org.

Since prokaryotes have no other option, it is rather obvious that the DNA is found in the cytoplasm, whereas in eukaryotes, the DNA is found inside the nucleus. Proteins that aid in organizing and stabilizing the overall chromosome structure are packaged into both prokaryotic and eukaryotic DNA, despite the differences in their locations, degrees of protection from outside forces, and most importantly, sizes. Although some of the proteins found in prokaryotic and eukaryotic chromosomes share structural similarities, relatively little is known about the packing of prokaryotic chromosomes. Consequently, the majority of beginning cell biology courses continue to focus on eukaryotic chromosomal packing (Figure 2.3).

Prokaryotic or eukaryotic naked DNA is a very thin strand of material with a diameter of about 11 nm. However, if the DNA was kept in the nucleus in that manner, it would become uncontrollably tangled due to the size of eukaryotic genomes. Imagine a bucket that you have filled with 100 meters of yarn, and you haven't tried to coil or bunch it in any way. Now think about whether you could reach into that bucket and pull on a single strand, in which case you would probably pull up a single strand, or if you would probably pull up at least a small tangle of yarn. The cell does for the yarn basically what you would do for it is carefully wound into more manageable, smaller skeins. The nucleosome, the first unit of chromosomal organization in DNA, is formed by each chromosome being looped around a histone complex (Figure 2.4).

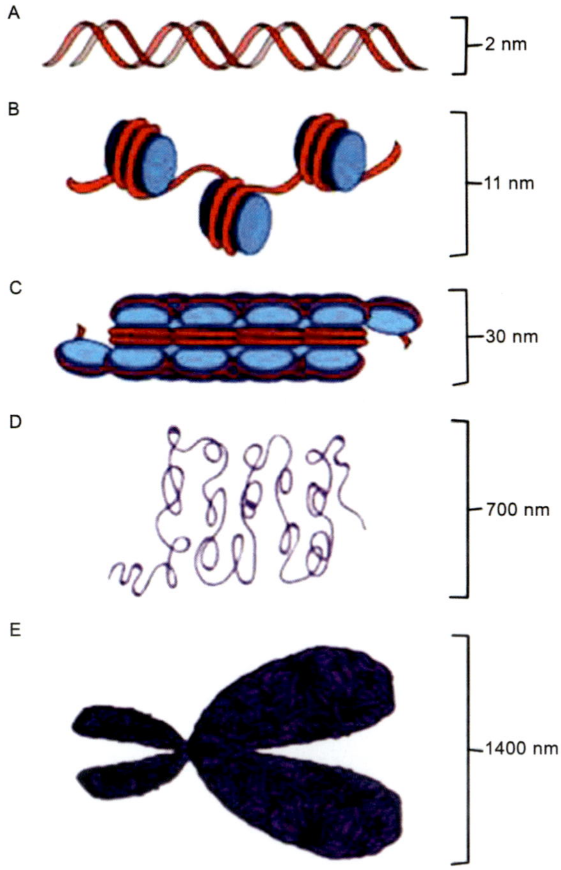

Figure 2.3. *DNA packaging.*

Note (A) A naked strand of DNA is approximately 2 nm in diameter. (B) Histones, which are octameric proteins depicted here as a roughly cylindrical protein, have positive charges distributed on the outer surface to interact with the negatively-charged DNA backbone. (C) Even the organization afforded by histone binding can leave an unmanageable tangle of DNA, especially with longer eukaryotic genomes, and therefore the histone-bound DNA is packaged into the "30-nm strand." This is held together, in part, by histone interactions. (D) The 30-nm fibers are looped into 700-nm fibers, which are themselves formed into the typical eukaryotic chromosome (E).

Source: E. V. Wong, DNA Structure, bio.libretexts.org.

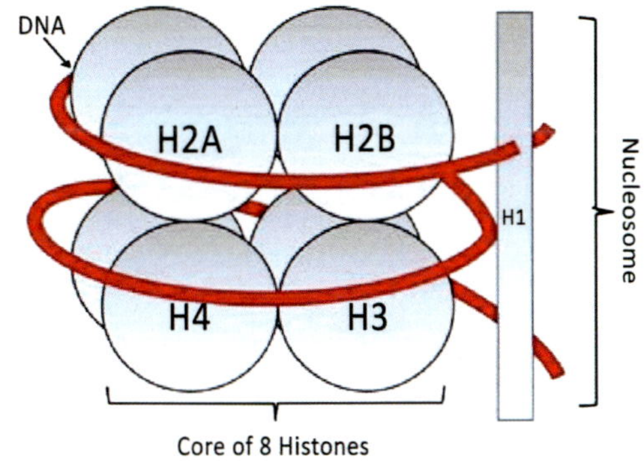

Figure 2.4. *The nucleosome is composed of slightly over two turns of DNA around a histone core containing two copies each of H2A, H2B, H3, and H4 histones. The H1 histone is not part of the core unit and functions in coordinating interaction between nucleosomes.*

Source: https://df0b18phdhzpx.cloudfront.net/ckeditor_assets/pictures/1272134/original_01.JPG.

A family of basic, positively-charged proteins is known as the histones. Their primary role is to organize DNA. The nucleosome is created when DNA wraps around a core of eight histones, two of each of H2A, H2B, H3, and H4. This process occurs a little more than twice. Since DNA is extremely negatively charged, as was previously mentioned, the quantity and orientation of the positive charges—mostly from lysines and arginines—are essential to their capacity to firmly bind DNA. It's not just a dating adage from the advice columns that "opposites attract."

2.1.4. Functions of DNA

The functions of DNA are central to the growth, development, functioning, and reproduction of cells and organisms. Here are the key functions of DNA:

- Genetic Storage:
  - **Function:** DNA serves as the

repository of genetic information. It contains the instructions for building and maintaining an organism. The genetic code is encoded in the sequence of nucleotide bases (adenine, thymine, cytosine, and guanine) along the DNA strands.

- Transmission of Genetic Information:
  - **Function:** DNA is passed from one generation to the next during reproduction. This ensures the continuity of genetic traits and characteristics from parents to offspring.
- Protein Synthesis:
  - **Function:** DNA provides the instructions for the synthesis of proteins, which are essential for the structure, function, and regulation of cells. The process involves the transcription of DNA into mRNA (messenger RNA) and translation of mRNA into proteins.
- Cellular Replication:
  - **Function:** DNA is duplicated during cell division (replication) to ensure that each daughter cell receives a complete set of genetic information. This process is crucial for growth, development, and tissue repair.
- Gene Expression Regulation:
  - **Function:** DNA plays a role in the regulation of gene expression. Various mechanisms, such as promoter regions and regulatory proteins, control when and to what extent specific genes are transcribed and translated into proteins.
- Epigenetic Information:
  - **Function:** DNA carries epigenetic information, including modifications such as DNA methylation and histone acetylation, which can influence gene expression without altering the underlying DNA sequence. Epigenetic modifications are crucial for cellular differentiation and the adaptation of cells to environmental factors.
- Chromosome Structure and Organization:
  - **Function:** DNA is organized into structures called chromosomes. Chromosomes provide a stable and organized way to package and segregate genetic material during cell division.
- Response to Environmental Stimuli:
  - **Function:** DNA can be influenced by environmental factors, and certain genes may be turned on or off in response to external stimuli. This adaptability allows organisms to respond to changes in their surroundings.
- Repair and Maintenance:
  - **Function:** DNA repair mechanisms exist to correct errors that may occur during DNA replication or damage caused by environmental factors (e.g., radiation, chemicals). Maintenance of DNA integrity is crucial for the prevention of mutations and the proper functioning of cells.

- Storage of Evolutionary Information:
  - **Function:** DNA serves as a record of evolutionary history. Changes in the DNA sequence over time contribute to the diversity of species and the adaptation of organisms to their environments.

2.1.5. Chemical Modification

Once a DNA molecule is put together, it can undergo chemical modification. This can happen inadvertently through oxidation, ionizing radiation, or the action of chemical carcinogens, or it can happen intentionally by specialized enzymes known as DNA methyltransferases. Enzymes known as nucleases are also capable of cleaving and breaking down DNA.

2.1.5.1. Methylation

There are three known forms of spontaneous methylation in DNA. N4-methylcytosine is created when cytosine is modified on the exocyclic amino group, or on the ring to create 5-methylcytosine. N^6-methyladenine can be created by modifying adenine. While 5-methylcytosine is widely distributed, N^4-methylcytosine and N^6-methyladenine are exclusively present in bacteria and archaea. This methylation is caused by specialized enzymes known as DNA methyltransferases, which identify particular DNA molecule sequences to modify only a portion of the bases. Sometimes carcinogens induce other methylations of the bases or of the deoxyribose. These must be eliminated if they are not to become mutagenic because they typically cause mispairing of the bases during replication.

There are numerous biological uses for natural methylation. Methylation plays a crucial role in the immune system of bacteria and archaea by shielding DNA molecules from restriction endonuclease fragmentation. Methylation aids in the removal of incorrect base sequences added during DNA replication in some organisms. A cellular mechanism called the mismatch repair system differentiates between the correctly sequenced portion on the template strand and the newly replicated strand where the errors occur by attaching a methyl group to the parental strand. By blocking DNA transcription, 5-methylcytosine regulates a variety of cellular processes in higher eukaryotes. Methylation is also thought to indicate imprinting, a process that results in the selective inactivation of certain genes inherited from one parent. Important genes that regulate embryonic development may also be repressed or activated by proper methylation. However, 5-methylcytosine may be mutagenic because the methylation process produces thymine, which changes C:G pairs into T:A pairs. Mammals exhibit selective methylation within the dinucleotide sequence CG, which is a rare sequence that has likely been lost through mutation. CG dinucleotide mutations are present in several malignancies in important genes.

2.1.5.2. Nucleases

The phosphodiester backbone of DNA is hydrolytically cleaved by enzymes called nucleases. Endonucleases cleave in the middle of chains, whereas exonucleases degrade from the end of the chain selectively. There are known nucleases that can function on single- or double-stranded DNA.

Restriction endonucleases are a special class that recognize and cleave specific sequences in DNA. Type II restriction endonucleases always cleave at or near their recognition sites. They produce small, well-defined fragments of DNA that help to characterize genes and genomes and that produce recombinant DNAs. Fragments of DNA produced by restriction endonucleases can be moved from one organism to another. In this way, it has been possible to express proteins such as human insulin in bacteria.

2.1.5.3. Mutation

Genetic material mutations can result from chemical modification of DNA. Anions that deaminate cytosine to form uracil, like bisulfite, can alter the genetic message by inducing C-to-T transitions. Purine residues are lost when exposed to acid, but cells have certain enzymes that can heal these wounds. While oxidative damage from free radicals or strong oxidizing agents can cause a variety of mutagenic lesions if not repaired, exposure to UV light can cause adjacent pyrimidines to dimerize. Direct reactions between uracil, adenine, and guanine and halogens like chlorine and bromine result in substituted bases, which are frequently mutagenic. In a similar way, primary amine groups interact with nitrous acid.

2.1.5.4. Supercoiling

The topologies of circular DNA molecules, like those in bacterial chromosomes and plasmids, can vary greatly. One is active supercoiling, in which a single DNA strand is cleaved, wound around the complementary strand one or more times, and then the molecule is sealed back up. The DNA can undergo one supercoiled turn for every full rotation, a process that can continue until the DNA is completely wound and collapses into a tight ball. It is also possible to reverse. Supercoiled DNA is wound and relaxed by specialized enzymes known as gyrases and topoisomerases. The DNA in eukaryotic linear chromosomes is typically tightly bound by proteins at different locations, which permits the supercoiling of the intervening stretches. The extreme compaction of DNA required to fit it inside the cell's walls is partly caused by this characteristic. Although the DNA in a single human cell can reach a length of two to three meters, it is packed extremely densely to fit inside the 10 micrometer-diameter nucleus of a human cell.

2.1.5.5. Sequence Determination

Frederick Sanger and Walter Gilbert invented techniques for figuring out DNA base sequences in the 1970s; for their contributions, they were awarded the 1980 Nobel Prize. The Sanger method is predicated on the *in vitro* enzymatic synthesis of DNA, whereas the Gilbert-Maxam method depends on the distinct chemical reactivities of the bases. Both techniques calculate the separation between a fixed DNA point and each instance of base—A, C, G, or T. Gel electrophoresis is used to divide DNA fragments from a sequence of reactions into four "lanes" based on their length. A distinct base is represented by each lane, and the sequence is read straight from the gel. Fluorescent dyes have made the Sanger method automated, allowing a single machine to generate tens of thousands of DNA base sequences in a single run.

2.1.6. DNA Translation

Transcription uses the genetic information found in DNA as a starting point to produce messenger RNA (mRNA). Afterward, single-stranded mRNA serves as a template for translation. By encouraging the binding of complementary transfer RNA (tRNA) anticodon sequences to the mRNA, ribosomes aid in the process of translation in the cytoplasm.

tRNAs carry particular amino acids, which are linked together by the ribosome. In this process, the mRNA is decoded to produce a specific amino acid chain, known as a polypeptide. Folding of the polypeptide creates an active protein, able to perform functions within the cell.

2.1.6.1. Cellular Components Involved in DNA Translation

The key components required for translation are mRNA, tRNA, ribosomes, and aminoacyl tRNA synthetases. These four structures are briefly explained below:

Ribosome

The complex organelle known as the ribosome is found in the cytoplasm and acts as the site of action for protein synthesis. It offers the enzymes required to form peptide bonds. Triplets, or codons, are recognized as the nucleotide sequence of mRNA. The amino acid is added to the chain when the ribosome forms a complementary codon sequence with an amino acid-bearing tRNA and bonds with the mRNA while moving along the single-strand mRNA.

A stop codon, which is a sequence of three nucleotides on the mRNA, signifies that translation is finished. When the ribosome reaches the stop codon, translation stops, releasing the newly synthesized polypeptide and mRNA.

Messenger RNA (mRNA)

Information from DNA is transferred to the ribosome via mRNA. Transcription produces this single-stranded molecule, which is complementary to the DNA template. Codons are the building blocks of mRNA strands; each one designates an amino acid that must be added to the polypeptide in a specific order. tRNA, which provides the amino acid once bound with a complementary mRNA codon, and ribosomal RNA (rRNA), the main part of the ribosomal machinery that recognizes the start and stop codons of mRNA, must interact with mRNA.

Transfer RNA (tRNA)

This RNA strand is a single strand made up of about 80 ribonucleotides. Every tRNA is read as an anticodon, which is a ribonucleotide triplet complementary to an mRNA codon. A specific amino acid carried by tRNA is added to the expanding polypeptide chain upon the formation of a complementary codon bond.

Aminoacyl tRNA Synthetases

These are enzymes that link each amino acid to their corresponding tRNA with the help of a two-step process. Each amino acid has a unique synthetase and the active site of each enzyme fits only one specific combination of the amino acid and tRNA.

2.1.6.2. Steps Involved in DNA Translation

There are three major steps in translation: initiation, elongation, and termination. These steps are briefly discussed below:

- **Initiation:** mRNA is bound by small ribosomal subunits. The start codon (AUG) of the mRNA is bound by the initiator tRNA, which is also fitted with the anticodon (UAC). The resultant massive complex starts the synthesis of proteins and forms a complete ribosome.
  - **Elongation:** A fresh tRNA-amino acid complex enters the codon adjacent to the AUG codon after initiation. Base pairing happens when the

new tRNA's anticodon aligns with the mRNA codon, forming a peptide bond that the ribosome uses to bind the two amino acids together.

Base pairing is impossible and the tRNA is rejected if the anticodon and codon do not match. The ribosome then advances one codon, creating room for the entrance of a fresh tRNA-amino acid complex. Up until the polypeptide is translated in its entirety, this process is repeated numerous times.

– **Termination:** One of the three stop codons for which there is no tRNA corresponding is encountered by the ribosome as it travels along the mRNA. The stop codon's terminator proteins attach to the ribosome and cause the freshly made polypeptide chain to be released. After that, the ribosome separates from the mRNA. The ribosome's small and large subunits split apart upon release from the mRNA and get ready for the subsequent translation cycle. Before becoming a fully functional protein, the polypeptide chains created during translation go through a few post-translational changes, such as folding.

2.2. RNA STRUCTURE AND FUNCTION

RNA is a complex, high-molecular-weight substance that is involved in the synthesis of proteins in cells and, in certain viruses, takes the place of DNA (deoxyribonucleic acid) as the carrier of genetic codes. Ribose nucleotides, which are nitrogenous bases joined to ribose sugars and joined by phosphodiester bonds to form strands of different lengths, are the building blocks of RNA. Adenine, guanine, cysteine, and uracil are the nitrogenous bases found in RNA; uracil takes the place of thymine in DNA.

Five carbons and one oxygen make up the cyclical structure of RNA's glucose sugar. RNA is susceptible to hydrolysis due to the presence of a chemically reactive hydroxyl (–OH) group linked to the second carbon group in the ribose sugar molecule. Since DNA lacks a reactive –OH group in the same location on the sugar moiety (deoxyribose), it is hypothesized that one of the reasons DNA evolved to be the preferred carrier of genetic information in the majority of organisms is because of its chemical lability. R provided a description of the RNA molecule's structure. W. Holley back in 1965.

2.2.1. RNA Structure

Phosphodiester bonds bind the mononucleotides that make up RNA, which is normally single-stranded. In the RNA chain, a ribonucleotide consists of a phosphate group, one of the four nitrogenous bases (A, U, G, and C), and ribose, or pentose sugar. While RNA is more suited for its shorter-term functions due to its relative instability, DNA is more stable due to the subtle structural differences between the sugars and is therefore better suited for storing genetic information. In place of the thymine used in DNA, RNA-specific pyrimidine uracil forms a complementary base pair with adenine. Notwithstanding the single-stranded nature, the majority of RNA molecules exhibit extensive intramolecular base pairing between complementary sequences, which forms a predictable three-dimensional structure that is vital to the molecules' functionality (Figures 2.5 and 2.6).

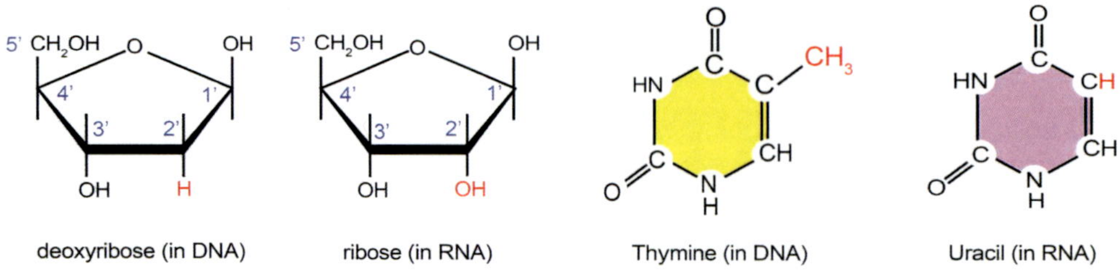

Figure 2.5. *(a) Ribonucleotides contain the pentose sugar ribose instead of the deoxyribose found in deoxyribonucleotides. (b) RNA contains the pyrimidine uracil in place of thymine found in DNA.*

Source: openstax.org/books/microbiology/pages/10-3-structure-and-function-of-rna.

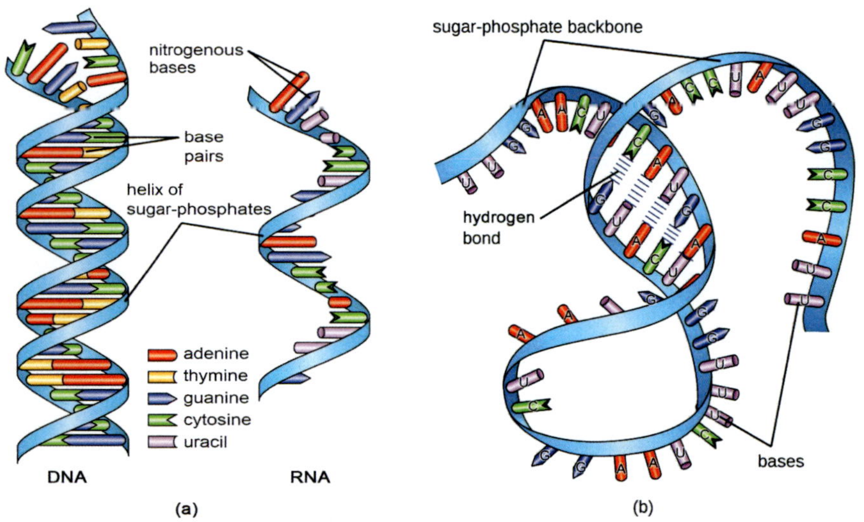

Figure 2.6. *(a) DNA is typically double-stranded, whereas RNA is typically single-stranded, (b) Although it is single-stranded, RNA can fold upon itself, with the folds stabilized by short areas of complementary base pairing within the molecule, forming a three-dimensional structure.*

Source: openstax.org/books/microbiology/pages/10-3-structure-and-function-of-rna.

2.2.2. Types of RNA

Of many types of RNA, the three well-known and most commonly discussed and found in almost all organisms. These three types of RNA are:

- mRNA (messenger RNA);
- rRNA (ribosomal RNA);
- tRNA (transfer RNA).

2.2.2.1. mRNA (Messenger RNA)

- It is a single-stranded RNA molecule that is complementary to one of the strands of DNA.
- mRNA is the version of the genetic materials that leave the nucleus and move to the cytoplasm where responsible proteins are synthesized.
- This RNA has utmost importance during protein synthesis, when the ribosome moves along this mRNA, it reads the base sequences and uses the genetic code to translate them into specific proteins.
- These codes are in the form of triplet sequences of nitrogenous bases and are often referred to as codons.
- As we now know that mRNA is responsible for transferring the genetic information into ribosomes where by reading the base sequences on mRNA, the translation of proteins is made possible, thus the name resembles its functions i.e., messenger RNA.
- We can even say mRNA is the molecule that uses genetic code present on a portion of DNA and makes proteins. If mRNA didn't exist then the information on DNA could never be used by our body (Figure 2.7).
- It is a single-stranded RNA molecule found in cells that forms part of the protein-synthesizing organelle, the ribosome (Figure 2.8).
  - It is synthesized inside the nucleus, particularly in the nucleolus where rRNA coding genes are present. The synthesized rRNA can be of varying sizes, commonly distinguished as small and large.

Figure 2.7. *mRNA (messenger RNA).*

Source: Created with BioRender.com.

2.2.2.2. rRNA (ribosomal RNA)

Figure 2.8. *rRNA (ribosomal RNA).*

Source: Created with BioRender.com.

- These newly synthesized rRNAs combine with ribosomal proteins and form smaller subunits and larger subunits of ribosomes respectively.
- These rRNAs are vital in recognizing conserved regions of incoming mRNAs and tRNA thus facilitating their binding and carrying out protein synthesis. Additionally,

rRNA also has enzymatic activity (peptidyl transferase) and catalyzes the formation of the peptide bond between two aligned proteins/amino acids during protein synthesis.

2.2.2.3. tRNA (Transfer RNA)

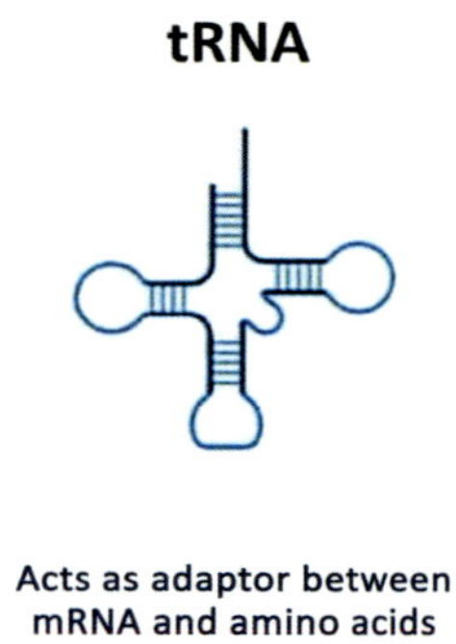

Figure 2.9. *tRNA (transfer RNA).*

Source: Created with BioRender.com.

- It is a type of RNA molecule that helps to decode information present in mRNA sequences into specific proteins (Figure 2.9).
- It is encoded by DNA in the cell nucleus and transcribed with the help of RNA polymerase ⊠⊠⊠.
- The structure of tRNA folds upon itself and creates an intra complementary base pairing which gives rise to hydrogen-bonded stems and associated loops that contain nucleotides with modified bases.
- The structure in two-dimensional resembles a cloverleaf having three loops and an open end. are usually 75–90 ribonucleotides in length.
- Each of these loops consisting of arms has a distinct name and function. The three-loop consisting arms are namely: DHU or D arm, which has a recognition site for specific enzyme amino-acyl tRNA synthetase; the T arm that consists of the ribosome recognition site and the anticodon arm that recognizes and binds to mRNA present in the ribosome.
- The open end with no loop is the site for attachment of amino acid, via 3' OH bonding with COOH-group of the amino acid.
- In general, tRNA reads the code on the mRNA sequence in the ribosome and translates specific amino acids, it does so along the length of the mRNA and gives out a polypeptide chain of amino acids (proteins) in association with other important enzymes like aminoacyl tRNA synthetase and peptidyl transferase.

2.2.2.4. Some other Types of RNA

- Ribozymes
  - These types of RNAs refer to those RNAs that are capable of showing enzymatic activities. They were first discovered in introns of precursor ribosomal RNA of *Tetrahymena thermophilus*, where it was found that these noncoding sequences were capable of excising themselves without any protein or external source.
  - These RNAs play vital roles in major reactions like RNA splicing, viral replication, and tRNA biosynthesis.
  - rRNAs also show enzymatic activities and thus can be termed as Ribozymes.

- Antisense RNAs
  - Antisense RNAs are those RNAs that contain sequences that are complementary to protein-coding sequences of mRNA.
  - These are single-stranded like mRNA but cannot code for proteins.
  - However, they can interfere and inactivate their complementary mRNA sequences thus inhibiting protein synthesis.
  - This ability of antisense RNAs has led researchers to create artificial antisense RNAs that can inhibit the protein synthesis of potential disease-causing organisms or of infected cells which then end up killing unwanted cells.

2.2.3. Functions of RNA

There are different types of RNA, each with specific functions. Here are some of the key functions of RNA:

- mRNA carries genetic information from the DNA in the cell nucleus to the ribosomes in the cytoplasm, where protein synthesis occurs. It serves as a template for protein production during the process of translation.
- rRNA is a structural component of ribosomes, which are cellular organelles responsible for protein synthesis. It helps in the binding of mRNA and tRNA (transfer RNA) during translation.
- tRNA plays a crucial role in protein synthesis by bringing amino acids to the ribosomes based on the information encoded in mRNA. Each tRNA molecule is specific to a particular amino acid.
- snRNA is involved in the splicing of pre-mRNA during the process of RNA maturation. It forms complexes with proteins to create small nuclear ribonucleoproteins (snRNPs) that help remove introns from the pre-mRNA.
- snoRNA is found in the nucleolus and is involved in the processing and modification of rRNA. It guides the chemical modifications, such as methylation and pseudouridylation, that occur in rRNA molecules.
- miRNAs are small RNA molecules that regulate gene expression by binding to target mRNA molecules. This binding can result in mRNA degradation or inhibition of translation, thereby controlling the production of specific proteins.
- While the exact functions of many lncRNAs are still being discovered, some have been implicated in the regulation of gene expression, chromatin organization, and other cellular processes. They may act as scaffolds for protein complexes or regulators of transcription.
- siRNAs are involved in the regulation of gene expression by inducing the degradation of specific mRNA molecules. They are often used in RNA interference (RNAi) experiments for gene silencing.

2.3. THE GENETIC CODE AND THE CONCEPT OF A GENE

DNA is the genetic material found in cells. Through DNA replication, the information contained in a cell's genetic material can be copied and passed on to offspring. A significant amount of DNA data is translated into mRNA, which is subsequently converted into proteins. However, some RNAs also serve vital purposes even though they are never translated into proteins. In everyday speech, expressions like "it's in my genes" or "in my DNA" denote that something is fundamental to a person's identity.

During the early 1900s, the fundamental unit of heredity was named a gene. It was once believed that genes were discrete loci arranged in a line on chromosomes. This theory was supported by breeding studies using the fruitfly Drosophila, which demonstrated that genes that are close to one another on a chromosome have a higher probability of being inherited together. The "one gene, one polypeptide" theory originated from the observation that gene mutations could result in changed phenotypes. Upon realizing that genes were composed of DNA, the term "central dogma of molecular biology" was established. This explains a two-step process whereby the DNA's genes are translated into RNA and then into the amino acid sequence that forms proteins. From DNA to RNA to protein is the information flow (Figure 2.10).

Figure 2.10. *The flow of genetic information.*

Source: Steve Minchin and Julia Lodge: Understanding biochemistry: structure and function of nucleic acids: doi: 10.1042/EBC20180038.

However, there are exceptions to this, firstly some viruses have RNA genomes and in some cases, these are reverse-transcribed into DNA before the genes can be expressed. The retrovirus HIV is an example of this. The other exception is that not all functional RNAs are translated into proteins.

2.3.1. The Genetic Code

The set of instructions that living cells use to convert genetic material information into proteins is known as the genetic code. Because nucleic acids are relatively simple molecules, many scientists were skeptical that they carried genetic information when DNA and RNA were first discovered. The question was how DNA, which only contains four different types of bases, could code for 20 different amino acids. DNA could only encode four amino acids if there was a 1:1 correlation between bases and amino acids. There would still be an insufficient number of combinations (16) if bases were paired. However, there are 64 possibilities if you consider a triplet code, which is more than sufficient. This is the code that is known to us, in which each codon—a string of three nucleotides—designates a specific amino acid. Even so, the triplet code did not make sense because there are now far more codons than required.

The genetic code as it is currently understood is shown in Table 2.1. Because RNA is what cells translate into amino acids, it is written as RNA with a U rather than a T. Because multiple codons frequently code for the same amino acid, the code is referred to as redundant or degenerate. The codon's third nucleotide, sometimes known as the degenerate position, is typically the one that differs.

Table 2.1. *The Genetic Code*

UUU	phe	UCE	Ser	UAU	Tyr	UGU	Cys
UUC		UCC		UAC		UGU	
UUA	Leu	UCA		UAA	STOP	UGA	STOP
UUG		UCG		UAG	STOP	UGG	Trp
CUU	Leu	CCU	Pro	CAU	His	CGU	Arg
CUC		CCC		CAC		CGC	
CUA		CCA		CAA	Gln	CGA	
CUG		CCG		CAG		CGG	
AUU	ILe	ACU	Thr	AAU	Asn	AGU	Ser
AUC		ACC		AAC		AGC	
AUA		ACA		AAA	Lys	AGA	Arg
AUG	Met	ACG		AAG		AGG	
GUU	Val	GCU	Ala	GAU	Asp	GGU	Gly
GUC		GCC		GAC		GGC	
GUA		GCA		GAA	Glu	GGA	
GUG		GCG		GAG		GGG	

Source: Steve Minchin and Julia Lodge: Understanding biochemistry: structure and function of nucleic acids: doi: 10.1042/EBC20180038.

2.3.2. Evidence for the Triplet Code

Long before we knew the sequence of DNA, scientists conducted the experiments that led to the decipherment of the genetic code. At that point, it was feasible to ascertain the relative amounts of every amino acid in a protein, but not the sequence in which they occurred. Sydney Brenner and Francis Crick used bacteriophage virus mutants in an experiment to provide important answers to some important questions. The normal or wild-type phage will infect *E. coli* and grow. Crick and Brenner investigated mutants that would not grow on some strains of *E. coli*.

Frameshifts are caused by mutants, which are insertions or deletions. A single adenine base inserted into the DNA sequence modifies not only the amino acid at the insertion site but also all other amino acids translated from that sequence (See Figures 2.11A and B). This causes a one-base shift in the reading frame and produces a non-functional protein. Nevertheless, a wild-type or almost wild-type phenotype is frequently obtained by inserting three nucleotides. This is because a complete triplet codon has been inserted; as a result, one or two amino acids that were not in the original sequence will now be present, but the reading frame will remain unchanged (Figure 2.11C), and the remainder of the sequence will read normally.

```
(A) atgtacaatattatatgcctgtccaaaaac
    MetTyrAsnIleIleCysLeuSerLysAsn

(B) atgtacagatattatatgcctgtccaaaaac
    MetTyrArgTyrTyrMetProValGlnLys

(C) atgtacagacatattatatgcctgtccaaaaac
    MetTyrArgHisIleIleCysLeuSerLysAsn
```

Figure 2.11. *DNA sequence showing the amino acid translation underneath.*

Source: Minchin S, Lodge J. (2019). Understanding biochemistry: structure and function of nucleic acids. Essays Biochem., 63(4), 433-456. doi: 10.1042/EBC20180038. PMID: 31652314; PMCID: PMC6822018.

Crick and Brenner searched for mutations known as suppressors, which could save the mutant and enable normal growth. They demonstrated that their suppressor mutants frequently added or subtracted one or more bases, rather than just reversing the original mutation. They discovered that a mutant phenotype results from the insertion or deletion of one, two, or four nucleotides. Nevertheless, there is little to no impact if three nucleotides are added or removed. This provided compelling evidence in favor of a triplet code. This further supports the idea of redundant coding, in which an amino acid can have multiple codes. There would be 20 codons that code for amino acids and 44 "nonsense" codons if the code were non-redundant. In this instance, adding three nucleotides would probably result in the introduction of a nonsense codon rather than the restoration of the wild-type. It was correctly proposed by Crick and Brenner that bases are read in groups of three and that the genetic code is read from a fixed starting point.

2.3.3. Cracking the Code

Approximately concurrently, two American scientists, Marshall Nirenberg and Heinrich Matthaei, created a cell-free apparatus capable of producing proteins in a test tube upon receiving an RNA molecule. They demonstrated how the system produced a polypeptide made completely of phenylalanine residues when given an artificial RNA chain made only of uracil (polyuracil). With this newfound tool, they could now decipher the genetic code. Polyproline was synthesized by RNA made of cytosine (C) residues, while polylysine was made by RNA made of adenosine (A) residues. Nucleotide combinations were used in experiments to show that, for instance, RNA made from A and C can only contain six amino acids: asparagine, glutamine, histidine, lysine, proline, and threonine. Eight triplet codons can be formed from A and C; two of these are known to encode cysteine and proline. Other combinations of A and C must be used to encode the remaining four amino acids. Naturally, this offers more proof of the genetic code's redundancy.

These efforts, which employed RNA molecules made up of arbitrary pairings of two or three bases, were insufficient to completely decipher the genetic code. Additional significant information was added through the use of chemically synthesized RNA molecules with known repeating sequences. For instance, two alternating codons, CAC and ACA, can be read from a synthetic RNA consisting of alternating A and G residues (AGAGAGAGAG). It produces a protein with threonine and histidine residues alternating.

2.3.4. Start and Stop Codons

Amino acids are encoded by 61 of the 64 potential codons. The three codons that are left, UAA, UAG, and UGA, are sometimes referred to as nonsense codons because they do not code for an amino acid. These are known as stop codons, and the ribosome halts protein synthesis in response to them. The most prevalent start codon is AUG, which also codes for the amino acid methionine.

2.3.5. Expanding the Genetic Code

As one of the goals of synthetic biology, it would be possible to create novel materials, some of which could have beneficial therapeutic properties, if we could engineer cells to use a wider range of building blocks with different physical and chemical properties than the small set of amino acids that nature uses to make proteins. Reprogramming the genetic code and designing the translation machinery to use these new combinations are necessary for this to be accomplished successfully. Some progress has been made, such as the ability to code for amino acids that are not typically found in proteins using both the AAG codon and the UAG stop codon for arginine.

2.3.6. Current Concept of the Gene

A gene is a sequence of bases on a DNA molecule that codes for a sequence of amino acids in a polypeptide chain or for an RNA molecule with a specific function, as was discovered after the genetic code was deciphered. Finding patterns typical of genes was made possible by the availability of DNA sequences of individual genes (see 'Recombinant DNA Technology and DNA Sequencing' in this issue of Essays in Biochemistry). An open reading frame is formed by a start codon, a string of codons encoding the amino acid sequence, and a stop codon in a protein-coding gene.

Unprecedented amounts of biological data have been made available through whole genome sequencing. The field of bioinformatics has developed as a result of the necessity to analyze sequence data to provide answers to biological questions. Homology is a fundamental idea in bioinformatics. A structure or a gene is considered homologous if it is shared by two organisms, and the same is true for a gene. For instance, pentadactyl limbs, which have five digits, are present in birds, amphibians, reptiles, and humans in addition to other mammals. The homology of the limbs indicates that these animal groups shared a common evolutionary ancestor. Genes work in the same way.

Haemoglobin is found in the red blood cells of all vertebrates; two ⊠ and two ⊠ globin molecules combine to form adult human hemoglobin. Since all vertebrate globin gene sequences are similar to one another, you can determine how long ago two animals shared an ancestor by comparing the DNA sequences of their globin genes. If a novel DNA sequence contains a segment that resembles a kn, this concept can also be utilized to locate genes in it.

A gene is more than just the sequence that codes for a protein; it also contains sequences that control how the gene is expressed, such as promoter sequences, which indicate the beginning of transcription and the locations where transcription-related proteins attach to DNA. Nearly every gene in bacteria is a single, continuous DNA sequence. The situation is more complex in eukaryotes because introns typically break the coding region. Precursor or pre-mRNA is the term used to describe the main transcript, which has both exons and introns. Exons are spliced together to form the mature mRNA, while the introns are eliminated during pre-mRNA processing before it exits the nucleus (Figure 2.12). The unique 5′ to 5′ linkage adds a methylated guanosine nucleotide to the 5′ end of eukaryotic mRNAs, known as the 5′ cap, which is crucial for starting translation.

The poly A tail, which is a chain of 100–250 adenine residues added to the mRNA to increase stability, is located at the 3′ end. Although there are many more distinct proteins than protein-coding genes, an analysis of the human genome sequence indicates that there are roughly 20,000–25,000 genes that code for proteins. This is due to the fact that many genes have the ability to encode different protein variants. Genes can have multiple alternative promoters and alternative poly A sites in addition to allowing different exon combinations to be included in the mature mRNA through alternative splicing. It is estimated that alternative splicing occurs in 95% of human genes.

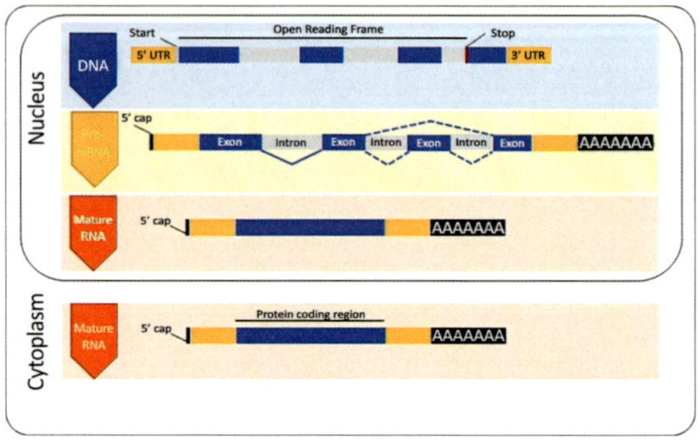

Figure 2.12. *The structure of a protein-coding eukaryotic gene.*

Source: Minchin S, Lodge J. (2019). Understanding biochemistry: structure and function of nucleic acids. Essays Biochem.. 63(4):433-456. doi: 10.1042/EBC20180038. PMID: 31652314; PMCID: PMC6822018.

2.3.7. Non-coding RNAs

Proteins are coded for in just 1.2% of the human genome. But when you look at the genomes of humans, mice, and dogs, you can see that since the species split, a lot more of the genome is subject to what is known as negative selection. Negative selection refers to the selection against unfavorable mutations. This implies that the fitness of the organism carrying the DNA is influenced by factors other than just the protein-coding regions. A portion of these are crucial DNA sequences that regulate the expression of certain genes (see the following section). On the other hand, a lot of processed RNA transcripts that do not encode proteins are being found through systematic screens. The most well-known are ribosomal and transfer RNAs, which are essential to protein synthesis as we shall see in a later section. On the other hand, we are starting to realize that other non-coding RNAs perform crucial biological functions. Two types of non-coding RNA, small inhibitory RNAs (siRNAs) and microRNAs (miRNAs) have a role in reducing gene expression after the mRNA has been transcribed from the DNA. They work by targeting a protein complex called RISC to specific mRNAs which it then degrades. Expression of the gene is specifically knocked out or reduced and the phenotypic effect of this can then be observed. Another group of non-coding RNAs plays an important role in increasing the stability and correct folding of ribosomal RNAs. This process takes place in a compartment within the nucleus called the nucleolus; the RNAs are called small nucleolar RNAs (snoRNAs). These are mostly generated from intron RNA after it has been spliced out of the precursor mRNA and they function in association with proteins. (Minchin & Lodge, 2019)

2.4. TRANSLATION OF RNA INTO PROTEINS

The ribosome, a complex structure made of proteins and RNA, is essential to the synthesis of proteins. The framework that the ribosome offers guarantees that the mRNA and tRNA are positioned correctly, facilitating the interpretation of the genetic code. Numerous other proteins are crucial for the process of protein synthesis; some of these are found within the ribosome, while others are likewise properly positioned by the ribosome's structure. As we shall see, ribosomal RNA, the small subunit, is an RNA molecule with catalytic properties akin to those of an enzyme, known as a ribozyme. A peptide bond can be created between two amino acids by ribosomal RNA.

2.4.1. Transfer RNA

The tRNA is the additional nucleic acid required for protein synthesis. The single-stranded tRNA molecule folds into a distinctive structure through base pairing (Figure 2.13). These function as adaptor molecules, carrying the amino acid that is specified by a particular mRNA codon and possessing an anticodon for that codon. On the mRNA, the anticodon and codon have complementary sequences.

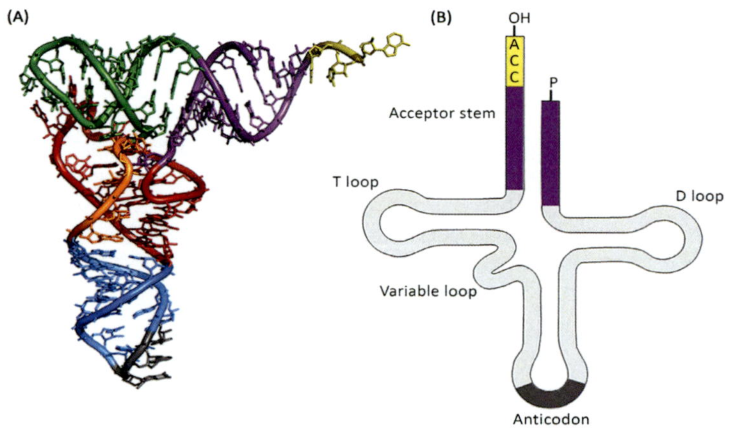

Figure 2.13. *Transfer RNA: (A) Tertiary structure of the phenylalanine tRNA from yeast showing the anticodon (grey), the acceptor stem (violet) with the nucleotides CAA at the 3 OH end (yellow). (B) Clover-leaf representation of the secondary structure of tRNA.*

Source: Steve Minchin and Julia Lodge: Understanding biochemistry: structure and function of nucleic acids: doi: 10.1042/EBC20180038.

A specific amino acid and its corresponding tRNA are recognized by the enzymes known as aminoacyl tRNA synthetases, which attach amino acids to tRNAs. The two-step reaction, which also needs ATP, proceeds as follows:

$$aa + ATP \rightarrow aa - AMP + PP$$

$$aa - AMP + tRNA \rightarrow aa - tRNA + AMP$$

The enzyme attaches the amino acid to the 3′ hydroxyl of the tRNA in the second step after hydrolyzing ATP to release pyrophosphate (PP) in the first. Aminoacyl tRNA synthetase enzymes are extremely selective; they will only bind to the appropriate tRNA after recognizing a given set of amino acids. This guarantees the proper coupling of tRNA molecules and amino acids, which is equally crucial for maintaining the integrity of protein synthesis as the ribosome's matching of the anticodon to the codon. Furthermore, because it generates the appropriate substrate for the ribosome and supplies a significant portion of the energy needed for the formation of peptide bonds during protein synthesis, this step is also known as activating the aminoacyl tRNA (Minchin & Lodge, 2019).

2.4.2. Structure of the Ribosome

Every living organism possesses ribosomes. Although ribosomes in bacteria are somewhat smaller than those in eukaryotic cells (Table 2.2), their general makeup and mode of operation are nearly identical. Three scientists—Ada Yonath, Thomas Steitz, and Venkatraman Ramakrishnan—were granted the 2009 Nobel Prize in Chemistry for their use of X-ray crystallography to solve the three-dimensional structure of the bacterial ribosome. The two subunits that make up the ribosome are the large subunit, which adds amino acids to the expanding polypeptide chain, and the small subunit, which reads the messenger RNA.

Table 2.2. *Properties of Ribosomes*

	Bacterial Ribosome	Eukaryotic Ribosome
Size	70S*, 2.3–2.6 MDa	80S, 3.3–4.5 MDa
Small subunit	30S	40S
	Approximately 20 proteins	32 proteins
	16S rRNA, ~1,600 nucleotides	18S rRNA
Large subunit	70S	60S
	Approximately 33 proteins	79–80 proteins
	23S rRNA, ~2,900 nucleotides	25S rRNA
	5S rRNA, ~120 nucleotides	5.8S & 5S

Source: Steve Minchin and Julia Lodge: Understanding biochemistry: structure and function of nucleic acids: doi: 10.1042/EBC20180038.

Three significant tRNA binding sites are located at the interface between the two subunits of the ribosome and are only formed when the two subunits come together. These locations, which are depicted in Figure 2.14, are known as the exit (E) site, where spent tRNAs exit the ribosome, the peptidyl (P) site, where peptide bonds between amino acids are formed, and the acceptor or aminoacyl (A) site.

The ribosome, mRNA, and tRNA are not the only small proteins needed for protein synthesis. Initiation factors, elongation factors, and termination factors are among them. These proteins are not a part of the ribosome's structure. The inherited disease Vanishing White Matter

Disease (VWM) serves as an example of the significance of these factors. Mutations in one of the initiation factors cause this neurodegenerative disease, which is a serious condition that causes lesions in the brain's white matter.

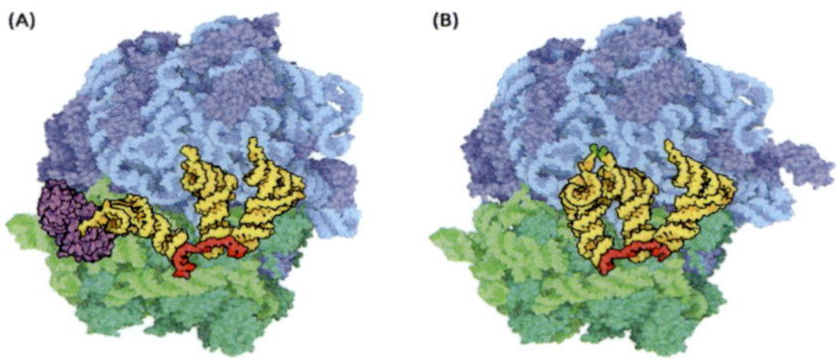

Figure 2.14. *The structure of the ribosome of Thermus thermophilus showing the small subunit (green), large subunit (blue), mRNA (red) and three tRNAs in the acceptor, peptidyl and exit sites (yellow). In (A) the new tRNA is delivered to the ribosome by elongation factor EF-Tu (purple). In (B) the amino acid on the incoming tRNA is brought close to the amino acid on the tRNA in the peptidly site to facilitate peptide bond formation (bright green).*

Source: Steve Minchin and Julia Lodge: Understanding biochemistry: structure and function of nucleic acids: doi: 10.1042/EBC20180038.

2.4.3. Protein Synthesis

The ribosome assembles the mRNA and the charged tRNA during protein synthesis; codon and anticodon are matched, and the amino acids are combined in the right order. This process consists of three stages: elongation, which reads the triplet code and adds amino acids to the expanding peptide chain, termination, which marks the end of protein synthesis, and initiation, which is when the ribosome assembles on the mRNA.

2.4.3.1. Initiation

The 5′ cap of the mRNA in the nucleus is bound by a complex of proteins known as the cap-binding complex. After that, the mRNA is exported to the cytoplasm, where it attracts the small (40S) ribosomal subunit, tRNA that has been charged with a methionine, and initiation factors. When the small subunit comes across the first AUG start codon, it continues to scan along the 5′ untranslated region of the mRNA in the presence of initiation factors (Figure 2.15A). The big subunit then docks to produce the translation complex after the initiator tRNA's anticodon codon identifies this. The subsequent tRNA can now be accepted by the 80S ribosome, which has the tRNA charged with methionine at the P site (Figure 2.15B).

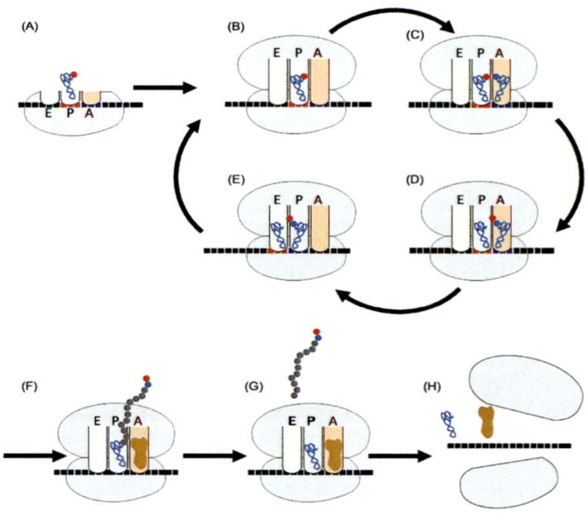

Figure 2.15. *Protein synthesis.*

Source: https://www.ncbi.nlm.nih.gov/pmc/articles/PMC6822018/pdf/ebc-63-ebc20180038.pdf.

Elongation

After initiation is finished, the mRNA is exposed to the next codon and is in the correct reading frame with the A site empty. An aminoacyl tRNA—a tRNA charged with an amino acid—enters the A site of the ribosome in a complex with an elongation factor during the elongation phase. It is properly positioned in the acceptor site and GTP is hydrolyzed on the elongation factor if the anticodon it carries is complementary to the exposed codon (Figure 2.15C). The C terminus of an amino acid in the P site and the N terminus of an amino acid in the A site subsequently form a peptide bond (Figure 2.16). The peptidyl transfer center of the large subunit of the ribosome catalyzes this reaction. As a result, an empty or spent tRNA is left in the P site and the growing peptide chain is transferred to the incoming aminoacyl tRNA in the A site (Figure 2.15D). Finally, the peptidyl tRNA with the growing peptide chain attached moves to the P site. This step is called translocation and the energy is provided by hydrolysis of GTP by the elongation factor EF-G. The spent tRNA moves to the exit site from where it can leave the ribosome. The mRNA moves so that the next codon is exposed in the A site (Figure 2.15E) ready to accept a new aminoacyl-tRNA charged with another amino acid. During the elongation phase the ribosome cycles through this process, adding amino acids to the growing peptide chain until a stop codon is exposed in the A site. The new protein emerges from the ribosome through an exit tunnel in the large subunit (Figure 2.16).

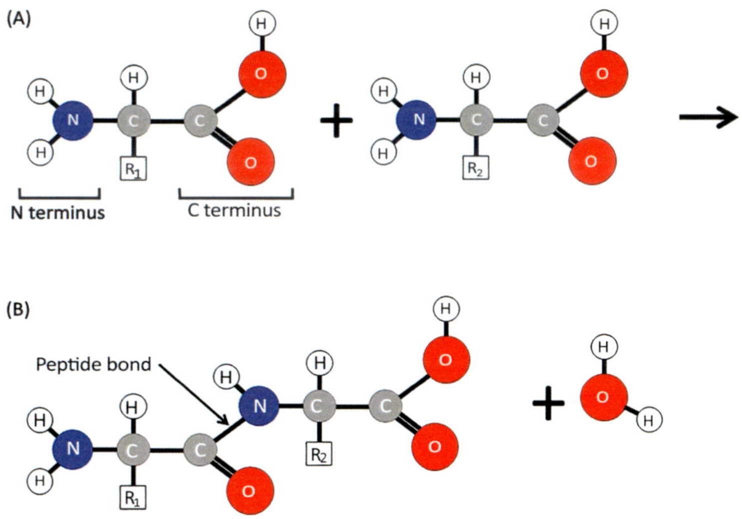

Figure 2.16. *Amino acids and peptide bonds.*

Source: Minchin S, Lodge J. Understanding biochemistry: structure and function of nucleic acids. Essays Biochem. 2019 Oct 16;63(4):433-456. doi: 10.1042/EBC20180038. PMID: 31652314; PMCID: PMC6822018.

Termination

An anticodon on a tRNA does not recognize the stop codon, hence it cannot be decoded. Rather, proteins known as release factors or termination factors detect it. All three stop codons that enter the A site are recognized by a single release factor (RF1) in eukaryotes

(Figure 2.15F). The peptide is released from the ribosome (Figure 2.15G) when the ester bond tying the peptide chain to the tRNA in the P site breaks (Figure 2.15H). The two ribosome subunits separate and are recycled.

With a significant core of structurally conserved proteins and rRNAs present in both prokaryotic and eukaryotic ribosomes, the structure and function of ribosomes are highly conserved. However, there are some variations in the extra proteins involved in translation as well as in the rRNAs. Although there are significant differences in the initiation of protein synthesis, the elongation phase is highly conserved. A particular sequence found in bacterial mRNAs is known as the Shine–Dalgarno sequence or ribosome binding site. Shine-Dalgarno sequence binds to a complementary sequence of the 16S rRNA in the small subunit to guarantee that the mRNA is positioned correctly in the ribosome. In bacteria, N-formylmethionine, a modified amino acid, charges the initiator tRNA.

Antibiotics, which are selective in that they influence protein synthesis in bacteria but not in mammalian cells, can take advantage of differences in the structure of bacterial and eukaryotic ribosomes. Protein synthesis is stopped by macrolide antibiotics such as erythromycin, which block the exit tunnel in the large subunit of bacterial ribosomes. Therefore, eukaryotic ribosomes are unaffected because their exit tunnel is marginally narrower. The antibiotic streptomycin binds to the 16S ribosome of bacteria, making it a valuable tool in the fight against tuberculosis. This leads to an incorrect reading of the mRNA by warping the structure of the decoding site. (Minchin & Lodge, 2019)

ROLE MODEL

RICHARD J. ROBERTS IS A NOBEL PRIZE-WINNING ENGLISH BIOCHEMIST AND MOLECULAR BIOLOGIST

Born Sept. 6, 1943, Derby, Eng., molecular biologist, the winner, with Phillip A. Sharp, of the 1993 Nobel Prize for Physiology or Medicine for his independent discovery of "split genes." Roberts received a Ph.D. in organic chemistry from the University of Sheffield, Eng., in 1968. After postdoctoral research at Harvard University, he took a post at Cold Spring Harbor Laboratory in New York in 1972. In 1992 he joined New England Biolabs, a biotechnology firm.

In 1977 Roberts and a team including Thomas Broker, Louise Chow, and Richard Gelinas established that the genes of the adenovirus—one of the viruses that cause the common cold—are discontinuous: the segments of DNA that code for proteins are interrupted by lengthy stretches of DNA that do not contain genetic information. The coding segments are called exons; the noncoding ones are called introns. A research team working under Sharp at the Massachusetts Institute of Technology produced the same finding that same year. Previously, based on studies of bacterial DNA, biologists believed that genes consisted of unbroken stretches of DNA, all of which encoded protein structure. It has since been established that the discontinuous gene structure discovered by Roberts and Sharp is the most common structure found in higher organisms (eukaryotes). In addition to having important implications for the study of genetic diseases, this structure is believed to drive evolution by allowing information from different parts of the gene to be brought together in new combinations.

Richard John Roberts is an English biochemist and molecular biologist who was felicitated with the Nobel Prize in Physiology or Medicine for his discovery of the mechanism of gene-splicing. He shared the prize with Philip Sharp who independently carried out similar work in the field. Roberts' fantasy with chemistry started as early as in high school. He was so interested in the subject that he chose to make it his career. Following his graduate studies and doctoral thesis, Roberts made his way to Harvard and eventually the Cold Harbor Laboratory. It was at the laboratory that Roberts first found success in restriction enzymes. By 1972, he discovered or characterized almost three-quarter of the world's first restriction enzymes. Later in 1977, Roberts demonstrated how RNA can be divided up into introns and exons, after which the exons can be joined together. The discovery was crucial as until then the scientific world believed that genes comprised unbroken sketches of DNA. The discovery had important implications for the study of genetic diseases.

Childhood and Early Life

- Richard John Roberts was born on September 6, 1943, in Derby, England, to Edna Allsop and John Roberts. His father was an auto mechanic by profession while his mother was a homemaker.
- When young Robert was just four, the family shifted to Bath. There, he attended the Christ Church infant school and later St Stephen's Boys' School. It was at Stephen's that Roberts developed a passion for logic and mathematics. He was encouraged by his headmaster, Mr Broakes who would spend hours producing problems and later helping Roberts solve them.
- Completing his elementary studies, he attended City of Bath Boys' School. Early on, Robert aimed at becoming a detective but when he was handed a chemistry play set, he changed his mind. His chemistry play set was transformed into a large in-house chemistry laboratory that comprised real apparatus and chemicals. He soon found his real calling in chemistry.
- Following his preliminary education, he enrolled at the University of Sheffield. He graduated from the same with a Bachelor of Science degree in Chemistry in 1965. Four years later, he attained his PhD in the subject, his thesis being on phytochemical studies of neoflavonoids and isoflavonoids.

Career

- While doing his PhD, Richard J. Roberts came across a book by John Kendrew that turned his focus to molecular biology. The book detailed the early history of crystallography and molecular biology focusing on the MRC Laboratory in Cambridge.
- For postdoctoral studies, Roberts looked for a laboratory that would provide him the pathway into molecular biology. At Harvard, Jack Strominger offered him an opportunity. He engaged himself in the research work from 1969 to 1972.
- In September 1972, he moved to Cold Spring Harbor Laboratory where he worked under James Dewey Watson, co-discoverer of the structure of DNA. The same year, he attended a seminar by Dan Nathans who described an enzyme Endonuclease R that could cleave DNA into specific pieces. Roberts realized that Nathans' restriction enzyme gave an immediate way to isolate small DNA molecules.
- At Cold Harbor Laboratory, he made preparations for Endonuclease R and the few other restriction enzymes known then. By 1973, he had found superlative success in restriction enzymes. Almost three-quarter of the world's first restriction enzymes were discovered or characterized under his guidance. He had compiled an enzyme collection that proved to be a valuable resource for scientists around the world.

- In 1974, Roberts was joined by Richard Gelinas to characterize the initiation and termination signals for an Adenovirus-2mRNA. The main purpose was to sequence the 5'-end of an mRNA, map its location on a restriction fragment, and then sequence the upstream region. Just before beginning the project, mRNA caps were discovered. The duo developed an assessment of the capped oligonucleotides.
- While working on the initiation and termination signals of Adenovirus-2mRNA, Roberts and Gelinas found that all late mRNAs seemed to begin with the same capped oligonucleotide, which was not encoded on the DNA next to the main body of the mRNA. Though they had biochemical evidence for the same, they did not have real proof.
- In March 1977, Roberts came up with the experiment that displayed the proposed split structure for Adenovirus-2mRNAs was accurate. Soon, the duo was joined by Louise Chow and Tom Broker. The four scientists aimed at visualizing the split structure by hybridizing an intact mRNA to its two different coding regions. Eventually, they came up with appropriate DNA fragments that were visible as split genes in the electron microscope.
- Before the discovery made by Roberts, scientists believed that genes consisted of unbroken stretches of DNA, all of which encoded protein structure. However, with his discovery, it was found that the discontinuous gene structure is the most common structure found in higher organisms (eukaryotes). In addition to having important implications for the study of genetic diseases, this structure is believed to drive evolution by allowing information from different parts of the gene to be brought together in new combinations.
- In 1992, Roberts moved to New England Biolabs. Don Cromb, founder and president of the establishment, had started the establishment mainly to produce restriction enzymes for commercial sale. Roberts started off as a chief consultant at the establishment and was no soon promoted as a joint Research Director.

Major Works

- His most distinguished work came in the 1970s when he successfully discovered split genes. The discovery was crucial and important in the scientific world as until then, scientists believed that the genes consisted of unbroken stretches of DNA, which encoded protein structure. It was Roberts who established the discontinuous gene structure. Through his experiments, he proved that the segments of DNA that code for proteins are interrupted by lengthy stretches of DNA that do not contain genetic information. He demonstrated how RNA can be divided up into introns and exons, after which the exons can be joined together. This can occur in different ways, giving a gene the potential to form many different proteins.

Awards and Achievements

- In 1993, Richard J. Roberts and Phillip A. Sharp were jointly awarded the Nobel Prize in Physiology or Medicine "for their discoveries of split genes." Sharp worked independently on the subject
- He was conferred with an Honorary Doctor of Science Degree by the University of Bath in 1994.
- In 1995, he was elected Fellow of the Royal Society.

Personal Life and Legacy

- Roberts married Jean and has four children, namely Alison, Andrew, Christopher and Amanda.
- In 2005, the chemistry department at his alma mater, University of Sheffield, was later named after him.

CLASS ACTIVITY

DNA EXTRACTION

Description: Students will extract DNA from strawberries using household materials like dish soap, salt, and rubbing alcohol. They will observe the extraction process and discuss the principles of molecular biology involved, such as cell lysis, DNA solubility, and precipitation.

Objective: To introduce students to basic molecular biology techniques and illustrate the concept of DNA extraction in a hands-on manner.

SUMMARY

- Molecular biology is a field of biology that explores the intricate mechanisms underlying the structure, function, and interactions of biological molecules within cells. Molecular biology delves into the molecular processes that drive life, examining the fundamental building blocks of living organisms at the smallest scales.
- DNA, also known as Deoxyribonucleic acid, is an organic chemical consisting of genetic instructions and information for protein synthesis. DNA can be seen in most cells of all living creatures. It is a vital part of reproduction, where genetic heredity occurs. Therefore, the main objective is to transfer the inheritance from the older generation to future generations.
- DNA is normally found as a double-stranded molecule in the cell whereas RNA is mostly single-stranded. It is important to understand though, that under the appropriate conditions, DNA could be made single-stranded, and RNA can be double-stranded.
- Nucleases are enzymes that hydrolytically cleave the phosphodiester backbone of DNA. Endonucleases cleave in the middle of chains, while exonucleases operate selectively by degrading from the end of the chain. Nucleases that act on both single- and double-stranded DNA are known.
- Chemical modification of DNA can lead to mutations in the genetic material. Anions such as bisulfite can deaminate cytosine to form uracil, changing the genetic message by causing C-to-T transitions.
- The genetic information in DNA is used as a basis to create messenger RNA (mRNA) by transcription. Single-stranded mRNA then acts as a template during translation. Ribosomes facilitate translation in the cytoplasm, by inducing the binding of complimentary transfer RNA (tRNA) anticodon sequences to the mRNA.
- mRNA is used to convey information from DNA to the ribosome. It is a single-strand molecule, complementary to the DNA template, and is generated through transcription. Strands of mRNA are made up of codons, each of which signifies a particular amino acid to be added to the polypeptide in a certain order.
- RNA is a complex compound of high molecular weight that functions in cellular protein synthesis and replaces DNA (deoxyribonucleic acid) as a carrier of genetic codes in some viruses. RNA consists of ribose nucleotides (nitrogenous bases appended to a ribose sugar) attached by phosphodiester bonds, forming strands of varying lengths. The nitrogenous bases in RNA are adenine, guanine, cytosine, and uracil, which replaces thymine in DNA.
- RNA is typically single-stranded and is made of ribonucleotides that are linked by phosphodiester bonds. A ribonucleotide in the RNA chain contains ribose (the pentose sugar), one of the four nitrogenous bases (A, U, G, and C), and a phosphate group.
- The genetic material in a cell is made of DNA and can be copied and passed on to progeny through DNA replication allowing for inheritance of the information that

it carries. A large proportion of the information on the DNA is first transcribed into mRNA and then translated into proteins.

- The genetic code is the set of rules used by living cells to translate the information encoded within genetic material into proteins. When DNA and RNA were first discovered, the relative simplicity of nucleic acids led many scientists to doubt that it carried the genetic information.

REVIEW QUESTIONS

- Elaborate on the structure and functions of DNA, highlighting its role in genetic information.
- Discuss the structure and functions of RNA, emphasizing its diverse roles in cellular processes.
- Explore the genetic code and provide insights into the concept of a gene.
- Evaluate the impact of mutations in DNA on gene expression and protein synthesis.
- Illustrate how the process of translation is regulated and its importance in cellular function.
- Illustrate how the process of translation is regulated and its importance in cellular function.

REFERENCES

1. Alberts, B., Johnson, A., Lewis, J., Morgan, D., Raff, M., Roberts, K., & Walter, P. (2014). *Molecular Biology of the Cell* (6th ed.). Garland Science.
2. Bellelli, R., & Boulton, S. J. (2021). Spotlight on the replisome: Aetiology of DNA replication-associated genetic diseases. *Trends in Genetics, 37*(4), 317–336. https://doi.org/10.1016/j.tig.2020.12.002.
3. Bentley, D. (2002). The mRNA assembly line: Transcription and processing machines in the same factory. *Current Opinion in Cell Biology, 14*(3), 336–342. https://doi.org/10.1016/S0955-0674(02)00318-1.
4. Bertone, P., Stolc, V., Royce, T. E., Rozowsky, J. S., Urban, A. E., Zhu, X., et al. (2004). Global identification of human transcribed sequences with genome tiling arrays. *Science, 306*(5705), 2242–2246. https://doi.org/10.1126/science.1103388.
5. Carninci, P., Kasukawa, T., Katayama, S., Gough, J., Frith, M. C., Maeda, N., et al. (2005). The transcriptional landscape of the mammalian genome. *Science, 309*(5740), 1559–1563. https://doi.org/10.1126/science.1112014.
6. Cheng, J., Kapranov, P., Drenkow, J., Dike, S., Brubaker, S., Patel, S., et al. (2005). Transcriptional maps of 10 human chromosomes at 5-nucleotide resolution. *Science, 308*(5725), 1149–1154. https://doi.org/10.1126/science.1116601.

7. Govindarajan, R., Duraiyan, J., Kaliyappan, K., & Palanisamy, M. (2012). Microarray and its applications. *Journal of Pharmacy & Bioallied Sciences, 4*(Suppl 2), S310–S312. https://doi.org/10.4103/0975-7406.100646.

8. Kokate, C., Jalalpure, S. S., & Hurakadle, P. J. (2016). *Textbook of Pharmaceutical Biotechnology: Expression Cloning*. Elsevier.

9. Lessard, J. C. (2013). Molecular cloning. In *Laboratory Methods in Enzymology: DNA* (Vol. 529, pp. 85–98). Academic Press.

10. Minchin, S. D., & Busby, S. J. W. (2013). Transcription factors. In S. Maloy & K. Hughes (Eds.), *Brenner's Encyclopedia of Genetics*. Elsevier.

11. Minchin, S., & Lodge, J. (2019). Understanding biochemistry: Structure and function of nucleic acids. *Essays in Biochemistry, 63*(4), 433–456. https://doi.org/10.1042/EBC20180038.

12. Roeder, R. G. (2003). The eukaryotic transcriptional machinery: Complexities and mechanisms unforeseen. *Nature Medicine, 9*, 1239–1244. https://doi.org/10.1038/nm938.

13. Simpson, B., Tupper, C., & Al Aboud, N. M. (2023). Genetics, DNA packaging. In *StatPearls* [Internet]. StatPearls Publishing. Retrieved from: https://www.statpearls.com (accessed on 3 September 2024).

14. Tian, J. (2013). *Molecular Imaging: Fundamentals and Applications*. Springer-Verlag Berlin & Heidelberg GmbH & Co. K. Retrieved from: https://doi.org/10.1007/978-3-642-34303-2.

CHAPTER

3

Structure and Functions of Cell

LEARNING OBJECTIVES

After studying this chapter, you will be able to:

- Understand the overview of the cell;
- Explain the types of cells;
- Define the molecular composition of cells;
- Discuss about cell cycle.

KEY TERMS FROM THIS CHAPTER

Cell cycle	Cell signaling	Cell wall
Centriole	Chromosome	Cytosol
Eukaryote	Flagellum	Microtubule
Nucleolus	Organelle	Peroxisome
Phospholipid bilayer	Plasma membrane	Prokaryote
Vacuole	Vesicle	

INTRODUCTION

The structure and functions of cells form the bedrock of biology, revealing the intricate organization and dynamic capabilities that sustain life. Cells are the smallest units of living organisms, each functioning as a complex and highly organized entity. They come in two primary types: prokaryotic and eukaryotic, each with distinct characteristics that influence their structure and function.

Prokaryotic cells, exemplified by bacteria, are simpler in structure. They lack a membrane-bound nucleus and other membrane-bound organelles found in eukaryotic cells. Instead, their genetic material floats freely within the cell, contained by a nucleoid region. Prokaryotes are adept at thriving in diverse environments and can exhibit remarkable metabolic versatility, making them foundational to ecological and industrial processes.

Eukaryotic cells, which include those of plants, animals, fungi, and protists, are more complex. They feature a membrane-bound nucleus that houses the cell's genetic material, providing compartmentalization that allows for more specialized functions. Eukaryotic cells also possess various organelles, such as mitochondria for energy production, the endoplasmic reticulum for protein synthesis, and the Golgi apparatus for processing and packaging molecules. These organelles work together in a coordinated manner to ensure the cell's survival and enable it to perform specialized tasks essential for the organism's overall function.

3.1. OVERVIEW OF THE CELL

The fundamental building block of all life are cells. Therefore, the smallest independent component that we would find if we dissected an organism down to its cellular level would be the cell.

Cell Definition

"A cell is defined as the smallest, basic unit of life that is responsible for all of life's processes."

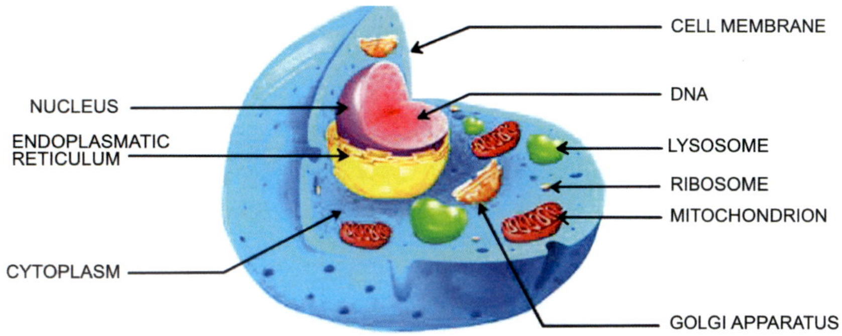

Figure 3.1. *Cells are the fundamental unit of life. They range in size from 0.0001 mm to nearly 150 mm across.*

Source: https://edurev.gumlet.io/ApplicationImages/Temp/14322571_0ffc2ecf-7784-4801-b69f-94e52be12437_lg.png?w=400&dpr=2.6.

The biological, structural, and functional components of every living thing are their cells. A cell is capable of autonomous replication. They are referred to as the building blocks of life as a result. The fluid known as cytoplasm, which is surrounded by a membrane, is found inside every cell. Numerous other biomolecules, including lipids, proteins, and nucleic acids, are also found in the cytoplasm. Moreover, the cytoplasm contains suspended cellular components known as cell organelles (Figure 3.1).

The basic and structural building block of life is a cell. Cell biology is the study of cells, from their fundamental structure to the roles played by each organelle. The first biologist to discover cells was Robert Hooke. Cells comprise all living things. They could consist of several cells (multicellular) or just one (unicellular). The smallest known cell is called a mycoplasma. The fundamental units of all living things are cells. They give the body structure and provide energy from the nutrients in food.

An organism's cells are complex structures, and each of their parts serves a variety of purposes. They are shaped and sized differently, much like the building's bricks. The various shaped and sized cells that comprise our body are diverse. The smallest unit of organization in all living forms is the cell. The number of cells varies from organism to organism. Compared to bacteria, humans are composed of a larger number of cells.

Numerous cell organelles that carry out specific tasks to carry out life processes make up cells. Each organelle possesses a unique structure. The cells contain the organisms' genetic material as well.

3.1.1. The Evolution of Cells

3.1.1.1. The Development of Genetic Information

The emergence of a group of catalysts that could encourage the synthesis of additional catalysts of the same kind was necessary for the existence of life on Earth. The early phases of the cell's evolutionary history are thought to have revolved around RNA molecules, which have the capacity to duplicate themselves by producing a complementary RNA molecule in addition to presenting particular catalytic surfaces. Eventually, it is thought, a tiny RNA molecule that could catalyze its own duplication emerged.

Numerous variant autocatalytic RNA molecules most likely originated from flaws in primitive RNA replication. RNA molecules that developed mutations that boosted their capacity for self-replication would have out multiplied less capable RNA molecules. Furthermore, natural selection was applied to other small RNA molecules that coexisted in symbiosis with autocatalytic RNA molecules because of their capacity to catalyze beneficial secondary reactions, like the synthesis of superior precursor molecules. Because cooperation between different molecules produced a system that was far more effective at self-replication than a collection of individual RNA catalysts, sophisticated families of RNA catalysts could have evolved together in this way.

The emergence of a primitive protein synthesis mechanism in one family of self-replicating RNA would have been another significant step in the evolution of the cell. Protein molecules are unable to supply the information needed to synthesize other protein molecules that are similar to them. The source of this data must ultimately be a sequence of nucleic acids. Compared to RNA synthesis, protein synthesis is far more complex, and it could not have developed before the evolution of a potent class of RNA catalysts. The RNA molecules that function in the current cell are thought to have counterparts for each of these catalysts: (1) an information RNA molecule, similar to messenger RNA (mRNA), whose nucleotide sequence was read to create an amino acid sequence; (2) a group of adaptor RNA molecules, similar to transfer RNA (tRNA), that could bind to both mRNA and a particular activated amino acid; and (3) finally, an RNA catalyst, similar to ribosomal RNA (rRNA), that helped join the amino acids aligned on the mRNA.

The first cell formed at some point during the evolution of biological catalysts. This would have required the division of the early biological catalysts into discrete membrane-enclosed units. Many amphiphilic molecules, which are half hydrophilic (loving water) and half hydrophobic (repelling water), aggregate to form bilayer sheets in which the hydrophilic portions of the molecules face the water while the hydrophobic portions line up in rows to form the interior of the sheet. This suggests that membrane formation may have happened quite simply. Just as modern cell phospholipid bilayer membranes can spontaneously close to form the walls of tiny, spherical vesicles, so too can such sheets (Figure 3.2).

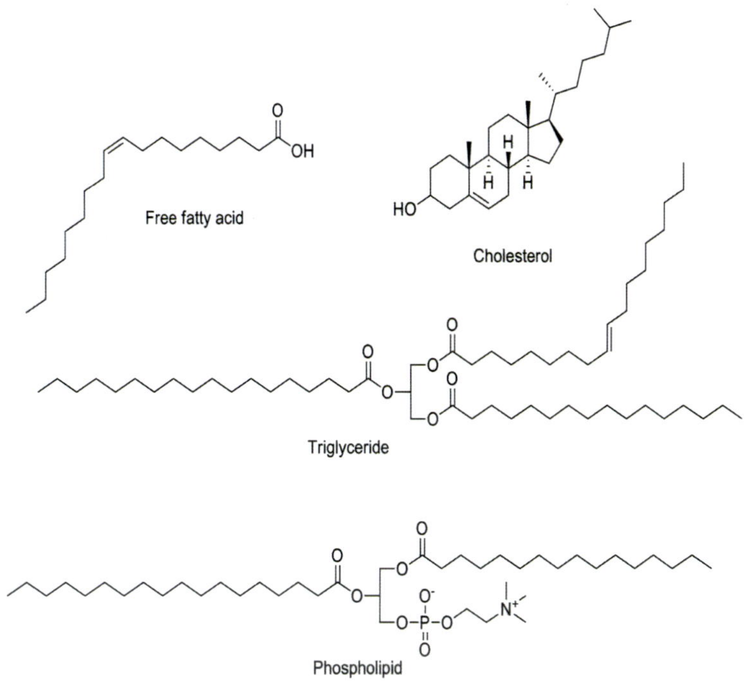

Figure 3.2. *Lipid structure.*

Source: By Lmaps at the English Wikipedia, CC BY-SA 3.0, https://commons.wikimedia.org/w/index.php?curid=4304232.

The biological catalysts would have started to compete with one another for the same resources as soon as they split up into tiny, independent units, or cells. The resultant intense competition must have sped up evolutionary change considerably, acting as a driving force behind the emergence of more effective cells. In this manner, new catalysts eventually led to the emergence of cells that could use simpler, more plentiful precursor molecules for growth. These cells were able to spread far beyond the small environments where the first primitive cells originated because they were no longer dependent on preformed ingredients for their survival.

A common belief is that the earliest cells didn't emerge until a crude form of protein synthesis evolved. It has been proposed that the first cells only contained RNA catalysts, but it is by no means certain that cells cannot exist without proteins. As a result, over time, cells emerged in which RNA acted primarily as genetic material, directly replicating in every generation and inherited by all progeny cells to specify proteins. In either scenario, protein molecules—with their chemically varied side chains—are more potent catalysts than RNA molecules.

With the increasing complexity of cells, a more stable method of storing genetic information than RNA would have been required. Though chemically more stable than RNA, DNA is related to RNA and most likely evolved relatively late in the history of cells. Direct replication of RNA molecules was lost as a result of the gradual transfer of genetic information from RNA sequences to DNA sequences. The fundamental process of biology—the sequential synthesis of DNA, RNA, and protein—did not emerge until this point.

3.1.1.2. The Development of Metabolism

The absence of nuclei and functional internal compartments, or organelles, is likely what made the earliest cells similar to prokaryotic cells. Additionally anaerobic (meaning they didn't need oxygen to function), these early cells got their energy from the fermentation of organic molecules that had previously accumulated over a long period of time on Earth. Eventually, more advanced cells developed the ability to perform rudimentary forms of photosynthesis, in which proteins bonded to membranes captured light energy to create organic molecules with chemical bonds rich in energy. A significant shift in the evolution of life occurred with the emergence of photosynthesizing prokaryotes that could produce molecular oxygen and only needed water as an electron donor. The blue-green algae, or cyanobacteria, are the prokaryotes' offspring and are still thriving life forms today. Their ancestors were so successful that they produced so much oxygen that the atmosphere grew rich in it. Other prokaryotes were able to evolve aerobic forms of metabolism—metabolism that was far more efficient at using organic molecules as a source of food—thanks to the free availability of this oxygen. It is estimated that bacteria made the transition from mostly aerobic metabolism about 2.5 billion years ago, or 1.5 billion years after the first cells formed. Since their lineage split off much earlier from that of the prokaryotes, aerobic eukaryotic cells—cells with nuclei and all other organelles—likely first appeared between 2.1 and 1.5 billion years ago. It's highly likely that eukaryotic cells developed into aerobic forms by consuming aerobic prokaryotes, with which they coexisted in harmony. Plants and animals both have mitochondria, which are descended from these prokaryotes. Later, an organism that resembled blue-green algae was ingested to carry out photosynthesis in branches of the eukaryotic lineage that gave rise to plants and algae. These organisms probably evolved into the chloroplasts over a very long time.

Thus, it appears that the eukaryotic cell originated as a mixture of various cells, evolving into an effective aerobic cell whose plasma membrane was released from energy metabolism—one of the primary roles of prokaryotic cell membranes. Because of this, the eukaryotic cell membrane was able to develop into a specialized structure for cell signaling and communication. This could be one of the reasons why eukaryotic cells eventually outperformed their simpler prokaryotic cousins in the formation of complex multicellular organisms.

3.1.2. The History of Cell Theory

Since early theories and predictions regarding the nature of the cell were typically misguided, the history of cell theory is really the history of the observation of cells themselves. The development of the microscope in the 16th century was a pivotal moment that made it possible to observe cells and sparked curiosity about the "invisible" world. The term "cell" was first used by English physicist Robert Hooke in 1665 when he described cork and other plant tissues. Hooke was reminded of monks' cell blocks by the cellulose walls of dead cork cells. No value was placed on the contents of plant cell walls, even after superb images of plant tissues were published in 1672. A study of the minute details of the contents of the cell was not possible due to the microscope's magnifying powers and the inadequate methods for preparing cells for observation. Beginning in 1673, the brilliant Dutch microscopist Antonie van Leeuwenhoek made the discoveries of spermatozoa, blood cells, and a vibrant world of "animalcules." It revealed a whole new universe of unicellular creatures. These discoveries added to the known diversity of living things but did not shed light on their fundamental similarity. Furthermore, Leeuwenhoek could only support the theory that his animalcules emerged spontaneously when he saw them swarming but overlooked their division.

It took nearly 200 years after the invention of microscopy for cell theory to be developed. There are several reasons for this delay, including the subpar quality of the microscopes and the continued use of antiquated notions about what constitutes a basic living unit. Numerous observations of cells were made, but it doesn't seem that any of the observers could firmly state that cells are the fundamental building blocks of biological structure and function.

Early cell theory development was greatly aided by three crucial discoveries made in the 1830s: better microscopes with appropriate lenses, higher magnification levels without aberration, and more adequate illumination. Scottish botanist Robert Brown made the first observation that the nucleus is a permanent part of plant cells in 1833. Then, it was discovered that certain animal cells had nuclei. Eventually, a living material known as protoplasm was identified inside of cells; its vitality was demonstrated by its vigorous streaming or flowing motions, particularly in plant cells. Following these three discoveries, cells—which were previously thought of as empty spaces in plant tissue—could no longer be regarded as such because they were found to contain living material.

German physiologist Theodor Schwann and German biologist Matthias Schleiden clearly stated in 1839 that cells are the "elementary particles of organisms" in both plants and animals and recognized that some organisms are unicellular and others multicellular. This statement was made in Schwann's Mikroskopische Untersuchungen über die Übereinstimmung in der Struktur und dem Wachstume der Tiere und Pflanzen (1839; Microscopical Researches into the Accordance in the Structure and Growth of Animals and Plants). Schleiden's contributions on plants were acknowledged by Schwann as the basis for his comparison of animal and plant structure.

The descriptive claims made by Schleiden and Schwann regarding the cellular foundation of biologic structure are clear-cut and consistent with contemporary theory. They identified the membrane, nucleus, and cell body as the three components that all cells share and described these elements through comparisons of different animal and plant tissues.

3.1.2.1. The Problem of the Origin of Cells

The work of Schwann and Schleiden's predecessors contained strong indications of the cell theory, suggesting that they were not the only ones who contributed to this great generalization of natural science. These early researchers recognized that the fundamental issue was how cells originated, so they developed the theory of "free cell formation," which postulated that cells formed from scratch from an unformed substance called a "cytoblastema" through a series of events that include the development of the nucleolus first, then the nucleus, the cell body, and finally the cell membrane. Their theory was based on the best physical model of the generation of formed bodies that was available at the time, which was crystallization. However, in retrospect, it would not appear that the hypothesis of free cell formation was justified, since earlier microscopists had regularly observed cell division, a feature not typical of crystallization processes, especially among single-celled organisms. The idea of free cell formation persisted for the majority of the 19th century, although cell division was repeatedly observed in the decades that followed. Nevertheless, it was increasingly considered as a potential exception to the general rule that cells reproduce by division. German pathologist and statesman Rudolph Virchow (Figure 3.3) declared in 1855 that "omnis cellula e cellula"—"all cells come from cells"—affirming the correct general principle.

Figure 3.3. *Rudolf Virchow.*

Source: https://upload.wikimedia.org/wikipedia/commons/thumb/8/80/Rudolf_Virchow.jpg/800px-Rudolf_Virchow.jpg.

Cell division's intrinsic complexity made it impossible to quickly resolve every change that takes place throughout the process. First, it was observed that a cell with a nucleus divides into two cells, each of which has a nucleus. As a result, it was determined that the nucleus must divide, and some people duly described the direct division of nuclei. Improved methods only contributed to the confusion, as it was discovered that the nucleus as such vanishes during cell division. Furthermore, faint masses that are now known to be chromosomes were briefly observed during the division process. German anatomist Walter Flemming's extremely accurate description and interpretation of cell division in 1882 was the result of observations made in the 1870s. By using his sophisticated methods for fixing and staining cells, he was able to observe that cell division marks the end of cell reproduction, which entails the transfer of chromosomes from parent to daughter cells during the process of mitosis.

The complete explanation of meiosis emerged from the finding that the number of chromosomes does not change between generations. The understanding of the physical foundation of reproduction and heredity resulted from the description of meiosis and the realization that fertilization is essentially the union of maternal and paternal sets of chromosomes. It was thus determined that meiosis and fertilization are complementary processes in the life cycle of organisms: meiosis reduces the number of chromosomes by half during the creation of spores (plants) or gametes (animals), while fertilization increases the number by uniting gametes. By the 1890s, "life" in all its forms could be understood as a cell's expression.

3.1.2.2. The Protoplasm Concept

During the 19th century, the notion that the cell is the fundamental unit of life progressed in tandem with the "protoplasm" concept, which postulated that life originates from the protoplasm inside the cell. According to the definition given in 1835, protoplasm is the fundamental component of all living things and is therefore in charge of all living processes. One perspective that contrasts this idea—that life is the activity of an elementary particle called the cell—is the idea that life is the manifestation of a living, complex substance, even a supermolecule, known as the protoplasm. Observations of the streaming motions of the apparently slimy contents of living cells provided evidence in favor of the protoplasm concept.

Proponents of the protoplasm theory suggested that cells were either protoplasmic fragments or containers. These researchers, suspicious and frequently disdainful of data derived from dead and stained cells, found the majority of fundamental data regarding the mechanical, optical, electrical, and contractile characteristics of the living cell.

It is challenging to evaluate the concept of protoplasm's utility. It was partially true; on the one hand, it promoted research into the mechanical and chemical characteristics of cell contents; however, it also led to resistance, which was visible as late as the 1930s, to the advancement of biochemical methods for cell fractionation and to the understanding that large molecules, or macromolecules, play a significant role in cell function. The term "protoplasm" has lost all meaning because the cell has been divided into its constituent parts. Nonetheless, the term protoplasm is still used to refer to the phenomenon known as protoplasmic streaming, which is where the word protoplasm first appeared.

3.1.2.3. Contribution of Other Sciences

Understanding of the cell as the building block of life has come from a number of significant sources, not the least of which is microbiology. Microorganisms could be grown quickly and easily, despite the fact that their small size made it difficult to observe much of their intricate structure until the development of electron microscopy. As a result, the 1861 publication of French chemist and microbiologist Louis Pasteur's studies on microbes contributed to the development of the theory of biogenesis, which holds that organisms can only arise through the reproduction of other organisms. Microbiology provided fundamental insights into the characteristics of cells' metabolism, namely their capacity to convert basic food components into cell substance and useable energy. Enzymatic activity is a fact that a large portion of cell chemistry is enzyme chemistry, and that enzymes are one characteristic that distinguishes cells. Pasteur may have overstated the relationship between catalysis and the living state of cells by viewing enzymatic action as a feature of the living cell rather than of the catalytic molecules (enzymes) contained in the cell. The methods of microbiology eventually made microbial genetics possible, which in turn gave rise to the means of resolving the basic issues of molecular biology that were initially impervious to direct attack by biochemical techniques.

If the cell theory were an oversimplified generalization, then molecular biology would be the most powerful science to refute it. Molecular biology, on the other hand, has emerged as the cornerstone of cell science because it has shown that fundamental processes like protein synthesis and the genetic code are common to all living systems and are enabled by the same cell components, e.g. membranes, ribosomes, and chromosomes (Figure 3.4).

Figure 3.4. *French chemist and microbiologist Louis Pasteur performing a scientific experiment. Pasteur's studies helped establish the principle of biogenesis, the development of new organisms from the reproduction of other organisms.*

Source: https://www.britannica.com/science/cell-biology/The-process-of-differentiation.

In the parallel histories of medicine and cell biology, two occasions stand out as particularly significant. One was the recognition of the ovum (unfertilized egg) as a cell in 1827 by Prussian-Estonian embryologist Karl Ernst Ritter von Baer. This was significant because of the numerous ways in which ova differ from other cells. Baer not only established the groundwork for reproductive biology, but at a crucial juncture, he also offered substantial support for the cell theory. The second significant development was Virchow's 1855 promotion of the idea of "cellular pathology." Cell theory gained credibility from his hypothesis that human diseases are diseases of cells and can be recognized and comprehended as such.

Cell theory and biochemistry may not have advanced as much, but they had an almost immediate influence on one another. Following the discovery that the majority of biochemical phenomena are universal to all cells, a cell's metabolism could be used to distinguish it from its structural characteristics. When Swiss biochemist Johann Friedrich Miescher proposed that the nucleus must have a distinctive chemistry in 1869, cytochemistry, also known as histochemistry, got off to a great start. Miescher later discovered nucleic acids, which have since been demonstrated to be the essential molecules of metabolism and heredity.

Cell theory is unable to fully explain how a multicellular organism develops and becomes one. An animal or plant is more than just a collection of individual cells, and a cell is not always an autonomously functioning unit. Thankfully, though, the protracted debate over the uniqueness and separation of cells is now over. These days, cell biology is primarily concerned with studying individual cells as well as how cells interact and communicate with one another. Although the impact of the surrounding environment on a cell has always been acknowledged, it is now known that other cells play a significant role in a cell's environment. Thus, the idea that an organism is more than the sum of its parts is not completely eliminated by cell theory. However, studying a given organism necessitates studying cells as individuals as well as groups. One example is the cancer problem: a plant or animal controls its own cell division; the appropriate cells must divide, differentiate, and then be integrated into the appropriate organ system at the appropriate time and location. Cancer is one of the abnormalities that arises from breakdown. The ultimate goal of a cell biologist's investigation into the regulation of cell division is to comprehend how an individual cell is affected by the actions of the entire organism. (Sturtevant & Beadle, 1939; Taylor and Wang, 1989).

3.2. TYPES OF CELLS

The nucleus is another fundamental cell structure that exists in many, but not all, living cells. DNA is found in the nucleus, a structure in the cytoplasm of a cell that is encircled by the nuclear membrane. There are two primary types of cells, distinguished by the presence or absence of a nucleus.

- Prokaryotic cells; and
- Eukaryotic cells.

3.2.1. Prokaryotic Cell

Cells without a nucleus are known as prokaryotic cells. In prokaryotic cells, the DNA is not encased in a nuclear membrane but rather resides in the cytoplasm. Bacteria and other single-celled organisms have prokaryotic cells (See Figure 3.5). Prokaryotes are defined as organisms possessing prokaryotic cells. They remain the most prevalent species even now because they were the first to evolve.

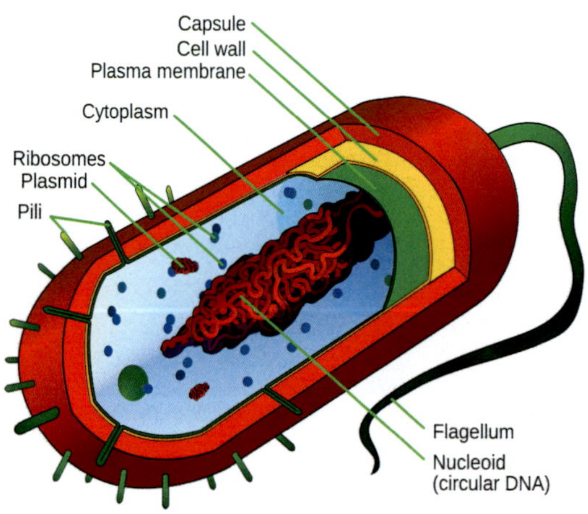

Figure 3.5. *Prokaryotic cell.*

Source: https://upload.wikimedia.org/wikipedia/commons/thumb/5/5a/Average_prokaryote_cell-_en.svg/1258px-Average_prokaryote_cell-_en.svg.png.

3.2.1.1. Components of Prokaryotic Cells

All cells share four common components:

- **A plasma membrane:** an outer covering that separates the cell's interior from its surrounding environment.
- **Cytoplasm:** a jelly-like cytosol within the cell in which other cellular components are found
- **DNA:** the genetic material of the cell
- **Ribosomes:** where protein synthesis occurs

A prokaryote is a simple, unicellular organism with one or more organized nuclei and no other organelles that are membrane-bound. We will soon discover that in eukaryotes, this is very different. The nucleoid, a central component of the cell, contains prokaryotic DNA.

Prokaryotes are characterized by their peptidoglycan cell walls and polysaccharide capsules in many cases. The cell wall serves as an additional barrier against injury, aids in shape maintenance, and keeps the cell from drying out. The cell can adhere to surfaces in its surroundings thanks to the capsule. Prokaryotes can have fimbriae, pili, or flagella. The mechanism of locomotion is flagella. Pili are used in conjugation, a kind of reproduction, to exchange genetic material. Bacteria use filaments to cling to a host cell.

Cell Size

Prokaryotic cells are much smaller than eukaryotic cells, which have diameters ranging from 10 to 100 µm, with a diameter of 0.1 to 5.0 µm. Owing to their small size, prokaryotes facilitate the rapid diffusion of ions and organic molecules into other areas of the cell. Similar to this, wastes generated in prokaryotic cells have a rapid diffusion outside of them. In contrast, eukaryotic cells have evolved distinct structural modifications to improve intracellular transport (Figure 3.6).

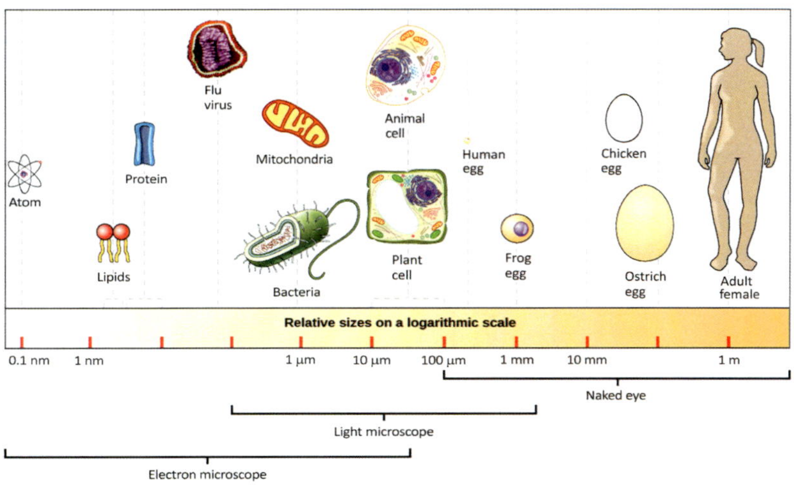

Figure 3.6. *Relative sizes of molecules, cells and organisms on a logarithmic scale.*

Source: https://courses.lumenlearning.com/suny-biology1/chapter/comparing-prokaryotic-and-eukaryotic-cells/.

For all cells, prokaryotic or eukaryotic, small size is generally required. Let's investigate this further. We'll start by thinking about the area and volume of a normal cell. Although not all cells have a spherical shape, the majority do. You may recall from your high school geometry class that a sphere's volume is equal to $4/3\pi r^3$, and its surface area is equal to $4\pi r^2$. Consequently, a cell's surface area grows as its radius squares, but its volume increases as its radius cube (much more quickly) as its radius increases. Consequently, a cell's surface area to volume ratio falls as it gets bigger. If the cell were shaped like a cube, the same idea would still hold true. The plasma membrane won't have enough surface area to support the rate of diffusion needed for the increased volume if the cell gets too big. Put another way, a cell loses efficiency as it grows. Separation is one method to increase efficiency; another is to create organelles with specialized functions. Due to these modifications, more advanced cells known as eukaryotic cells were created (Figure 3.7).

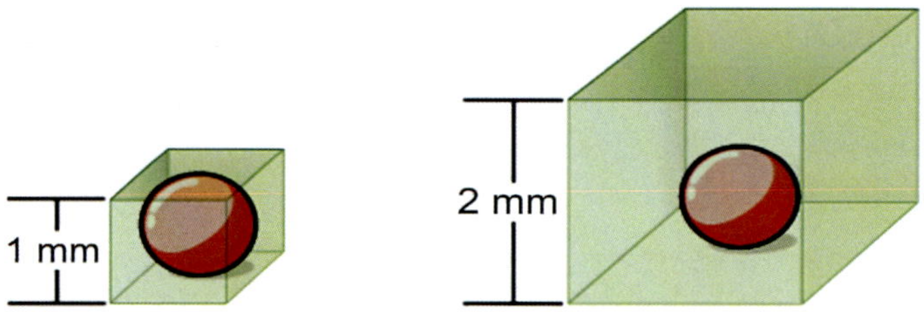

Figure 3.7. *Cell surface size.*

Source: https://bio.libretexts.org/Bookshelves/Introductory_and_General_Biology/Book%3A_General_Biology_%28Boundless%29/04%3A_Cell_Structure/4.04%3A_Studying_Cells_-_Cell_Size.

Different Features of Prokaryotes

The shoe rack, a framework to store and arrange all this essential stuff, would undoubtedly be the center of attention if you had a large collection of shoes and a walk-in closet. The most notable distinction between prokaryotes and eukaryotes is whether or not they have a nucleus. Keep in mind that the DNA in your eukaryotic cells is linear. The genome of prokaryotic cells is made of smaller, circular DNA. Prokaryotes essentially don't need a nucleus because of their smaller size and genome. Their nucleoid, which is essentially not a structure at all, contains their DNA. It is not a separate compartment because there are no membranes surrounding it. The nucleoid is just the part of the cell that contains DNA. This is how prokaryotic and eukaryotic cells differ from one another.

Prokaryotic cells are simpler than eukaryotic cells, but they are nevertheless capable of carrying out similar tasks. They just do them in a different way. This cytoplasmic nucleoid region is where processes like transcription and DNA replication occur, for instance. Even though ribosomes are still used for translation, they are not membrane-bound organelles; instead, every ribosome in a bacterium floats freely within the cytoplasm. The endoplasmic reticulum is absent (Figure 3.8).

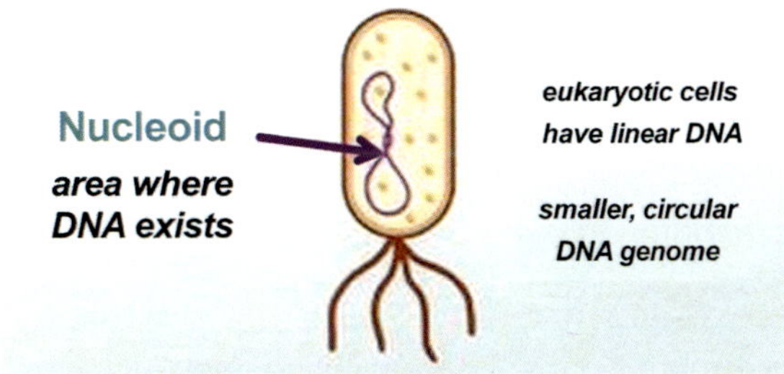

Figure 3.8. *Since prokaryotes do not have nuclei, their DNA is housed in a nucleoid.*

Citation: http://study.com/academy/lesson/eukaryotic-and-prokaryotic-cells-similarities-and-differences.html.

3.2.2. Eukaryotic Cells

Cells with a nucleus are known as eukaryotic cells. In Figure 3.9, a typical eukaryotic cell is displayed. Eukaryotic cells are primarily found in multicellular organisms and are typically larger than prokaryotic cells. Fungal species and humans are examples of eukaryotes, which are organisms having eukaryotic cells.

Aside from the nucleus, eukaryotic cells also have other organelles. An organelle is a cytoplasmic structure that serves a particular purpose for the cell. For instance, the cell's mitochondria supply energy, while the vacuoles are responsible for storing materials. Eukaryotic cells are capable of performing more tasks than prokaryotic cells due to the presence of organelles. As a result, prokaryotic cells cannot match the level of cell specificity exhibited by eukaryotic cells. In prokaryotic cells, ribosomes are the only organelles present. They are where proteins are made.

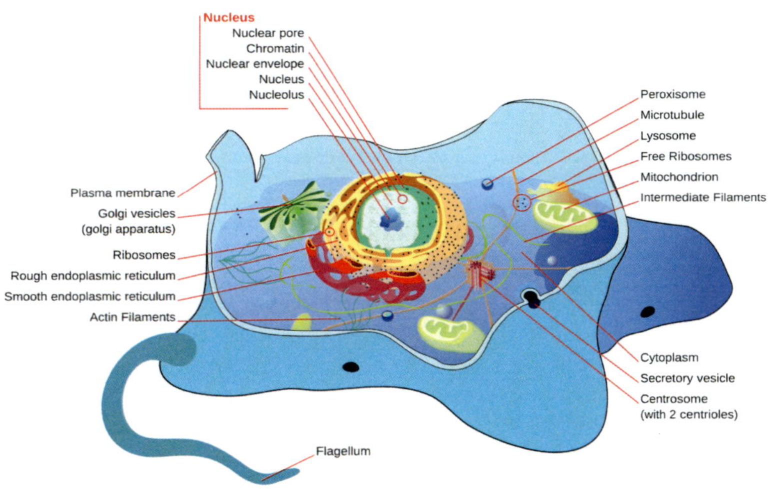

Figure 3.9. *Eukaryotic cell.*

Source: http://ailbay.weebly.com/uploads/3/7/9/6/37965993/prokaryotic-and-eukaryotic-cells-.pdf.

A cell is like a bag of Jell-O filled with plastic. Its fundamental structure is a cytoplasm-filled plasma membrane. Similar to mixed fruit that contains Jell-O, the cytoplasm of a cell also has different structures, including organelles and a nucleus.

3.2.2.1. Structure to Function in Eukaryotic Cells

The fluid mosaic model describes the basic structure shared by the membranes that form organelles in eukaryotic cells and the cell surface membrane. The phospholipid bilayer and membrane proteins of these membranes allow for the regulation of substances' entry and exit into cells and organelles.

These membranes separate the cytoplasm of eukaryotic cells into various compartments or organelles. Organelles enable various processes to take place effectively and concurrently in various areas of the cell. For instance, the outer double membrane of the mitochondrion distinguishes between reactions that take place in the general cytoplasm and those that happen within the mitochondria. Additionally, the electron transfer chain reactions (ETC) and the enzymatic reactions of the Krebs cycle can be kept apart thanks to the internal membranes of the mitochondria. This is crucial because the pH optima of the two sets of reactions differ due to the different enzymes involved. The membranes that make up the crista allow the mitochondria's cytoplasm to be divided, which concentrates enzymes and substrates and optimizes pH. Therefore, the interior of eukaryotic cells can be said to be compartmentalized by membranes.

It is possible to characterize the relationship between structure and function in terms of either the entire eukaryotic cell or its individual organelles. The description of eukaryotic organelles' structure and function is found below in Table 3.1.

Table 3.1. *The Structure and Function of Eukaryotic Organelles*

Major Eukaryotic Organelles				
Organelle	**Main Function**	**Structure**	**Organisms**	**Notes**
Cell membrane	Separates the interior of all cells from the outside environment (the extracellular space) which protects the cell from its environment.	double-layered, fluid sheet of phospholipids	All eukaryotes	
Cell wall	The cell wall is a rigid structure composed of cellulose that provides shape to the cell, helps keep the organelles inside the cell, and does not let the cell burst from osmotic pressure.	various	Plants, protists, rare kleptoplastic organisms	

Chloroplast (plastid)	photosynthesis, traps energy from sunlight	double-membrane compartment	Plants, algae, rare kleptoplastic organisms	Has own DNA; theorized to be engulfed by the ancestral archaeplastid cell (endosymbiosis)
Endoplasmic reticulum	Translation and folding of new proteins (rough endoplasmic reticulum), expression of lipids (smooth endoplasmic reticulum)	Single-membrane compartment	All eukaryotes	Rough endoplasmic reticulum is covered with ribosomes (which are bound to the ribosome membrane), has folds that are flat sacs; smooth endoplasmic reticulum has folds that are tubular
flagellum	locomotion, sensory	Protein	Some eukaryotes	
Golgi apparatus	Sorting, packaging, processing and modification of proteins	Single-membrane compartment	All eukaryotes	Cis-face (convex) nearest to rough endoplasmic reticulum; trans-face (concave) farthest from rough endoplasmic reticulum
Mitochondria	Energy production from the oxidation of glucose substances and the release of adenosine triphosphate	Double-membrane compartment	Most eukaryotes	Constituting element of Chondriome; has own DNA; theorized to have been engulfed by an ancestral eukaryotic cell (endosymbiosis) (Murakami, Otsuka, Sugimoto, & Mitsui, 2012)

Nucleus	DNA maintenance, controls all activities of the cell, RNA transcription	Double-membrane compartment	All eukaryotes	Contains bulk of genome
Vacuole	Storage, transportation, helps maintain homeostasis	Single-membrane compartment	All eukaryotes	

Minor Eukaryotic Organelles and Cell Components

Organelle/ Macromolecule	**Main function**	**Structure**	**Organisms**
Acrosome	Helps spermatozoa fuse with ovum	Single-membrane compartment	Most animals (including sponges)
Autophagosome	Vesicle that sequesters cytoplasmic material and organelles for degradation	Double-membrane compartment	All eukaryotes
Centriole	Anchor for cytoskeleton, organizes cell division by forming spindle fibers	Microtubule protein	Animals
Cilium	Movement in or of external medium; "critical developmental signaling pathway." (Ogushi, & Kori, 2013)	Microtubule protein	Animals, protists, few plants
Cnidocyst	Stinging	Coiled hollow tubule	Cnidarians
Eyespot apparatus	Detects light, allowing phototaxis to take place		Green algae and other unicellular photosynthetic organisms such as euglenids
Glycosome	Carries out glycolysis	Single-membrane compartment	Some protozoa, such as *Trypanosomes*.
Glyoxysome	Conversion of fat into sugars	Single-membrane compartment	Plants
Hydrogenosome	Energy & hydrogen production	Double-membrane compartment	A few unicellular eukaryotes
Lysosome	Breakdown of large molecules (e.g., proteins + polysaccharides)	Single-membrane compartment	Animals
Melanosome	Pigment storage	Single-membrane compartment	Animals

Mitosome	Probably plays a role in Iron–sulfur cluster (Fe–S) assembly	Double-membrane compartment	A few unicellular eukaryotes that lack mitochondria
Myofibril	Myocyte contraction	Bundled filaments	Animals
Nucleolus	Pre-ribosome production	Protein–DNA–RNA	Most eukaryotes
Ocelloid	Detects light and possibly shapes, allowing phototaxis to take place	Double-membrane compartment	Members of the family Warnowiaceae
Parenthesome	Not characterized	Not characterized	Fungi
Peroxisome	Breakdown of metabolic hydrogen peroxide	Single-membrane compartment	All eukaryotes
Porosome	Secretory portal	Single-membrane compartment	All eukaryotes
Proteasome	Degradation of unneeded or damaged proteins by proteolysis	Very large protein complex	All eukaryotes, all archaea, and some bacteria
Ribosome (80S)	Translation of RNA into proteins	RNA-protein	All eukaryotes
Stress granule	mRNA storage (Palmer, & Delwiche, (1996)	Membraneless (mRNP complexes)	Most eukaryotes
TIGER Domain	mRNA encoding proteins	Membraneless	Most organisms
Vesicle	Material transport	Single-membrane compartment	All eukaryotes

Source: https://en.wikipedia.org/wiki/Organelle.

3.2.3. Differences Between Eukaryotic and Prokaryotic Cells

Eukaryotic and prokaryotic cells represent two fundamentally distinct forms of cellular organization found in living organisms. These differences encompass various aspects, including their structural complexity, organization of genetic material, internal organelles, metabolic capabilities, and evolutionary origins. Below, I'll delve into these disparities, highlighting the unique features of each cell type.

3.2.3.1. Structural Complexity

Eukaryotic cells are typically larger and more complex than prokaryotic cells. They have a defined nucleus enclosed within a double membrane, while prokaryotic cells lack a true nucleus. Instead, their genetic material is concentrated in a region called the nucleoid.

3.2.3.2. Genetic Material Organization

In eukaryotic cells, DNA is organized into linear chromosomes within the nucleus, along with histone proteins that help in packaging and regulating gene expression. Prokaryotic cells contain a single, circular chromosome located in the nucleoid region, lacking histone proteins.

3.2.3.3. Membrane-Bound Organelles

Eukaryotic cells feature various membrane-bound organelles, such as mitochondria, endoplasmic reticulum, Golgi apparatus, lysosomes, and chloroplasts (in plant cells). These organelles compartmentalize cellular functions. Prokaryotic cells lack membrane-bound organelles, though they may contain simpler structures like ribosomes, cytoplasm, cell wall, plasma membrane, and sometimes flagella or pili.

3.2.3.4. Cell Size

Eukaryotic cells are generally larger, ranging from 10 to 100 micrometers in diameter, while prokaryotic cells are much smaller, typically ranging from 0.1 to 5 micrometers.

3.2.3.5. Reproduction

Eukaryotic cells reproduce via mitosis (for somatic cells) or meiosis (for gametes), involving complex processes of nuclear division and cytoplasmic division (cytokinesis). Prokaryotic cells reproduce through binary fission, where the single circular chromosome replicates, and the cell divides into two daughter cells.

3.2.3.6. Metabolic Diversity

Eukaryotic cells exhibit diverse metabolic pathways due to the presence of various organelles specialized for different functions. Prokaryotic cells have simpler metabolic pathways, but they can be incredibly diverse, allowing them to thrive in a wide range of environments, including extreme conditions.

3.2.3.7. Cell Wall Composition

In eukaryotic cells, if present, the cell wall is primarily composed of cellulose (in plants), chitin (in fungi), or peptidoglycan (in some protists). Prokaryotic cells (bacteria and archaea) have cell walls containing peptidoglycan, although the composition may vary between different bacterial species.

3.2.3.8. Evolutionary Origins

Eukaryotic cells are believed to have evolved from prokaryotic ancestors through a process called endosymbiosis, where a prokaryotic cell engulfed another prokaryotic cell, leading to the formation of mitochondria and chloroplasts.

3.2.3.9. Flagella Structure

Eukaryotic flagella are complex structures consisting of microtubules arranged in a "9+2" pattern, surrounded by the plasma membrane, and are involved in cell motility. Prokaryotic flagella are simpler, composed of a filament, hook, and basal body, and are not membrane-bound.

3.2.3.10. DNA Replication

Eukaryotic DNA replication is initiated at multiple points along the chromosome, and the process is more complex due to the larger genome size and presence of histones. Prokaryotic DNA replication typically starts at a single origin of replication and proceeds bidirectionally around the circular chromosome. (Sturtevant & Beadle, 1939; Taylor & Wang, 1989).

3.3. THE MOLECULAR COMPOSITION OF CELLS

Water, inorganic ions, and (organic) molecules containing carbon make up cells. At least 70% of the total mass of a cell is made up of water, making it the most prevalent molecule in cells. As a result, water's interactions with the other components of cells are crucial to understanding biological chemistry. In this regard, water's polarity—in which the oxygen atoms have a slight negative charge and the hydrogen atoms have a slight positive charge—is its key characteristic (Figure 3.10). Water molecules' polarity allows them to interact with positively or negatively charged ions, as well as form hydrogen bonds with other polar molecules. These interactions lead to the easy solubility of ions and polar molecules in water (hydrophilicity). On the other hand, nonpolar molecules—which are hydrophobic—are poorly soluble in an aqueous solution because they are unable to interact with water. As a result, nonpolar molecules prefer to form close relationships with one another to reduce their interaction with water.

The inorganic ions of the cell, including sodium (Na^+), potassium (K^+), magnesium (Mg^{2+}), calcium (Ca^{2+}), phosphate (HPO_4^{2-}), chloride (Cl^-), and bicarbonate (HCO_3^-), constitute 1% or less of the cell mass. These ions are involved in a number of aspects of cell metabolism and thus play critical roles in cell function.

The distinct components of cells are the organic molecules. The majority of these organic compounds fall into one of four molecular classes: nucleic acids, proteins, lipids, or carbohydrates. The majority of carbohydrates (polysaccharides), proteins, and nucleic acids are macromolecules that are created through the joining (polymerization) of hundreds or thousands of low-molecular-weight precursors, namely amino acids, nucleotides, and simple sugars. These macromolecules make up 80–90% of the dry weight of the majority of cells. The other main component of cells is lipids. A range of small organic molecules, including precursors of macromolecules, make up the remaining mass of the cell. Therefore, the structures and roles of four main classes of organic molecules can be used to understand the fundamental chemistry of cells.

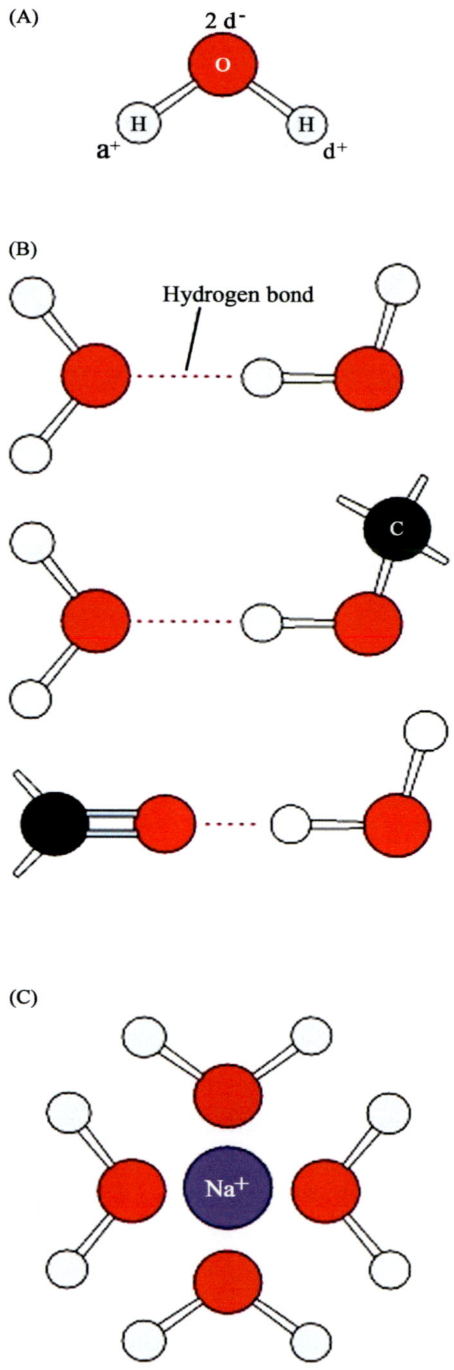

Figure 3.10. *Characteristics of water.*

Source: https://www.ncbi.nlm.nih.gov/books/NBK9879/.

Note: (A) Water is a polar molecule, with a slight negative charge (δ-) on the oxygen atom and a slight positive charge (δ+) on the hydrogen atoms. Because of this polarity, water molecules can form hydrogen bonds (dashed lines) either with each other or with other polar molecules (B), in addition to interacting with charged ions (C).

3.3.1. Carbohydrates

Polysaccharides and simple sugars are two types of carbohydrates. The main nutrients in cells are these simple sugars, like glucose. Their disintegration supplies the building blocks for the synthesis of other cell components as well as a source of cellular energy. Sugars can be stored as polysaccharides, which are structural elements of the cell. Furthermore, polysaccharides and shorter sugar polymers serve as indicators for several cell recognition mechanisms, such as cell adhesion to neighboring cells and protein transport to specific intracellular locations.

The basic formula for these molecules is $(CH_2O)_n$, from which the name carbohydrate is derived (C= "carbo" and H_2O= "hydrate"). The six-carbon (n= 6) sugar glucose ($C_6H_{12}O_6$) is especially important in cells since it provides the principal source of cellular energy. Other simple sugars have between three and seven carbons, with three- and five-carbon sugars being the most common. Sugars containing five or more carbons can cyclize to form ring structures, which are the predominant forms of these molecules within cells. The cyclized sugars exist in two alternative forms (called α or β), depending on the configuration of carbon 1 (Figure 3.11).

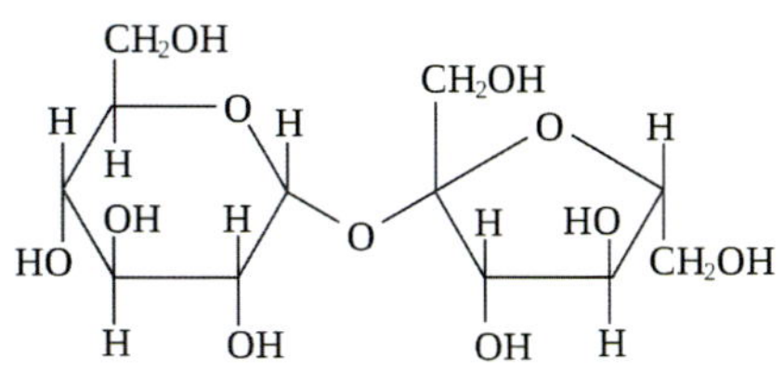

Figure 3.11. *Structure of simple sugars.*

Source: Public Domain, https://commons.wikimedia.org/w/index.php?curid=791528.

Through dehydration reactions, which remove H_2O and link the sugars by a glycoside bond between two of their carbons, monosaccharides can be joined together. An oligosaccharide is the name given to the polymer formed when a small number of sugars are combined. Polymers that result from the involvement of a significant quantity of sugars, such as hundreds or thousands, are known as polysaccharides (Figure 3.12).

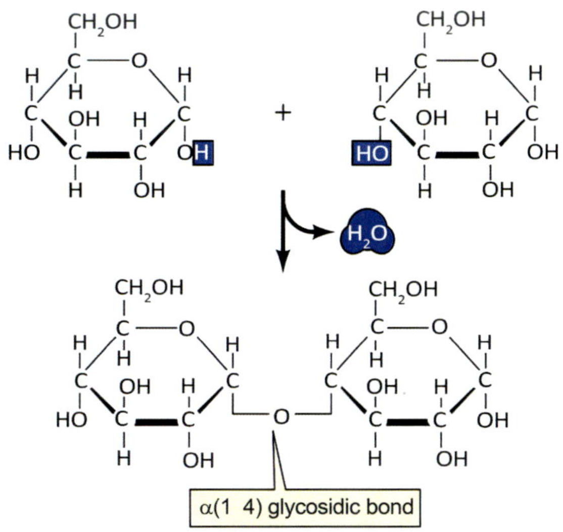

Figure 3.12. *Formation of a glycosidic bond.*

Source: https://courses.lumenlearning.com/suny-orgbiochemistry/chapter/disaccharides/.

In plant and animal cells, respectively, the storage forms of carbohydrates are two common polysaccharides: starch and glycogen. Glycogen molecules in the α configuration make up the entirety of both starch and glycogen (Figure 3.13). The main connection is made between the carbons 1 and 4 of one glucose and the second. Furthermore, there are sporadic α (1→6) linkages in glycogen and amylopectin, which connect the carbon 1 of one glucose to the carbon 6 of another. When two independent α (1→4) linked chains are joined, these linkages cause branches to form. While amylopectin and glycogen both contain these branches, amylose, a different type of starch, is an unbranched molecule.

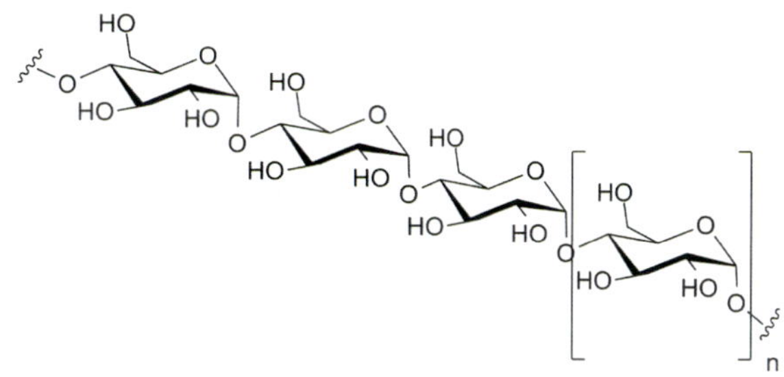

Figure 3.13. *Structure of polysaccharides.*

Citation: https://upload.wikimedia.org/wikipedia/commons/7/71/Amylose_3Dprojection.corrected.png.

Because they both store glucose, glycogen and starch essentially have similar structures. On the other hand, cellulose serves a very specific purpose as the main structural element of the plant cell wall. It may come as a surprise to learn that all of the molecules in cellulose are glucose molecules. However, cellulose is an unbranched polysaccharide, and its glucose residues are arranged in the β configuration as opposed to the α configuration. When glucose residues are linked by β (1→4) bonds instead of α (1→4) bonds, cellulose forms long, extended chains that stack one on top of the other to create fibers with exceptional mechanical strength.

Apart from their functions in energy storage and cellular architecture, oligosaccharides and polysaccharides play a crucial role in numerous cell signaling pathways. For instance, oligosaccharides are commonly attached to proteins, where they act as markers to direct proteins toward the cell surface or various subcellular organelles for incorporation. In addition to acting as surface markers, oligosaccharides and polysaccharides are crucial for cell recognition and intercellular communication in the tissues of multicellular organisms.

3.3.2. Lipids

In cells, lipids play three main functions. They offer a significant means of storing energy, to start. Lipids are the main building blocks of cell membranes, which makes them extremely significant in the field of cell biology. Third, lipids are crucial for cell signaling because they function as messenger molecules, which transfer signals from cell surface receptors to specific locations within the cell, as well as steroid hormones.

Fatty acids are the most basic type of lipids. They are made up of long hydrocarbon chains with a carboxyl group (COO-) at one end, usually comprising 16 or 18 carbon atoms (Figure 3.14). Saturated fatty acids have all of the carbon atoms bonded to as many hydrogen atoms as possible, while unsaturated fatty acids have one or more double bonds between the carbon atoms. The only C–H bonds found in the lengthy hydrocarbon chains of fatty acids are nonpolar, meaning they cannot interact with water. Much of the behavior of complex lipids, especially in the formation of biological membranes, is due to the hydrophobic nature of these fatty acid chains.

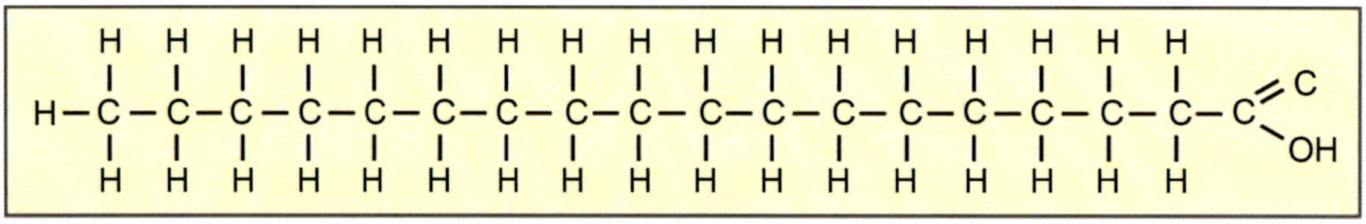

Figure 3.14. *Structure of fatty acids.*

Source: https://courses.lumenlearning.com/suny-wmopen-biology1/chapter/lipids/.

Fatty acids are stored in the form of triacylglycerols, or fats, which consist of three fatty acids linked to a glycerol molecule. Triacylglycerols are insoluble in water and therefore accumulate as fat droplets in the cytoplasm. It is noteworthy that fats are a more efficient form of energy storage than carbohydrates, yielding more than twice as much energy per weight of material broken down. Fats therefore allow energy to be stored in less than half the body weight that would be required to store the same amount of energy in carbohydrates a particularly important consideration for animals because of their mobility (Figure 3.15).

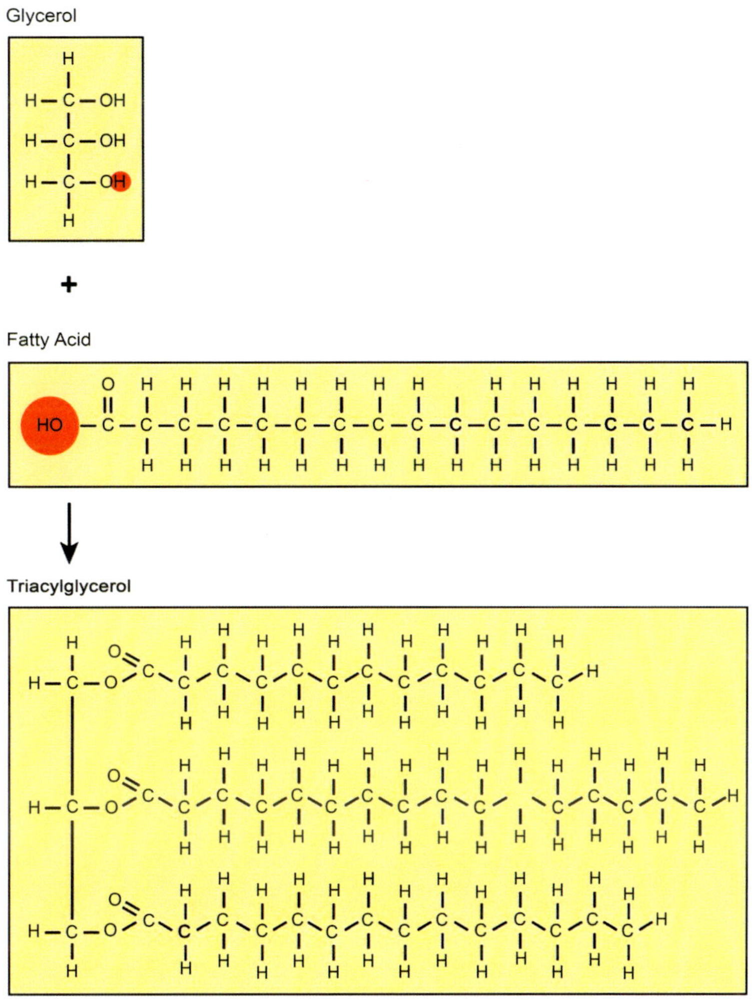

Figure 3.15. *Structure of triacylglycerols.*

Source: https://courses.lumenlearning.com/suny-wmopen-biology1/chapter/lipids/.

Phospholipids, which are the main components of cell membranes, are made up of two fatty acids connected to a polar head group. In glycerol phospholipids, the fatty acids are linked to carbon atoms in glycerol, similar to triacylglycerols. However, the third carbon of glycerol is connected to a phosphate group, which is often linked to another small polar molecule like choline, serine, inositol, or ethanolamine. Sphingomyelin, the only phospholipid in cell membranes that is not based on glycerol, consists of two hydrocarbon chains attached to a polar head group made from serine. All phospholipids have hydrophobic tails made up of the two hydrocarbon chains and hydrophilic head groups consisting of the phosphate group and its polar attachments. As a result, phospholipids are considered amphipathic molecules, being partially soluble in water and partially insoluble. This unique property of phospholipids is essential for the creation of biological membranes.

3.3.3. Nucleic Acids

The main informational molecules in a cell are called nucleic acids, and they consist of DNA and RNA. Since DNA is the genetic material found in the nucleus of eukaryotic cells, it plays a special role in these cells. Numerous cellular processes involve the participation of various forms of ribonucleic acid (RNA). Information from DNA is transported by messenger RNA (mRNA) to the ribosomes, where it acts as a template for the synthesis of proteins. Protein synthesis involves two additional forms of RNA: transfer RNA and ribosomal RNA. The processing and transport of both RNAs and proteins are carried out by still other types of RNAs. RNA can catalyze a variety of chemical reactions in addition to serving as an informational molecule. These include processes that are involved in both protein synthesis and RNA processing in modern cells.

Nucleotides, the building blocks of DNA and RNA, are polymers made up of purine and pyrimidine bases connected to phosphorylated sugars. DNA is made up of two pyrimidines, thymine and cytosine, and two purines, adenine and guanine. RNA also contains adenine, guanine, and cytosine, but instead of thymine, uracil is found in RNA. Nucleosides are formed when bases are joined to sugars, either ribose in RNA or 2′-deoxyribose in DNA. In addition, nucleotides have one or more phosphate groups attached to the nucleoside sugars' 5′ carbon (Figure 3.16).

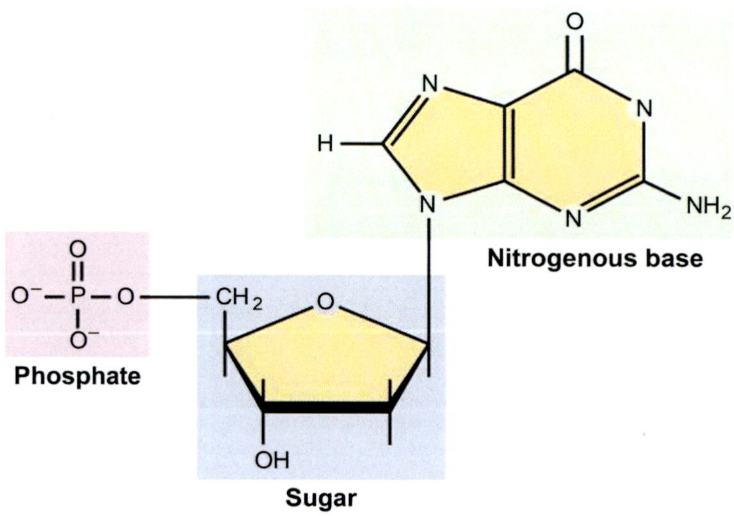

Figure 3.16. *Components of nucleic acids.*

Source: https://courses.lumenlearning.com/wm-nmbiology1/chapter/reading-structure-of-nucleic-acids/.

Phosphodiester bonds are created between the 3′ hydroxyl and 5′ phosphate of one nucleotide during the polymerization of nucleotides to create nucleic acids. Small polymers called oligonucleotides have only a few nucleotides; in contrast, large polynucleotides, such as those found in cellular RNA and DNA, can have thousands or even millions of nucleotides. It is crucial to understand that a polynucleotide chain has a sense of direction, ending in a 3′ hydroxyl group at one end and a 5′ phosphate group at the other. A free nucleotide is added to the 3′ OH group of a growing chain during the 5′ to 3′ direction of polynucleotide synthesis. It is customary for the base sequences in DNA and RNA to be written from 5′ to 3′ (Figure 3.17).

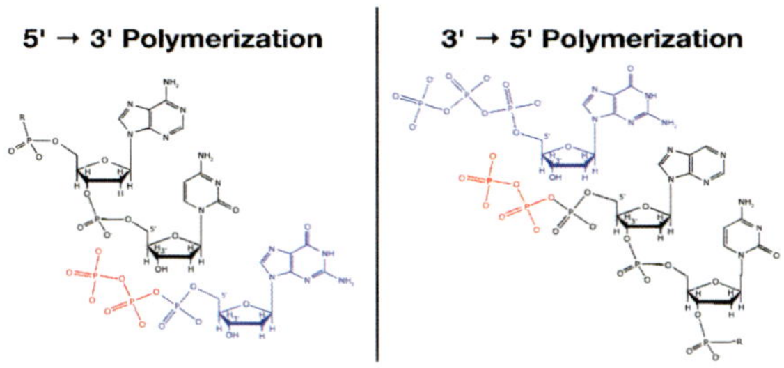

Figure 3.17. *Polymerization of nucleotides.*

Source: http://www.manhattanmetric.com/Ph.D.-Dissertation/reactions.png.

The arrangement of bases in a polynucleotide chain encodes information in both DNA and RNA. DNA is a double-stranded molecule made up of two chains of polynucleotides that run counter to one another. The bases are inside the molecule, and hydrogen bonds between complementary base pairs—adenine pairing with thymine and guanine pairing with cytosine—join the two chains. One strand of DNA (or RNA) can function as a template to guide the synthesis of a complementary strand, which is an important outcome of such complementary base pairing. Because of their exceptional ability to control their own self-replication, nucleic acids serve as the basic information molecules in cells. Specific proteins are synthesized in response to instructions carried by DNA and RNA, and these proteins regulate the majority of cellular functions (Figure 3.18).

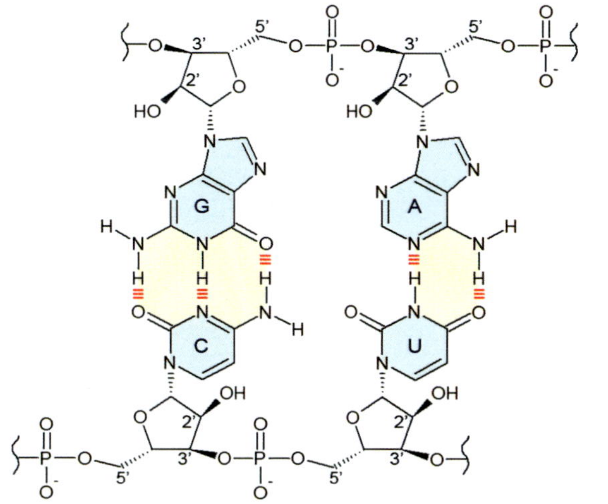

Figure 3.18. *Complementary pairing between nucleic acid bases.*

Source: By Iquo – Illustrated by the uploader, CC BY-SA 3.0, https://commons.wikimedia.org/w/index.php?curid=1042812.

In addition to serving as the building blocks of nucleic acids, nucleotides are essential for several other cellular functions. The most well-known example is probably adenosine 5′-triphosphate (ATP), which is the main chemical energy source in cells. In a wide range of metabolic reactions, other nucleotides serve a similar role as carriers of energy or reactive chemical groups. Furthermore, certain nucleotides—like cyclic AMP, for instance—are crucial signaling molecules in cells.

3.3.4. Proteins

Although the genetic information of the cell is carried by nucleic acids, proteins are primarily responsible for carrying out the tasks that are directed by that information. Of all macromolecules, proteins are the most diverse. There are thousands of distinct proteins found in each cell, and they all serve a variety of purposes. The structural elements of cells and tissues, the movement and storage of small molecules, information transmission between cells, and infection protection are just a few of the functions of proteins. But the most fundamental characteristic of proteins is their capacity to function as enzymes, which enables them to catalyze almost all chemical reactions in biological systems. As a result, almost every cell activity is controlled by proteins.

20 distinct amino acid polymers make up proteins. The components of every amino acid are a carbon atom, also known as the α carbon, bound to a hydrogen atom, an amino group (NH_3^+), a carboxyl group (COO–), and a characteristic side chain (Figure 3.19). The roles that each amino acid plays in the structure and function of proteins are determined by the unique chemical characteristics of the various side chains.

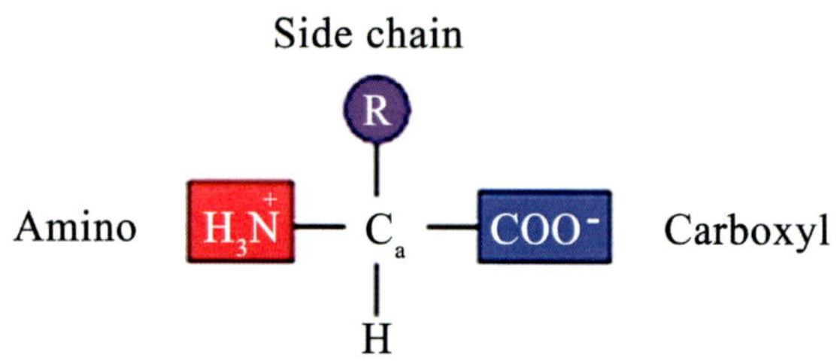

Figure 3.19. *Structure of amino acids.*

Source: https://www.ncbi.nlm.nih.gov/books/NBK9879/.

The characteristics of the side chains of the amino acids allow them to be divided into four major groups. The nonpolar side chains of 10 amino acids don't interact with water. The most basic amino acid is glycine, which has just one hydrogen atom in its side chain. The hydrocarbon side chains of aspartic acid, valine, leucine, and isoleucine can contain up to four carbon atoms. Due to their hydrophobic nature, these amino acids' side chains are more likely to be found inside proteins, away from water. Similar to proline, proline has a hydrocarbon side chain, but it differs from other hydrocarbon side chains in that it is bonded to both the α carbon and the amino group's nitrogen, creating a cyclic structure. Sulfur atoms can be found in the side chains of two amino acids: cysteine and methionine. Because of its sulfhydryl (SH) group, cysteine is less hydrophobic than methionine. Disulfide bonds between the side chains of various cysteine residues can form, which is why the sulfhydryl group of cysteine is important for protein structure. Lastly, the side chains of two nonpolar amino acids—phenylalanine and tryptophan—contain extremely hydrophobic aromatic rings.

Side chains containing charged basic groups are present in the amino acids histidine, arginine, and lysine. Extremely basic amino acids like arginine and lysine have positively charged side chains inside of cells. They are highly hydrophilic and can be found on the surface of proteins in contact with water. Histidine often participates actively in enzymatic reactions involving the exchange of hydrogen ions because it can be either positively charged or uncharged at physiological pH.

Lastly, the acidic side chains of two amino acids—aspartic acid and glutamic acid—end in carboxyl groups. These amino acids are commonly referred to as aspartate and glutamate because they are negatively charged within the cell. These acidic amino acids are highly hydrophilic, just like the basic amino acids, and are typically found on the surface of proteins.

Peptide bonds, which connect the α amino group of one amino acid to the α carboxyl group of another, bind amino acids together. Polypeptides are linear sequences of amino acids that range in length from hundreds to thousands of amino acids. Every polypeptide chain has two unique ends: the amino, or N, terminus, which ends in an α amino group, and the carboxy, or C, terminus, which ends in an α carboxyl group. The amino acid sequence in a polypeptide is written (by convention) in the same order as the polypeptide is synthesized from the amino to the carboxy terminus (Figure 3.20).

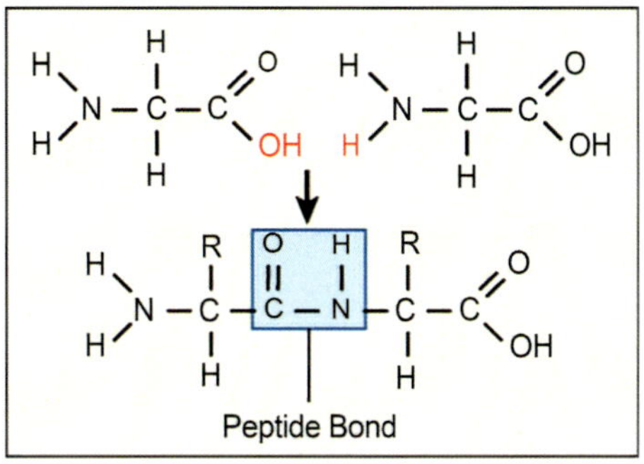

Figure 3.20. *Formation of a peptide bond.*

Source: https://courses.lumenlearning.com/wm-biology1/chapter/reading-amino-acids/.

The fact that proteins are polypeptides with particular amino acid sequences is what makes them unique. It was discovered that insulin is made up of two polypeptide chains connected by disulfide bonds made up of cysteine residues. The most significant finding of Sanger's experiment was that every protein has a unique amino acid sequence. Currently, automated techniques are used to sequence proteins, and over 100,000 proteins' entire amino acid sequences are known. The arrangement of nucleotides in a gene determines the distinct sequence of amino acids that make up each.

The folding of the polypeptide chain that results from interactions between the side chains of amino acids located in various regions of the primary sequence is known as the tertiary structure (Figure 3.21). The fundamental units of tertiary structure in most proteins are called domains, which are compact globular structures made up of combinations of α helices and β sheets joined by loop regions of the polypeptide chain. Larger proteins can have several different domains, many of which are linked to specific functions. Small proteins, like ribonuclease or myoglobin, only have one domain.

The location of hydrophilic amino acids on the protein's surface, where they interact with water, and hydrophobic amino acids inside the protein is a crucial factor in determining the tertiary structure. Hydrophobic amino acids arranged in α helices and β sheets make up the majority of the interiors of folded proteins. These secondary structures are present in the hydrophobic cores of proteins because hydrogen bonding balances the polar nature of the CO and NH groups of the polypeptide backbone. On the surface of folded proteins, where the polar peptide bond components form hydrogen bonds with water or the polar side chains of hydrophilic amino acids, are the loop regions that connect the elements of secondary structure. Important factors that determine tertiary structure also include interactions between polar amino acid side chains (ionic and hydrogen bonds) on the protein surface. Furthermore, the folded structures of numerous cell-surface or secreted proteins are stabilized by the covalent disulfide bonds that form between the sulfhydryl groups of cysteine residues.

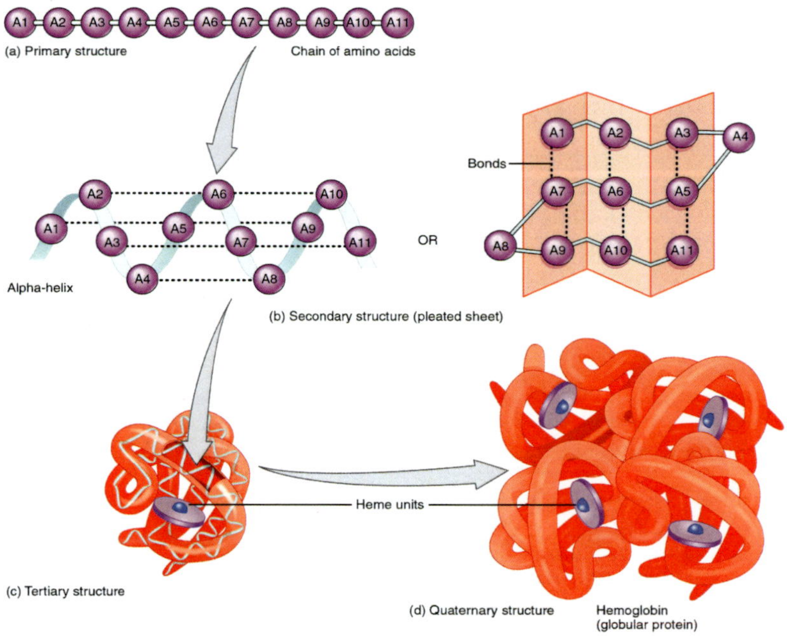

Figure 3.21. *Tertiary structure of ribonuclease.*

Source: https://en.m.wikipedia.org/wiki/File:225_Peptide_Bond-03.jpg. /.

In proteins made up of multiple polypeptides, the interactions between various polypeptide chains make up the fourth level of protein structure, known as quaternary structure. The four polypeptide chains that makeup hemoglobin, for instance, are bound together by the same kinds of interactions that preserve the tertiary structure.

Thus, there is a great deal of variation in the three-dimensional conformations of folded proteins due to the unique chemical properties of the 20 different amino acids.

3.4. CELL CYCLE

The primary purpose of the cell cycle is to precisely duplicate the enormous amount of DNA found in chromosomes, and then to separate the copies into two genetically identical daughter cells. These mechanisms delineate the two principal stages of the cell cycle. In a typical mammalian cell, DNA duplication takes place during the S phase (S for synthesis), which takes 10–12 hours and accounts for roughly half of the cell cycle. After the S phase, the M phase (M for mitosis) is when chromosome segregation and cell division take place.

In a mammalian cell, this phase lasts for less than an hour. Nuclear division, or mitosis, is the first of several dramatic events that make up the M phase. The process of chromosome condensation, which starts mitosis, causes duplicated DNA strands that were previously packaged into elongated chromosomes to condense into the much more compact chromosomes needed for their segregation. The replicated chromosomes, each made up of two sister chromatids, attach to the microtubules of the mitotic spindle when the nuclear envelope disintegrates.

The chromosomes are positioned at the mitotic spindle's equator, ready for segregation, during a brief period of time known as metaphase, which occurs as mitosis advances. The chromosomes travel to the opposite poles of the spindle during anaphase, when they decondense and reform intact nuclei. This process is initiated by the abrupt separation of sister chromatids. Following cytoplasmic division, also known as cytokinesis, the cell is split in half, marking the completion of cell division (Figure 3.22).

The majority of cells take a lot longer to grow and double the mass of their organelles and proteins than they do to divide and replicate their DNA. Most cell cycles have two extra gap phases added, a G_1 phase between the M phase and S phase and a G_2 phase between the S phase and mitosis, in part to give the cells more time for growth. The G_1, S, G_2, and M phases are the four sequential phases into which the eucaryotic cell cycle is typically divided (Figure 3.23). We refer to G_1, S, and G_2 collectively as interphase. A normal human cell growing in culture may spend 23 hours of a 24-hour cycle in interphase and 1 hour in the M phase.

The two gap phases do more for cell growth than just act as temporal delays. They also give the cell time to observe its surroundings, both internal and external, to make sure everything is ready and conditions are met before the cell commits to the major events of the S phase and mitosis. The G_1 phase is particularly significant in this regard. Its length can change significantly based on outside factors and extracellular cues from other cells. For example, in unfavorable extracellular environments, cells postpone going through G_1 and may even enter a specialized resting state called G_0 (G zero), where they can stay for days, weeks, or even years before starting to proliferate again. In fact, a large number of cells never leave G_0 until either the organism or they pass away.

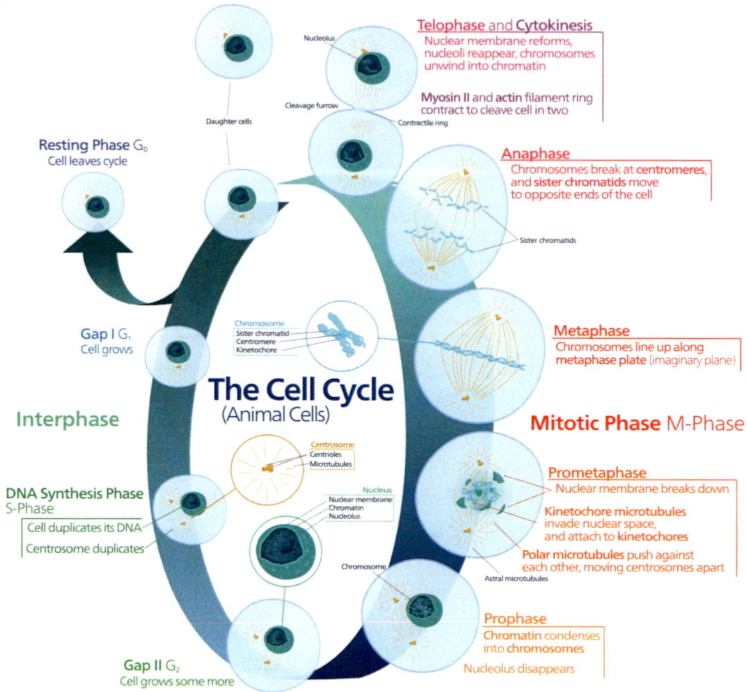

Figure 3.22. *The events of eucaryotic cell division as seen under a microscope.*

Source: By Kelvinsong – Own work, CC0, https://commons.wikimedia.org/w/index.php?curid=22965076.

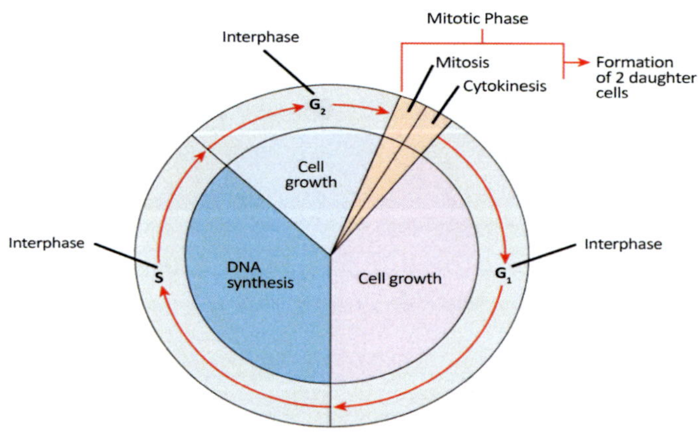

Figure 3.23. *The phases of the cell cycle. The cell grows continuously in interphase, which consists of three phases: DNA replication is confined to the S phase; G1 is the gap between the M phase and the S phase, while G2 is the gap between the S phase and the M phase. In the M phase, the nucleus and then the cytoplasm divide.*

Source: https://courses.lumenlearning.com/suny-biology1/chapter/the-cell-cycle/.

Beginning at the end of G_0, cells in early G_1 or G_0 proceed through a commitment point called the restriction point (in mammalian cells) or Start (in yeasts) if extracellular conditions are favorable and signals to grow and divide are present. Cells are dedicated to DNA replication once they have passed this stage, even in the absence of extracellular cues that promote cell division and growth.

3.4.1. The Cell-Cycle Control System Is Similar in All Eukaryotes

Even within the same organism, there are significant differences in certain aspects of the cell cycle between cell types, such as the duration of specific events. All eucaryotic cells, however, share a nearly identical basic cycle organization and control system. The control system proteins initially emerged more than a billion years ago. Surprisingly, many of them work flawlessly when transferred from a human cell to a yeast cell because they have been so well-conserved throughout evolution. Thus, by examining the cell cycle and its control in a range of organisms, we can combine the knowledge from each to create a comprehensive understanding of the process by which eukaryotic cells divide.

3.4.2. The Cell-Cycle Control System Can Be Dissected Genetically in Yeasts

The cell-cycle regulation mechanisms of yeasts, which are microscopic, single-celled fungi, are strikingly similar to those of humans. Cell cycle research typically involves two species. The African beer that is made with the fission yeast *Schizosaccharomyces pombe* is the source of the yeast's name. The growth of this rod-shaped cell occurs through elongation at the ends. The process of division involves the creation of a cell plate, or septum, in the rod's center (Figure 3.24A). Both bakers and brewers use *Saccharomyces cerevisiae*, a budding yeast. It is an oval cell that undergoes mitosis to divide by forming a bud, which initially emerges during G_1 and grows gradually until it splits off from the mother cell (Figure 3.24B).

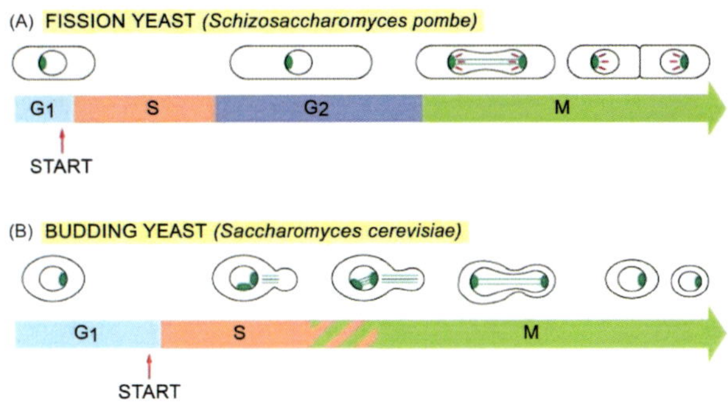

Figure 3.24. *A comparison of the cell cycles of fission yeasts and budding yeasts.*

Source: https://image.slideserve.com/793935/a-comparison-of-the-cell-cycles-of-fission-yeasts-and-budding-yeasts-l.jpg.

The two yeast species are different in appearance, but they are similar in a lot of ways that make them great for genetic research. Their genome is smaller than 1% of a mammal's, and they reproduce nearly as quickly as bacteria. They are quickly manipulable through molecular genetics; genes can be added, removed, or changed. The ability to proliferate in a haploid state, where each gene is only found in one copy in the cell, is perhaps their most remarkable trait. Since there is no hassle of having two copies of a gene in the cell, it is simple to identify and investigate mutations that render a gene inactive in haploid cells.

Systematic searches for yeast mutations that inactivate genes encoding crucial elements of the cell-cycle control system have yielded many significant discoveries about cell-cycle control. These mutations impact a class of genes called cell-division-cycle genes, or cdc genes. Many mutations result in cell arrest at a particular stage of the cell cycle, indicating that the normal gene product is necessary for the cell to continue beyond this point.

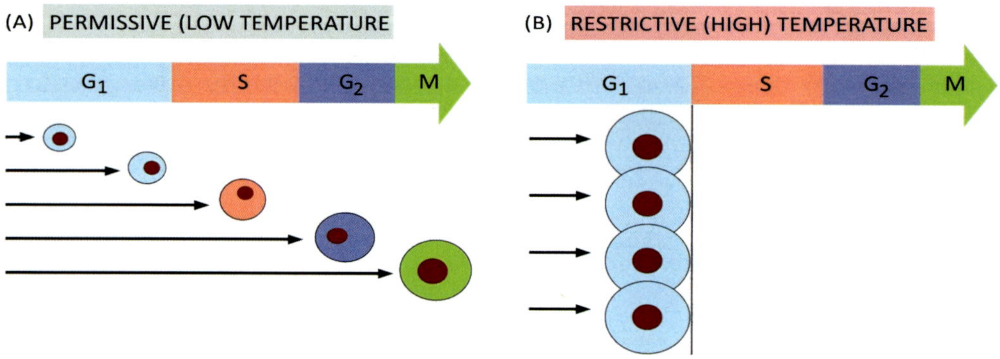

Figure 3.25. *The behavior of a temperature-sensitive CDC mutant.*

Source: https://image.slideserve.com/793935/the-behavior-of-a-temperature-sensitive-cdc-mutant-l.jpg.

A mutant that is unable to finish the cell cycle cannot proliferate. Therefore, only if the phenotype of CDC mutants is conditional—that is, if the gene product only stops working under particular conditions—can they be selected for and kept in the population. Temperature-sensitive mutations, in which the mutant protein fails to function at high temperatures but functions adequately to permit cell division at low temperatures, account for the majority of conditional cell-cycle mutations. To turn off the function of the mutant gene, a temperature-sensitive CDC mutant can be raised to a higher temperature (the restrictive condition) after being propagated at a lower temperature (the permissive condition). The cells proceed through the cell cycle at the higher temperature until they reach the point where the mutant gene's function is necessary for them to go forward; at this point, they stop (Figure 3.25). A uniform cell-cycle arrest of this kind can be identified in budding yeasts simply by observing the cells; the size of the bud and its presence or absence indicate the stage of the cycle at which the mutant is arrested (Figure 3.26).

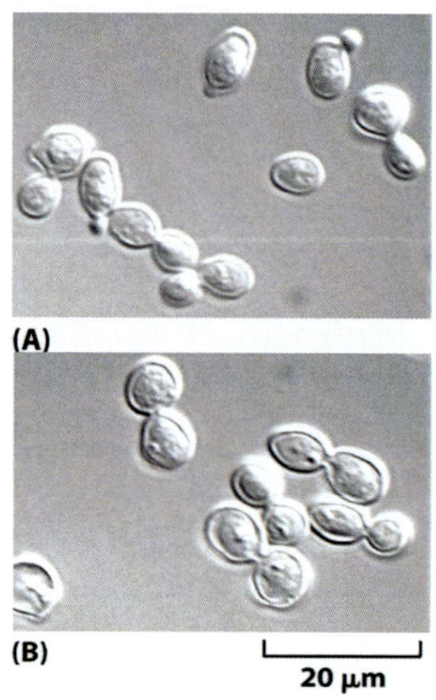

Figure 3.26. *The morphology of budding yeast cells arrested by a cdc mutation.*

Source: https://images.slideplayer.com/16/5214002/slides/slide_3.jpg.

3.4.3. The Cell-Cycle Control System Can Be Analyzed Biochemically in Animal Embryos

The biochemistry of the cell cycle can be most easily studied in the massive fertilized eggs of many animals, which carry enormous stores of the proteins required for cell division. Yeasts are great for studying the genetics of the cell cycle. For instance, the Xenopus frog egg has a diameter of more than 1 mm and contains 100,000 times the amount of cytoplasm as a typical human cell (Figure 3.27). The Xenopus egg undergoes an incredibly quick series of cell divisions known as cleavage divisions upon fertilization. During these divisions, a single giant cell divides without proliferating to produce an embryo made up of thousands of smaller cells. A tiny quantity of protein and DNA, which are needed to create thousands of new nuclei, are the only macromolecules that are synthesized during this process. Following an initial division that takes roughly 90 minutes, the remaining 11 divisions happen roughly simultaneously at 30-minute intervals, resulting in the production of 4096 (2^{12}) cells in 7 hours. Every cycle has two distinct phases, S and M, lasting roughly 15 minutes each. G_1 or G_2 phases are not discernible.

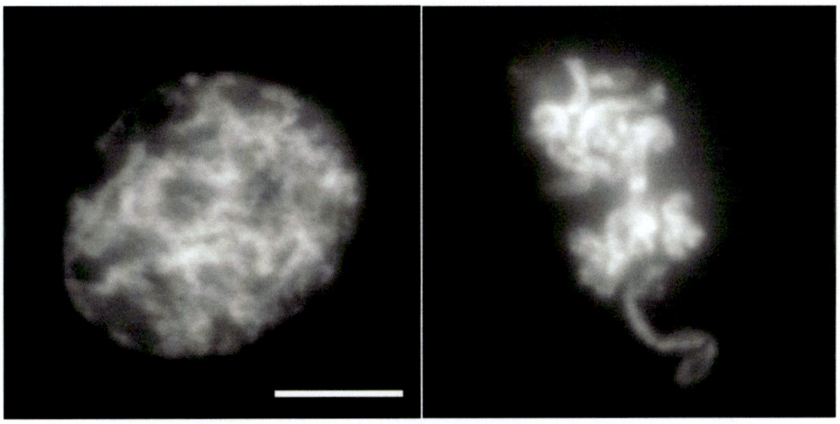

Figure 3.27. *A mature Xenopus egg, ready for fertilization.*

Source: By Bravy – Own work, CC BY-SA 4.0, https://commons.wikimedia.org/w/index.php?curid=37797928.

Thus, in the absence of growth or many of the regulatory mechanisms that function in more complex cell cycles, the cells in early Xenopus embryos, as well as those of the clam Spisula and the fruit fly Drosophila, are capable of dividing extremely quickly. Therefore, these early embryonic cell cycles show how the cell-cycle control system functions, reduced to the bare minimum required to fulfill the most basic needs—the duplication of the genome and its segregation into two daughter cells. Their large size is another advantage of these early embryos for cell-cycle analysis. Testing chemicals to see how they affect the development of the cell cycle in an egg is a reasonably simple process. Additionally, it is feasible to create nearly pure cytoplasm from Xenopus eggs and recreate several cell cycle events in a test tube (Figure 3.28). Such cell extracts allow for the extremely simplified and controlled observation and manipulation of cell-cycle events.

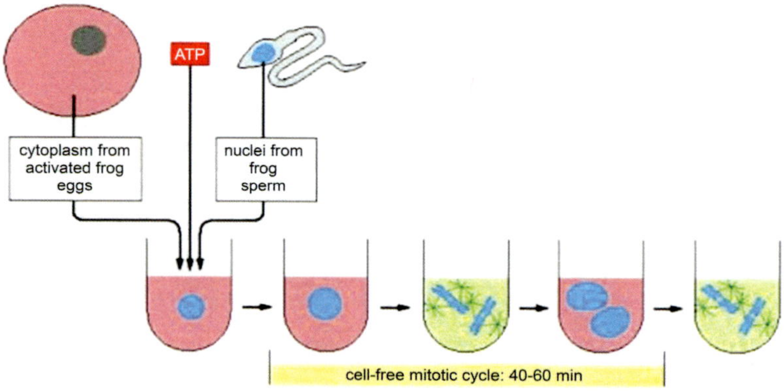

Figure 3.28. *Studying the cell cycle in a cell-free system.*

Source: Bruce, A., Alexander, J., Julian, L., Martin, R., Keith, R., & Peter, W., (2002). Molecular Biology of the Cell. New York: Garland Science. ISBN-10: 0–8153–3218–1.

3.4.4. The Cell-Cycle Control System of Mammals Can Be Studied in Culture

Individual cells in an intact mammal are difficult to observe. Thus, the majority of research on mammalian cell-cycle regulation employs cells that have been separated from healthy tissues or malignancies and cultured in plastic plates with necessary nutrients and additional elements (Figure 3.29). However, there is a problem. Under standard culture conditions, cells derived from normal mammalian tissues frequently cease to proliferate after a certain number of division cycles. Replicative cell senescence is the process by which human fibroblasts, for instance, permanently stop dividing after 25–40 divisions.

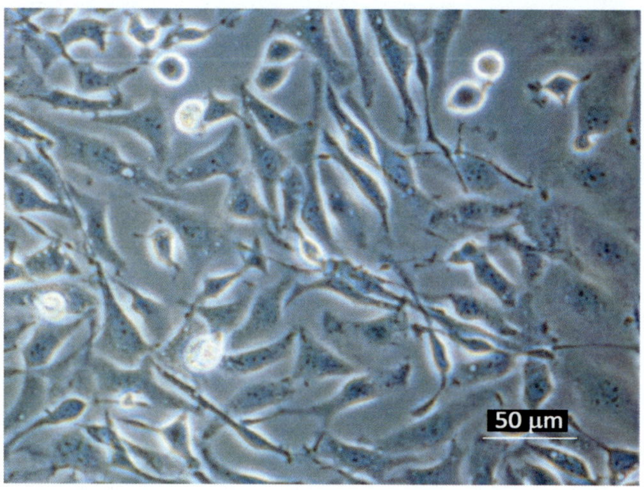

Figure 3.29. *Mammalian cells in culture.*

Source: By User: Alcibiades – Self-made during work, public domain, https://commons.wikimedia.org/w/index.php?curid=591443.

Sometimes, mutations occur in mammalian cells that enable them to multiply easily and endlessly in culture as "immortalized" cell lines. Such cell lines are widely used for cell-cycle studies and for cell biology in general, even though they are not normal. This is because they offer an endless supply of genetically homogeneous cells. These cells can also be subjected to biochemical analysis of the proteins involved in cell-cycle control because they are large enough to allow for detailed cytological observations of cell-cycle events.

Examining the molecular mechanisms controlling the regulation of cell proliferation in multicellular organisms has been made particularly easy by studies conducted on cultured mammalian cells. These investigations are crucial for comprehending both the loss of these controls in cancer and the typical controls of cell numbers in tissues. (Sturtevant & Beadle, 1939; Taylor and Wang, 1989).

3.4.5. Cell-Cycle Progression Can Be Studied in Various Ways

One method of determining the stage of the cell cycle at which an animal cell is in is to use a microscope to observe living cells. A portion of the population of mammalian cells growing in culture can be seen to have rounded up and be in mitosis. Others are visible during the cytokinesis process. However, simple observation cannot identify the S-phase cells.

However, by providing them with visible molecules integrated into freshly synthesized DNA, such as ^{3}H-thymidine or the synthetic thymidine analog bromo-deoxyuridine (BrdU), they can be identified. Autoradiography is used to visualize cell nuclei that have incorporated ^{3}H-thymidine, while anti-BrdU antibody staining is used to visualize cell nuclei that have incorporated BrdU (Figure 3.30).

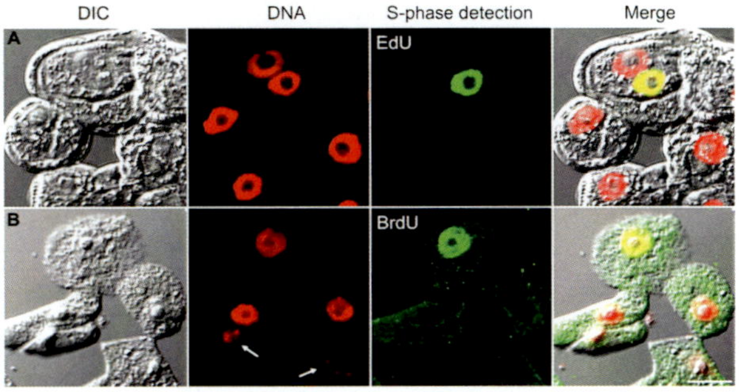

Figure 3.30. *Labeling S-phase cells.*

Source: Kotogány, E., Dudits, D., Horváth, G. V., et al., (2010). A rapid and robust assay for detection of S-phase cell cycle progression in plant cells and tissues by using ethynyl deoxyuridine. Plant Methods, 6, 5. https://doi.org/10.1186/1746-4811-6-5.

Generally, 30–40% of the cells in a population that is all proliferating quickly but asynchronously will be in the S phase at any given time and will be labeled by a brief pulse of BrdU or ^{3}H-thymidine. One can calculate the length of the S phase as a percentage of the total cell cycle duration by looking at the labeling index, which is the percentage of labeled cells in a population. Similarly, the length of the M phase can be inferred from the percentage of these cells that are in mitosis (the mitotic index). Furthermore, the duration of time it takes for an S-phase cell to advance from G_2 into the M phase, from the M phase into G_1, and lastly from G_1 back into the S phase can be measured by delivering a pulse of ^{3}H-thymidine or BrdU and letting the cells continue around the cycle for predetermined amounts of time.

Measuring a cell's DNA content, which doubles during the S phase of the cell cycle, is another technique to determine where it is in the cycle. The use of flow cytometers and DNA-binding fluorescent dyes, which enable the quick and automated analysis of large numbers of cells, greatly facilitates this approach (Figure 3.31). By tracking a population of cells that has been preselected to be in a specific phase of the cell cycle over time, DNA content measurements on this synchronized population of cells reveal how the cells progress through the cycle. This method can also be used to determine the lengths of the G_1, S, and G_2 + M phases in flow cytometry. (Sturtevant & Beadle, 1939; Taylor and Wang, 1989).

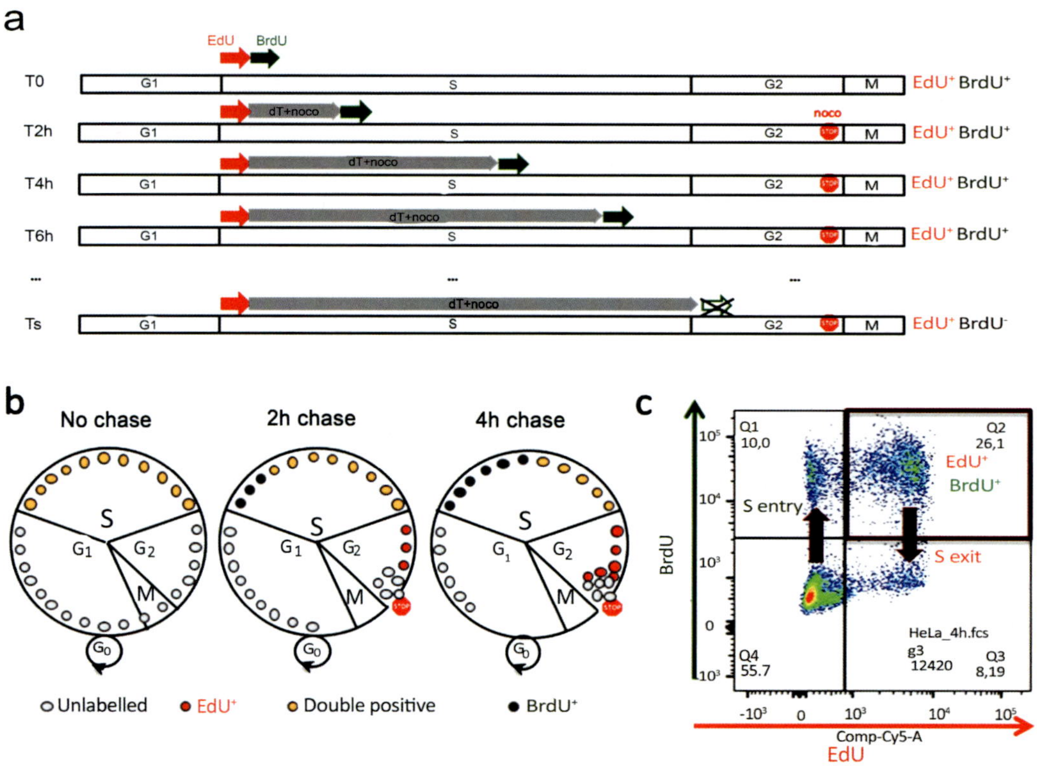

Figure 3.31. *Analysis of DNA content with a flow cytometer.*

Source:Bialic, M., Al Ahmad, N. B., Ko☒lak, M., Coulon, V., Schwob, E., (2022). Measuring S-phase duration from asynchronous cells using dual EdU-BrdU pulse-chase labeling flow cytometry. Genes, 13(3), 408. https://doi.org/10.3390/genes13030408.

CASE STUDY

APPLICATION OF CELL BIOTECHNOLOGY IN TISSUE ENGINEERING

Background

Cell biotechnology plays a pivotal role in tissue engineering, a multidisciplinary field aiming to create functional biological substitutes that restore, maintain, or improve tissue function. This case study focuses on how cell biotechnology has advanced tissue engineering for regenerative medicine applications.

Challenges

- **Cell Sourcing and Characterization:** Identifying suitable cell sources and ensuring their viability, functionality, and compatibility for tissue engineering.
- **Biocompatibility and Integration:** Developing biomaterials and scaffolds that promote cell adhesion, proliferation, differentiation, and integration into host tissues.
- **Clinical Translation:** Moving from laboratory-scale research to scalable manufacturing and regulatory approval for clinical use.

Solutions Implemented

- Cell Sourcing and Characterization:
  - **Action:** Researchers explore various cell sources, including stem cells (e.g., mesenchymal stem cells, induced pluripotent stem cells) and differentiated cells (e.g., chondrocytes, osteoblasts).
  - **Rationale:** Stem cells offer versatility in differentiation potential, while differentiated cells provide tissue-specific functionality.
  - **Implementation:** Utilizing advanced cell culture techniques, genetic engineering (if applicable), and quality control measures to ensure cell quality and reproducibility.
- Biomaterials and Scaffolds:
  - **Action:** Engineers design biomaterials (e.g., natural polymers, synthetic materials, hydrogels) and three-dimensional scaffolds that mimic the native tissue microenvironment.
  - **Rationale:** Biomaterials provide structural support and biochemical cues to guide cell behavior and tissue regeneration.

- **Implementation:** Optimizing scaffold properties (e.g., porosity, mechanical strength, biodegradability) through computational modeling, additive manufacturing (3D printing), and surface modification techniques.

- Clinical Translation:
  - **Action:** Collaborating with clinicians, regulatory agencies, and industry partners to translate laboratory research into clinical applications.
  - **Rationale:** Ensuring safety, efficacy, and regulatory compliance for tissue-engineered products before clinical trials and commercialization.
  - **Implementation:** Conducting preclinical studies, optimizing manufacturing processes for scalability, and navigating regulatory pathways (e.g., FDA approval in the United States) to obtain market authorization.
- Results and Future Considerations
  - **Clinical Success:** Tissue-engineered products such as skin substitutes, cartilage implants, and bone grafts have shown promising outcomes in clinical trials, offering alternatives to conventional therapies.
  - **Technological Advances:** Advances in cell biotechnology continue to drive innovation in tissue engineering, including biofabrication techniques for complex tissues and organs.
  - **Challenges Ahead:** Addressing long-term safety, scalability, and cost-effectiveness remain critical for widespread adoption and commercialization of tissue-engineered products.

CLASS ACTIVITY

OBJECTIVE: UNDERSTAND THE DIVERSITY AND IMPORTANCE OF BIOMOLECULES IN CELLS.

Assessment:

Divide students into small groups and assign each group one type of biomolecule (carbohydrates, lipids, proteins, nucleic acids).

Ask students to research and list examples of their assigned biomolecule, its functions in cells, and where it is commonly found.

Have each group present their findings briefly to the class, emphasizing the roles and significance of their assigned biomolecule in cellular processes.

Discuss as a class how these biomolecules work together to maintain cellular functions and support life processes.

SUMMARY

- A cell is the basic structural and functional unit of all living organisms. Cells are the smallest units of life capable of carrying out essential processes necessary for survival. They are highly organized entities enclosed by a cell membrane that separates their internal environment from the external surroundings.
- Prokaryotic cells are typically smaller and simpler in structure compared to eukaryotic cells. They lack membrane-bound organelles and a distinct nucleus. Instead, their genetic material, usually a single circular DNA molecule, is found in the cytoplasm in an area called the nucleoid. Prokaryotic cells are represented by organisms within the domains Bacteria and Archaea. They have a cell wall made of peptidoglycan (in bacteria) or other materials (in archaea), which provides structural support and protection.
- Eukaryotic cells are larger and more complex. They have a well-defined nucleus that houses the genetic material (DNA), which is organized into chromosomes. The nucleus is surrounded by a double membrane called the nuclear envelope. Eukaryotic cells also contain membrane-bound organelles such as mitochondria, endoplasmic reticulum, Golgi apparatus, and lysosomes, each with specific functions that contribute to cellular processes.
- The molecular composition of cells is incredibly diverse and reflects the intricate balance of various biomolecules that perform essential functions in cellular processes. These molecules can be broadly categorized into carbohydrates, lipids, proteins, nucleic acids, and smaller molecules like ions and vitamins.
- The cell cycle is a highly regulated process that describes the sequence of events a cell undergoes from its formation to division into two daughter cells. It consists of interphase, which includes three phases (G1, S, and G2), followed by mitosis or meiosis, depending on the cell type. The cell cycle is crucial for growth, development, tissue repair, and reproduction in organisms.

REVIEW QUESTIONS

1. What are the fundamental characteristics that define a cell?
2. Briefly describe the major stages in the evolution of cells.
3. Describe the structure of a typical prokaryotic cell.
4. List the main organelles found in eukaryotic cells and their functions.
5. What are the main differences between prokaryotic and eukaryotic cells?
6. What are the major biomolecules found in cells?
7. Describe the phases of the cell cycle.

REFERENCES

1. Agard, D. A., Hiraoka, Y., Shaw, P., & Sedat, J. W. (1989). Fluorescence microscopy in three dimensions. In D. L. Taylor & Y. L. Wang (Eds.), *Methods in Cell Biology* (Vol. 30, pp. 1–10). Academic Press.
2. Allen, R. D. (1985). New observations on cell architecture and dynamics by video-enhanced contrast optical microscopy. *Annual Review of Biophysics and Biophysical Chemistry, 14*, 265–290. https://doi.org/10.1146/annurev.bb.14.060185.001405.
3. Böttcher, B., Wynne, S. A., & Crowther, R. A. (1997). Determination of the fold of the core protein of hepatitis B virus by electron cryomicroscopy. *Nature, 386*, 88–91. https://doi.org/10.1038/386088a0.
4. Celis, J. E. (1998). *Cell Biology: A Laboratory Handbook* (2nd ed., Vol. 3, Part 1—Histology and histochemistry, pp. 219–248). Academic Press.
5. Celis, J. E. (1998). *Cell Biology: A Laboratory Handbook* (2nd ed., Vol. 2, Part 6—Immunocytochemistry, pp. 457–494). Academic Press.
6. Celis, J. E. (1998). *Cell Biology: A Laboratory Handbook* (2nd ed., Vol. 3, Part 9—Electron microscopy, pp. 249–362). Academic Press.
7. Dubochet, J., Adrian, M., Chang, J. J., et al. (1988). Cryoelectron microscopy of vitrified specimens. *Quarterly Reviews of Biophysics, 21*, 129–228. https://doi.org/10.1017/S0033583500003344.
8. Fowler, W. E., & Aebi, U. (1983). Preparation of single molecules and supramolecular complexes for high resolution metal shadowing. *Journal of Ultrastructure Research, 83*, 319–334. https://doi.org/10.1016/S0022-5320(83)80034-2.
9. Gabashvili, I. S., Agrawal, R. K., Spahn, C. M. T., et al. (2000). Solution structure of the *E. coli* 70S ribosome at 11.5 Å resolution. *Cell, 100*, 537–549. https://doi.org/10.1016/S0092-8674(00)80613-6.
10. Harlow, E., & Lane, D. (1988). *Antibodies: A Laboratory Manual*. Cold Spring Harbor Laboratory Press.
11. Haugland, R. P. (1996). *Handbook of Fluorescent Probes and Research Chemicals* (8th ed.). Molecular Probes, Inc. Available online at: http://www.probes.com/ (accessed on 3 September 2024).
12. Hayat, M. A. (1978). *Introduction to Biological Scanning Electron Microscopy*. University Park Press.
13. Hayat, M. A. (2000). *Principles and Techniques of Electron Microscopy* (4th ed.). Cambridge University Press.
14. Heitlinger, E., Peter, M., Haner, M., et al. (1991). Expression of chicken lamin B2 in *Escherichia coli*: Characterization of its structure, assembly and molecular interactions. *Journal of Cell Biology, 113*, 485–495. https://doi.org/10.1083/jcb.113.3.485.
15. Heuser, J. (1981). Quick-freeze, deep-etch preparation of samples for 3-D electron microscopy. *Trends in Biochemical Sciences, 6*(2), 64–68. https://doi.org/10.1016/0968-0004(81)90064-0.

16. Hoenger, A., & Aebi, U. (1996). Three-dimensional reconstructions from ice-embedded and negatively stained macromolecular assemblies: A critical comparison. *Journal of Structural Biology, 117*(1), 99–116. https://doi.org/10.1006/jsbi.1996.0069.

17. Kuhlbrandt, W., Wang, D. N., & Fujiyoshi, Y. (1994). Atomic model of plant light-harvesting complex by electron crystallography. *Nature, 367*, 614–621. https://doi.org/10.1038/367614a0.

18. Mancini, E. J., Clarke, M., Gowen, B. E., Rutten, T., & Fuller, S. D. (2000). Cryo-electron microscopy reveals the functional organization of an enveloped virus, Semliki Forest virus. *Molecular Cell, 5*, 255–266. https://doi.org/10.1016/S1097-2765(00)80437-7.

19. Maunsbach, A. B. (1998). Immunolabelling and staining of ultrathin sections in biological electron microscopy. In J. E. Celis (Ed.), *Cell Biology: A Laboratory Handbook* (Vol. 3, pp. 268–276). Academic Press.

20. Murakami, S., Otsuka, Y., Sugimoto, M., & Mitsui, T. (2012). 3H1010 Controlled cell migration with ultrasound (Cell Biology III: Cytoskeleton & Motility, Oral Presentation). *Seibutsu Butsuri, 52*(Supplement), S70.

21. Ogushi, F., & Kori, H. (2013). Dependence of cell differentiation ratio on cell-cell interaction and noise (24. Mathematical biology, Poster). *Seibutsu Butsuri, 53*(Supplement 1–2), S257.

22. Palmer, J. D., & Delwiche, C. F. (1996). Second-hand chloroplasts and the case of the disappearing nucleus. *Proceedings of the National Academy of Sciences, 93*(15), 7432–7435. https://doi.org/10.1073/pnas.93.15.7432.

23. Petran, M., Hadravsky, M., Egger, M. D., & Galambos, R. (1968). Tandem-scanning reflected light microscope. *Journal of the Optical Society of America, 58*(5), 661–664. https://doi.org/10.1364/JOSA.58.000661.

24. Philippe, H., Germot, A., & Moreira, D. (2000). The new phylogeny of eukaryotes. *Current Opinion in Genetics & Development, 10*(6), 596–601. https://doi.org/10.1016/S0959-437X(00)00163-2.

25. Ploem, J. S., & Tanke, H. J. (1987). *Introduction to Fluorescence Microscopy*. Royal Microscopical Society Microscopy Handbook, No. 10. Oxford University Press.

26. Rubin, G. M., Yandell, M. D., Wortman, J. R., et al. (2000). Comparative genomics of the eukaryotes. *Science, 287*(5461), 2204–2215. https://doi.org/10.1126/science.287.5461.2204.

27. Shotton, D. M. (1990). *Electronic Light Microscopy: Techniques in Modern Biomedical Microscopy*. Wiley-Liss.

28. Shotton, D. M. (1998). Freeze fracture and freeze etching. In J. E. Celis (Ed.), *Cell Biology: A Laboratory Handbook* (Vol. 3, pp. 310–322). Academic Press.

29. Sturtevant, A. H., & Beadle, G. W. (1939). *An Introduction to Genetics*. Saunders. Reprinted 1988 by Garland.

30. Taylor, D. L., & Wang, Y. L. (1989). *Methods in Cell Biology* (Vols. 29 & 30: Fluorescence microscopy of living cells in culture, Parts A & B). Academic Press.

31. The, C. (1998). Genome sequence of the nematode *C. elegans*: A platform for investigating biology. *Science, 282*(5396), 2012–2018. https://doi.org/10.1126/science.282.5396.2012.

32. Tinsley, R. C., & Kobel, H. R. (1996). *The Biology of Xenopus* (pp. 16, 379, 391–401). Clarendon Press.

33. Unwin, P. N. T., & Henderson, R. (1975). Molecular structure determination by electron microscopy of unstained crystal specimens. *Journal of Molecular Biology, 94*(2), 425–440. https://doi.org/10.1016/0022-2836(75)90270-2.

34. Weiss, D. G., Maile, W., Wick, R. A., & Steffen, W. (1999). Video microscopy. In A. J. Lacey (Ed.), *Light Microscopy in Biology: A Practical Approach* (2nd ed.). Oxford University Press.

35. White, J. G., Amos, W. B., & Fordham, M. (1987). An evaluation of confocal versus conventional imaging of biological structures by fluorescence light microscopy. *Journal of Cell Biology, 105*(1), 41–48. https://doi.org/10.1083/jcb.105.1.41.

36. Wigge, P. A., Jensen, O. N., & Holmes, S., et al. (1998). Analysis of the *Saccharomyces* spindle pole by matrix-assisted laser desorption/ionization (MALDI) mass spectrometry. *Journal of Cell Biology, 141*(4), 967–977. https://doi.org/10.1083/jcb.141.4.967.

37. Wilson, L., & Matsudaira, P. (1993). *Methods in Cell Biology* (Vol. 37: Antibodies in cell biology). Academic Press.

38. Yamashita, Y. M. (2018). Cell biology of stem cells: Studying stem cells at the level of cell biology and studying cell biology using stem cells. *Molecular Biology of the Cell, 29*(24), 2912. https://doi.org/10.1091/mbc.E18-06-0351.

CHAPTER 4

Genomics and Proteomics

LEARNING OBJECTIVES

After studying this chapter, you will be able to:

- Describe the functional genomics;
- Understand the proteomics.

KEY TERMS FROM THIS CHAPTER

Biotinylation	Data analysis	DNA polymerase
Drug targets	Metabolomics	Mutations
Phycoerythrin	Phylogenetic	Polyadenylation
Polymorphisms	Post-translational modifications	
Protein electrophoresis	Proteolysis	
Qualitative expression	RNA extraction	

INTRODUCTION

Genomics and proteomics are two interconnected fields at the forefront of modern biological research, revolutionizing our understanding of life at the molecular level. Genomics focuses on the study of an organism's entire genetic material, encompassing its DNA sequence, organization, function, and evolution. With the advent of high-throughput sequencing technologies, scientists can now decipher the entire genetic blueprint of an organism, unlocking vast amounts of information encoded within its DNA. This wealth of genomic data not only provides insights into an organism's traits, behaviors, and susceptibility to diseases but also enables researchers to explore the intricate mechanisms governing gene expression, regulation, and inheritance.

Proteomics, on the other hand, delves into the complex world of proteins – the molecular machines that execute the instructions encoded in the genome. Unlike genomics, which focuses on the static blueprint, proteomics investigates the dynamic interplay of proteins within cells, tissues, and organisms. By analyzing the entire complement of proteins (the proteome) and their functions, proteomics sheds light on crucial biological processes such as metabolism, signaling, and disease pathways. Advanced techniques such as mass spectrometry and protein microarrays allow researchers to identify, quantify, and characterize thousands of proteins simultaneously, offering unprecedented insights into the molecular mechanisms underlying health and disease.

Together, genomics and proteomics form the cornerstone of systems biology, an interdisciplinary approach that aims to decipher the complex networks of interactions governing living organisms. By integrating genomic and proteomic data with other omics disciplines such as transcriptomics and metabolomics, scientists can unravel the intricate web of molecular interactions driving biological phenomena. This holistic understanding not only enhances our knowledge of fundamental biological processes but also holds immense promise for personalized medicine, agriculture, biotechnology, and beyond. As technology continues to advance and our understanding deepens, genomics and proteomics will undoubtedly remain at the forefront of biological research, shaping the future of science and medicine.

4.1. FUNCTIONAL GENOMICS

The study of genomes and individual genes at the DNA level, including sequence differences-based phylogenetic relationships, polymorphisms, and mutations, is referred to as genomics. Functional genomics, also known as transcriptomics, is a branch of genomics that focuses on transcription patterns. It can be applied qualitatively to identify genes that are expressed or quantitatively to quantify changes in gene transcription levels. Whole genome transcription is evaluated in relation to various clinical settings, mutations, natural or toxic agents, distinct cell or tissue types, or distinct points in time during biological processes like cell division or organism development. Finding the genes that are up- and down-regulated under particular circumstances is one of the goals of gene expression research (the transcriptome). One or a small number of genes' transcription could be monitored at a time in the past. Thousands of genes can be tracked simultaneously in transcription (gene expression profiling) of a cell or tissue sample using functional genomics methodology. Serial analysis of gene expression (SAGE) and DNA microarrays are the two primary experimental methods for identifying gene expression profiles. Owing to the substantial volume of data produced by these experiments, specialized computational tools are needed to gather, store, and evaluate the data.

4.1.1. DNA Microarray Technology

The process of hybridizing a nucleic acid sample (target) derived from a cell or tissue's messenger RNAs (mRNAs) to single-stranded DNA sequences (probes) bound in an ordered arrangement to a solid platform is known as a DNA microarray (also known as a DNA chip or gene chip) experiment. A particular kind of DNA microarray is made by depositing amplified cDNA sequences from mRNAs of a single cell or all or particular sets of an organism's coding sequences onto a glass slide or nylon membrane using the polymerase chain reaction (PCR). In this instance, a 1-cm^2 area can hold roughly 10,000 distinct probe arrays.

An alternative microarray technology uses sets of oligonucleotides, which typically correspond to thousands of genes, as probes. One widely used platform involves the direct (in situ) synthesis of the probes using photolithography, a light-directed technique, on a quartz wafer. In a predetermined location (probe cell or feature) on the array surface, thousands of copies of an oligonucleotide with the same precise nucleotide sequence are synthesized. The probes for this kind of microarray usually consist of 10–40 nucleotides, and multiple probes with various sequences will be created on the microarray for each gene. One can also use longer oligonucleotides, up to 100 nucleotides in length. More than 500,000 probes that correspond to up to 30,000 genes can be found on a full whole-genome oligonucleotide array. In general, the experiment's goal and the necessary level of resolution determine how the probes (or probe set) for a microarray are designed. In order for all target sequences to bind to their complementary probe sequences under the same circumstances, computer programs identify probe sequences that are specific for their target sequences, are least likely to hybridize with nontarget sequences (cross-hybridize), have no secondary structure (foldback) that would prevent hybridization with the target sequence, and have similar melting (annealing) temperatures. Probes representing an entire genome, a single chromosome, specific genomic regions, or specific coding regions from one or more different organisms can be included in oligonucleotide microarrays. Genomic DNA microarrays do not include repetitive sequences.

Reverse transcriptase is used to synthesize cDNA using the extracted mRNA as a template after mRNA is extracted and purified from cells or tissues for the majority of gene expression profiling experiments that use microarrays. mRNA is typically extracted from two or more sources, such as cells grown in various environments or diseased versus normal tissue, whose expression profiles are compared. During cDNA synthesis, fluorescently labeled nucleotides are added to each source's cDNA to give it a unique fluorophore. For instance, the test sample uses a red-emitting fluorescent dye (Cy5), while the normal (reference) sample uses a green-emitting fluorescent dye (Cy3) (Figure 4.1). The cDNA samples are labeled, combined, and hybridized to the same microarray. The intensities of Cy5 and Cy3 are measured for every probe cell using a laser scanner. The amounts of Cy3 and Cy5 in the probe cells determine their color, and the ratio of a probe cell's red (Cy5) to green (Cy3) fluorescence intensity reveals the relative expression levels of the represented gene in the two samples (Figure 4.2).

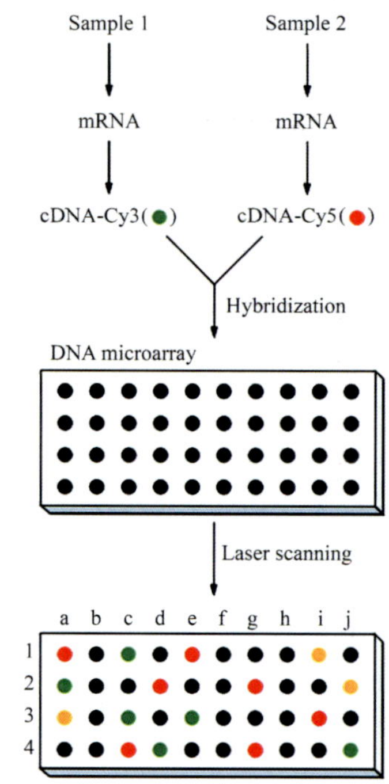

Figure 4.1. *Gene expression profiling with a DNA microarray. mRNA is extracted from two samples (sample 1 and sample 2), and during reverse transcription, the first cDNA strands are labeled with the fluorescent dyes Cy3 and Cy5, respectively. The cDNA samples are mixed and hybridized to an ordered array of either gene sequences or gene-specific oligonucleotides. After the hybridization reaction, each probe cell is scanned for both fluorescent dyes and the separate emissions are recorded. Probe cells that produce only a green or red emission represent genes that are transcribed only in samples 1 and 2, respectively; yellow emissions denote genes that are active in both samples; and no emissions (black) represent genes that are not transcribed in either sample.*

Source: Bernard R. Glick, Jack J. Pasternak, and Cheryl L. Patten.; Molecular biotechnology: principles and applications of recombinant DNA; ISBN 978-1-55581-498-4.

Reverse labeling and hybridization of reference and test samples to a different microarray are common methods used to prevent variation caused by intrinsic and sequence-specific discrepancies in labeling efficiencies between Cy3 and Cy5. Labeling the reference sample with Cy5 and the test sample with Cy3 would be known as reverse labeling, or dye swapping, in the example above. As an alternative, on some microarray platforms, reference and test samples are hybridized to separate microarrays while the target sequences from each source are labeled with the same fluorescent dye.

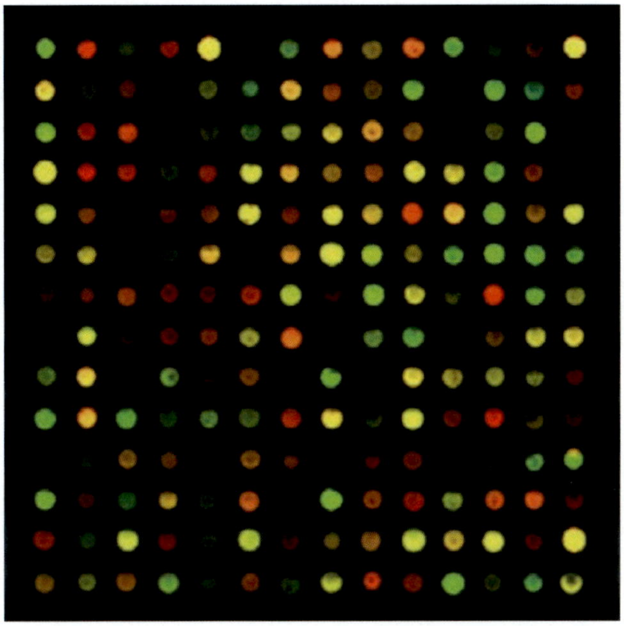

Figure 4.2. *Fluorescence image of a DNA microarray hybridized with Cy3- and Cy5-labeled cDNA.*

Source: Bernard R. Glick, Jack J. Pasternak, and Cheryl L. Patten.; Molecular biotechnology: principles and applications of recombinant DNA; ISBN 978-1-55581-498-4.

An alternative method involves purifying mRNA using an oligo(dT) sequence, which attaches to the poly(A) tail of eukaryotic mRNA. The T7 primer, a short extension, contains the sequence for the promoter of the bacteriophage T7 RNA polymerase (Figure 4.3). Reverse transcriptase is primed to synthesize cDNA from mRNA by the oligo(dT)-T7 sequence. Double-stranded cDNA containing the T7 RNA polymerase promoter is produced when DNA polymerase is used to synthesize the second DNA strand. Next, in the presence of biotin-labeled ribonucleotides, T7 RNA polymerase is employed to create RNA copies (complementary RNA [cRNA] or antisense RNA) from the second cDNA strand as a template. Following linear amplification and biotinylation by this reaction, cRNA is broken up into fragments that range in length from 50 to 100 nucleotides and are best suited for hybridization. Following hybridization, streptavidin, which is attached to the fluorescent protein phycoerythrin, is applied to the microarray. The specific biotin residues of hybridized cRNA are bound by streptavidin, and the emissions of phycoerythrin during laser scanning serve as a marker for hybridization.

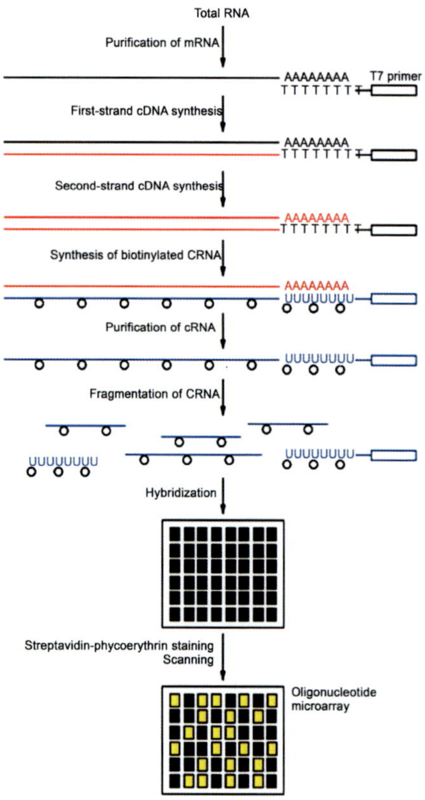

Figure 4.3. *Gene expression profiling with an oligonucleotide microarray. mRNA is purified with a poly(dT) sequence that has a T7 RNA polymerase primer sequence extension. After two-stranded cDNA synthesis, the second cDNA strand acts as a template for the synthesis of cRNA by T7 RNA polymerase in the presence of ATP, cytidine triphosphate (CTP), guanosine triphosphate (GTP), uridine triphosphate (UTP), biotinylated CTP, and biotinylated UTP. The gray circles represent incorporated biotinylated nucleotides. The biotinylated cRNA is purified, fragmented into pieces from 50 to 100 nucleotides in length, and hybridized to an oligonucleotide microarray. The microarray is treated with streptavidin–phycoerythrin, and the probe cells (black squares) are scanned for emission (yellow) from the biotin-bound streptavidin–phycoerythrin.*

Source: Bernard R. Glick, Jack J. Pasternak, and Cheryl L. Patten.; Molecular biotechnology: principles and applications of recombinant DNA; ISBN 978-1-55581-498-4.

Owing to the enormous volume of data produced by microarray experiments, specialized software has been created to optimize information output. One-dye and two-dye hybridized microarray analyses are comparable. This is a summary of a widely used technique for information extraction from two-dye hybridized microarrays. A confocal scanning microscope is used to scan each probe cell in a two-dye hybridized microarray. Fluorescence emitted from each probe cell, detected at 532 and 635 nm for Cy3 and Cy5, respectively, after the dye is excited by laser is collected through the objective lens of the microscope and transformed into an electrical signal using a photomultiplier tube. Each probe cell's fluorescence intensities from both dyes are measured and stored, along with the microarray's background readings. To calculate background fluorescence, one must subtract the measured fluorescence intensities for each probe cell from the fluorescence measured in blank areas where no probe cells have been spotted. Internal controls, or particular probes, are built into microarray designs to assess the dependability of the hybridization process and guarantee that the laser scanner operates as intended. As of now, the microarray data (i.e., the set of fluorescence intensities of every probe cell) is normalized to account for variances (systematic errors) brought about by technical elements that affect a probe cell's fluorescence intensities and allow comparisons between the microarrays in an experiment. Replica samples are independently prepared under the same conditions and hybridized to different microarrays to reduce errors. Multiple probe cells for each gene are included on a single microarray. Variations in sample preparation, including growth conditions, RNA extraction, cDNA synthesis, and labeling efficiencies, can lead to misleading interpretations. Other sample preparation variations include variations in the concentrations of probes on different microarrays, unequal amounts of target sequences applied to different microarrays, or unequal distribution of targets on a single microarray. The fluorescence intensity of a gene that is not differentially expressed under different conditions is used as a reference point in many normalization techniques, such as spiked control sequences that are sufficiently different from the target sequences to bind only to a corresponding control probe cell, and adjusting the total fluorescence intensity for each microarray to a similar value based on the theory that a relatively small number of genes are expected to change under different conditions.

A microarray experiment's primary goal is to find the genes whose expression varies in response to a specific biological condition. By comparing the average fluorescence intensity of each gene (i.e., probe cell) under two distinct conditions and computing the ratio—often represented as an n-fold change—it is possible to ascertain the response to a biological condition. The raw data of each probe cell's dye emissions from a microarray are frequently converted to log2 ratios for useful comparisons (Table 4.1). The dye with the higher intensity is indicated by the sign. Positive log ratios typically indicate higher levels of Cy5 than Cy3, which indicates higher levels of gene expression in the test sample compared to the reference sample. Positive values (more Cy3 than Cy5) suggest that the test sample's expression level is lower than the reference sample's. An expression matrix is a table that is created by compiling the log ratios for every probe cell.

Table 4.1. *Converting Cy3 and Cy5 Intensities to log_2 Ratios*

Feature (gene)	Cy3 intensity	Cy5 intensity	Cy5/Cy3	log_2(Cy5/Cy3)
1	180	10,000	56	+5.81
2	5,400	5,400	1	0
3	8,400	400	0.05	–4.39

Source: Bernard R. Glick, Jack J. Pasternak, and Cheryl L. Patten.; Molecular biotechnology: principles and applications of recombinant DNA; ISBN 978-1-55581-498-4.

Moreover, co-expressed genes under various circumstances or over time can be found using microarray analysis, which helps identify which gene products are involved in a particular pathway. There are several computational techniques that group the data into clusters, or related groups. A uniform display of the clustered data is achieved by arbitrarily color-coding ranges of log ratio values. Generally, a log ratio of zero is represented by black, increasing positive log ratios by dark to bright red, and decreasing negative log ratios by dark to bright green. Put differently, red is commonly used to indicate overexpression of a gene, and green to indicate under expression. Gene expression profiles, where the rows represent the rearranged genes and the columns represent either conditions or samples, are a visual depiction of a clustered microarray (Figure 4.4).

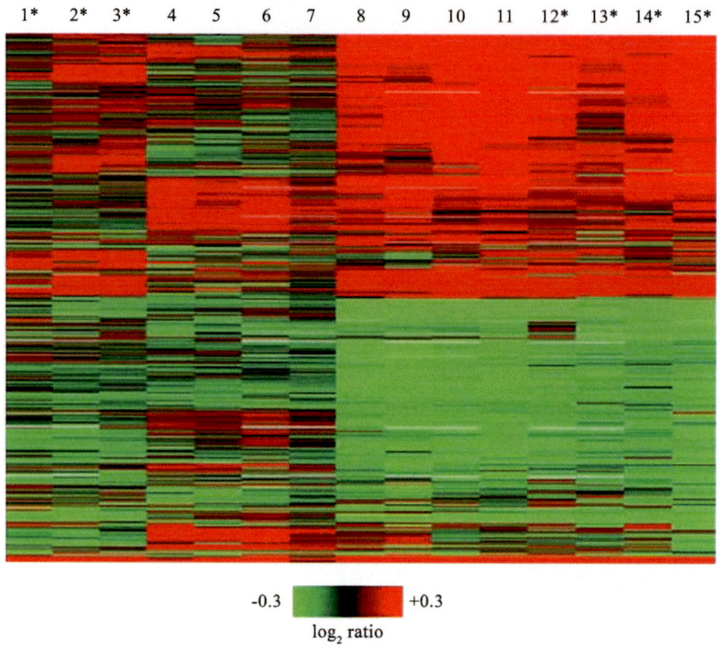

Figure 4.4. *Gene expression profile of cirrhotic liver tissue. Columns 1 to 7 and 8 to 15 are liver samples from patients with ethanol- and hepatitis C virus-induced cirrhosis of the liver, respectively. Each patient's sample was compared to normal liver tissue. The cluster consists of 2,965 genes. The asterisks denote patients with severe cirrhosis of the liver.*

Source: Lederer et al., Virol. J. 3:98, 2006.

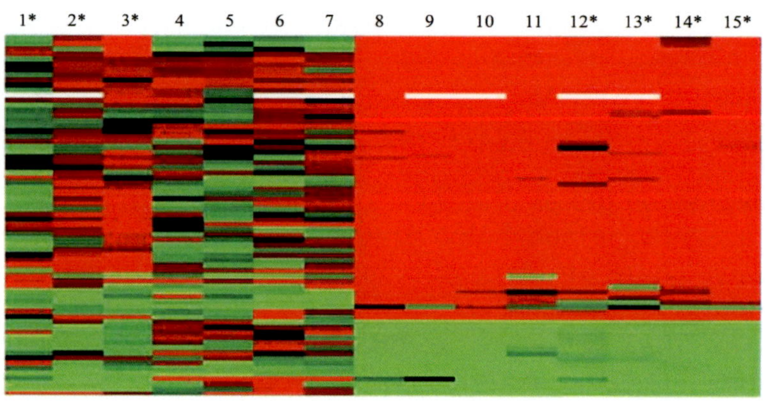

Figure 4.5. *Gene expression profile of lymphocyte-specific genes from cirrhotic liver tissue. Columns 1 to 7 and 8 to 15 are liver samples from the patients shown in Figure 4.4 with ethanol- and hepatitis C virus-induced cirrhosis of the liver, respectively. Each patient's sample was compared to normal liver tissue. The cluster consists of about 70 genes. The asterisks denote patients with severe cirrhosis of the liver.*

Source: Lederer et al. (2006). Virol. J. 3:98.

The gene expression profile in Figure 4.4, determined by microarray analysis, clearly shows that different genes are transcribed in patients with cirrhosis of the liver than in healthy individuals and in patients with ethanol-induced cirrhosis than in those with cirrhosis induced by the hepatitis C virus. Moreover, there is a difference between the genes that are turned on during advanced ethanol-induced liver damage and those in patients with less severe ethanol-induced cirrhosis (Figure 4.4). No such distinction is evident among individuals with different severities of virus-induced cirrhosis (Figure 4.4). In addition, information about the transcription of genes that contribute to a particular pathway or cellular activity can be extracted from a gene expression profile. For example, genes that are transcribed during lymphocyte proliferation and activation are highly expressed in virus-induced liver cirrhosis and to a much lesser extent in ethanol-associated cirrhotic samples (Figure 4.5).

It is impossible to overestimate the significance and widespread use of DNA microarrays. DNA microarrays are being developed for clinical use. The U.S. Permission for the first commercial diagnostic assay based on a DNA microarray was given by the Food and Drug Administration in 2007. Here, a 70-gene expression profile (MammaPrint) is used to differentiate between patients whose cancer has a low risk of metastasizing and those whose cancer is likely to migrate to other sites. One of the main concerns has been the reproducibility and dependability of microarrays from different laboratories and in different formats. However, several international groups have proposed standards that should mitigate these issues for both conducting microarray experiments and analyzing the data. Furthermore, it should be mentioned that DNA microarrays are utilized in addition to gene expression studies to identify the binding sites for DNA-binding proteins e.g. chromatin immunoprecipitation [ChIP], which is used in ChIP-on-chip assays to identify proteins bound to a DNA microarray, the locations of gene transcription initiation and termination, and numerous other features of genome architecture. The eukaryotic genome is much more extensively transcribed than previously believed, according to this research; many genes have multiple start and termination sites, some of which are hundreds of kilobases from known sites for many genes; splicing occurs between RNA molecules; transcription occurs on both strands of many genomic regions; and some transcription factors bind to dozens of sites scattered throughout the genome. To put it briefly, larger complexity in the transcription-controlling mechanisms of a eukaryotic organism has been uncovered by whole-genome microarray analysis than was anticipated based on smaller-scale transcriptional analysis.

4.1.2. Serial Analysis of Gene Expression

SAGE uses recombinant DNA techniques to clone randomly linked short sequences of cDNA prepared from extracted cellular mRNA, which can be efficiently sequenced to identify expressed genes, in contrast to DNA microarrays that rely on hybridization and signal detection (Figure 4.6). An oligo(dT) sequence labeled with biotin is used to capture polyadenylated mRNA and is attached to a streptavidin-coated magnetic bead. Reverse transcriptase is used to create the first strand of double-stranded cDNA from the purified mRNA template and oligo(dT) primer. DNA polymerase is then used to create the second, complementary strand of cDNA. A strong magnet is used to retain the magnetic beads with attached cDNAs during successive treatments and washings. The cDNAs are cut with the restriction endonuclease NlaIII, which recognizes the sequence $\begin{matrix}\text{CATG}\\\text{GTAC}\end{matrix}$ and cuts outside the G:C base pairs, leaving a 3′ GTAC extension. Since NlaIII cuts, on average, 1 in 256 base pairs (bp), there is a high probability that each cDNA will have at least one NlaIII recognition site. Because the cDNAs are bound to beads, each NlaIII cut site that is closest to the 3′ end of a cDNA is retained, and all unbound fragments are washed away. The SAGE nomenclature refers to NlaIII as an anchoring enzyme. The cDNA sample that was digested with NlaIII is then split in half. Adaptor A is used to ligate one aliquot, and adaptor B is used for the other. Both adaptors contain a 5-bp recognition site for the restriction endonuclease BsmFI, a sequence for priming a PCR, and a CATG extension that is complementary to the extension generated by NlaIII digestion. The adaptors have different primer sequences to prevent intrastrand base pairing (snap back) during subsequent PCR steps.

The products are treated with BsmFI after the adaptors are ligated to NlaIII-digested cDNA. This type II restriction endonuclease, in contrast to NlaIII and other type II restriction endonucleases, cuts regardless of the intervening nucleotide sequence 10 nucleotides downstream from its recognition site in one DNA strand and 14 nucleotides in the other strand. BsmFI is referred to as a tagging enzyme in SAGE, and the cDNA segment that results from the BsmFI treatment is known as a tag. The adaptor-tag molecules are liberated from the beads by BsmFI digestion and can subsequently be extracted and recovered from solution. After filling in the BsmFI-cut DNA's 4-nucleotide extension to create a blunt-end molecule, the pools of adaptor A- and adaptor B-tags are combined, and T4 DNA ligase is added. In these circumstances, primer sequences surround a two-tag molecule, or ditag, which is created by joining the blunt ends of two tags. Ditags are easily amplified during PCR using primer sequences found in the adaptors because ditag formation is entirely random and joins tags from distinct cDNAs during ligation.

NlaIII treatment of the amplified ditags releases the adaptor sequences and results in ditags with an NlaIII extension at both ends. To create numerous, randomly joined combinations (concatemers) of ditags, the NlaIIIdigested ditags are ligated. Concatemers that are about 500 bp in length are isolated and cloned in an *E. coli* plasmid. Concatemers are sequenced, each tag's sequence is noted, and a dedicated "tag to gene" database is searched to find the matching gene. The 3′ end of the gene corresponds to the sequenced tag since it is derived from the 3′ end of the mRNA. Each tag's sequence count, which indicates how abundant it was in the original sample, is counted. It is possible to identify mRNAs that are up- or down-regulated by comparing the tag frequencies across various samples. Typically, one experiment yields more than 10,000 distinct tags. 30-plus million SAGE tags from humans and an additional 35 million from other organisms have been assembled. SAGE has the potential to identify every transcript in a sample, while a DNA microarray can only identify sequences that match probes on the array.

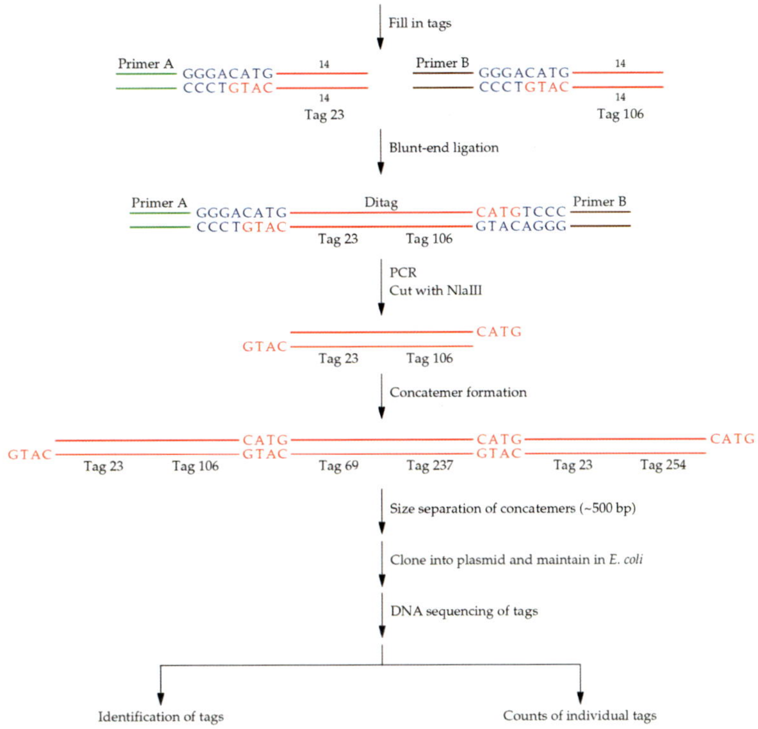

Figure 4.6. *SAGE. Poly(A)+ mRNA transcripts are isolated by poly(dT) hybridization. The poly(dT) sequence is biotinylated (B) and bound to streptavidin (SA)- coated magnetic beads (yellow circles), or the poly(dT) is attached directly to magnetic beads. Double-stranded cDNA is synthesized from the captured mRNAs and then cut with the restriction endonuclease NlaIII. The fragments of cDNA that are not bound to the magnetic beads are eluted. The NlaIII-cut cDNA sample that is attached to the magnetic beads is divided in two, and one sample is ligated with adaptor A and the other with adaptor B. Each adaptor contains a 4-base extension that is complementary to the 4-base extension produced by cleavage with NlaIII, a recognition site for the restriction endonuclease BsmFI, and its specific primer sequence (primer A and primer B). The ligated "adaptor–NlaIII-cut" molecules are treated with BsmFI, which cleaves the DNA 10 and 14 nucleotides downstream from its recognition site (open arrows). The extensions of the BsmFI fragments are filled in by DNA synthesis, and the mixture is blunt-end ligated. Some of the ligated molecules consist of two joined segments from different cDNAs (ditags) that are flanked by sequences for primers A and B.*

Source: Bernard R. Glick, Jack J. Pasternak, and Cheryl L. Patten.; Molecular biotechnology: principles and applications of recombinant DNA; ISBN 978-1-55581-498-4.

To create longer tags, more SAGE protocols have been developed, including LongSAGE and SuperSAGE (Table 4.2). Also, other anchoring enzymes, e. g. Sau3AI and additional restriction endonucleases, which identify and cleave at particular 4-bp sequences known as 4-base cutters, have been utilized for the purpose of identifying transcribed genes lacking an easily accessible NlaIII site. SAGE Genie is an online tool that can be used to match tags to putative genes, find the frequency of a tag across different SAGE libraries, and provide other relevant data.

Table 4.2. *Different SAGE Systems*

System	Tagging enzyme	Recognition site	Tag (bp)
SAGE	BsmFI	GGGAC (N) $_{10}$	10–14
		CCCTG (N) $_{14}$	
LongSAGE	MmeI	TCCRAC (N) $_{21}$	17–21
		AGGYTG (N) $_{19}$	
SuperSAGE	EcoP15I	CAGCAG (N) $_{25}$	26
		GTCGTC (N) $_{27}$	

Source: Bernard R. Glick, Jack J. Pasternak, and Cheryl L. Patten.; Molecular biotechnology: principles and applications of recombinant DNA; ISBN 978-1-55581-498-4.

4.2. PROTEOMICS

The extensive characterization of a cell line, tissue, or organism's whole protein complement was the definition of "proteomics" when the term was first used in 1995. There are currently two definitions of proteomics in use. The first is the traditional definition, which limits large-scale gene product analysis to protein-only studies. The second and broader definition combines analyses with a genetic readout, like genomics, mRNA analysis, and yeast two-hybrid analysis, with protein studies. But proteomics still has the same objective, i.e., instead of examining each protein separately, to gain a more comprehensive and worldwide understanding of biology through the study of all the proteins in a cell. (Anderson, N. G., and N. L. Anderson, 1996)

Now that proteomics has been defined more broadly, a wide range of research fields are included under its umbrella (Figure 4.7). These comprise, among other things, studies on protein-protein interactions, protein modifications, protein function, and protein localization. Proteomics seeks to map out the location of all proteins within a cell in three dimensions (3-D) in addition to identifying every protein in the cell. These lofty objectives will undoubtedly necessitate the collaboration of numerous academic fields, including molecular biology, biochemistry, and bioinformatics. More potent computers will probably need to be developed in the field of bioinformatics alone to arrange the vast amounts of data produced by these initiatives.

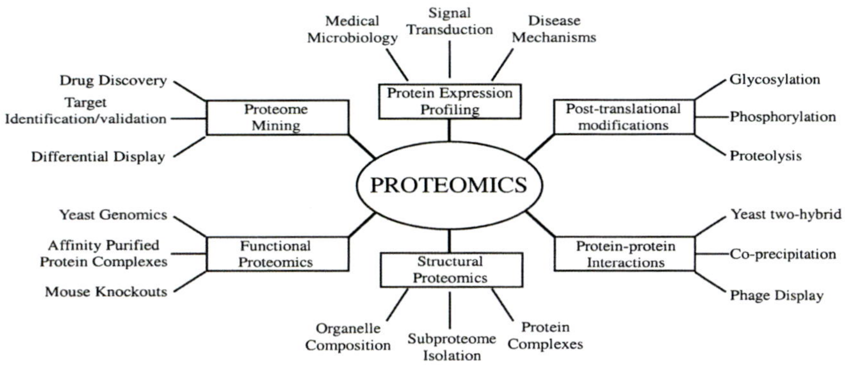

Figure 4.7. Types of proteomics and their applications to biology.

Source: Paul R. Graves and Timothy A. J. Haystead; Molecular Biologist's Guide to Proteomics; doi: 10.1128/MMBR.66.1.39-63.2002.

It is important to keep in mind that the proteome is dynamic when attempting to characterize the proteome of a particular cell or organism. A cell's proteome will mirror the immediate environment it is being studied. Proteins can be synthesized, degraded, or undergo posttranslational modifications in response to internal or external stimuli.

They can also undergo translocations within the cell. Analyzing a cell's proteome is akin to taking a "snapshot" of the protein environment at any particular moment. Taking everything into account, it is most likely possible for any given genome to give rise to an infinite number of proteomes.

4.2.1. Proteomics Origins

With the development of the two-dimensional gel and the mapping of proteins from *Escherichia coli*, mice, and guinea pigs, respectively, O'Farrell, Klose, and Scheele initiated the first protein studies that can be referred to as proteomics in 1975. Numerous proteins could be divided and seen, but they could not all be identified. Notwithstanding these drawbacks, a large-scale examination of every human protein was suggested soon after. This project, known as the human protein index, aimed to catalog all human proteins through the use of two-dimensional protein electrophoresis (2-DE) and other techniques. Unfortunately, financial constraints and technological issues made it impossible for this project to proceed.

The discovery of 2-DE represented a significant advancement, but proteomics research would have to wait until the proteins that 2-DE displayed could be identified. Sensitive protein-sequencing technology was one issue that needed to be solved. Because biological samples are frequently limited and because the capacity of both one-dimensional (1-D) and two-dimensional (2-D) gels to load proteins is limited, increasing sensitivity was essential to the experiment's success. The sequencing of proteins by Edman degradation was the first significant technological advancement for protein identification. The creation of methods for electroblotted protein microsequencing was a significant advance. The first 2-D databases were made using this method of protein identification from 2-D gels. In the 1990s, Edman sequencing became more sensitive to high picomole amounts due to advancements in microsequencing technology.

The advancement of MS technology has been one of the most significant developments in the field of protein identification. Over the past 10 years, there has been a significant increase in both the sensitivity and accuracy of protein identification by mass spectrometry. Proteins in the femtomolar range are thought to be detectable in gels at this point. Edman sequencing has virtually been superseded by mass spectrometry (MS) as the preferred method for protein identification due to its increased sensitivity, tolerance for protein mixtures, and ease of use with large-scale operations.

4.2.2. Why Proteomics?

The study of genes by itself is not sufficient to obtain many types of information. For instance, the phenotypes of cells are caused by proteins rather than genes. By looking only at the genome, it is impossible to understand the mechanisms underlying aging, disease, and environmental effects. Protein modifications can only be described and drug targets found through protein research.

4.2.2.1. Annotation of the Genome

Finding the total number of genes in a given genome will be one of the first uses of proteomics. It is still challenging to correctly predict genes from genomic data, which makes this "functional annotation" of a genome necessary. One issue is that bioinformatics cannot reliably predict the exon-intron structure of the majority of genes. To accomplish this, data from protein studies and genomic information will need to be combined to validate the presence of a specific gene.

4.2.2.2. Protein Expression Studies

The use of diverse techniques for analyzing mRNA expression has grown in popularity in recent years. These techniques include DNA microarray technology and serial analysis of gene expression (SAGE). Nevertheless, the amount of protein in a cell cannot be directly inferred from an analysis of its mRNA. As a result, numerous investigations have now demonstrated a weak relationship between the levels of mRNA and protein expression. The synthesis of a protein is the outcome of a protracted series of events that begins with the formation of mRNA (Figure 4.8). First, polyadenylation, mRNA editing, and alternative splicing are examples of posttranscriptional control mechanisms that affect mRNA. This stage allows for the generation of numerous distinct protein isoforms from a single gene. Second, regulation of mRNA may occur at the translational level of proteins. After they are formed, proteins can undergo posttranslational modification. Up to 200 distinct kinds of posttranslational protein modification are thought to exist. Proteolysis and compartmentalization are additional mechanisms for controlling proteins. It was estimated that there are one or two protein forms on average per gene in bacteria, three in yeast, and more in humans. It is evident from this that the idea of "one gene, one protein" is oversimplified. Furthermore, mRNA analysis is not possible for some body fluids, such as serum or urine, because they lack an mRNA source.

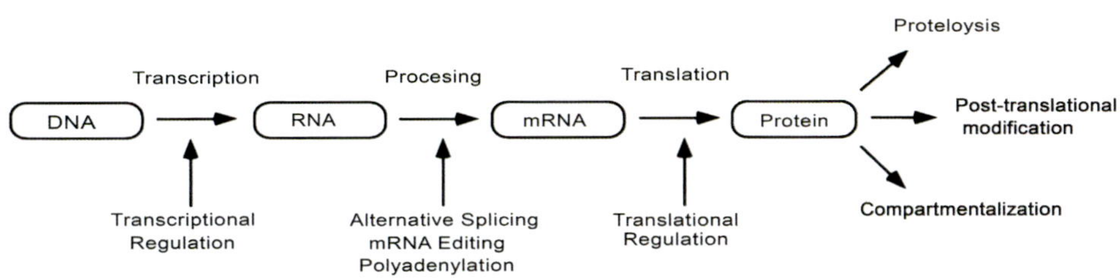

Figure 4.8. Mechanisms by which a single gene can give rise to multiple gene products. Multiple protein isoforms can be generated by RNA processing when RNA is alternatively spliced or edited to form mature mRNA. mRNA, in turn, can be regulated by stability and efficiency of translation. Proteins can be regulated by additional mechanisms, including posttranslational modification, proteolysis, or compartmentalization.

Source: Paul R. Graves and Timothy A. J. Haystead; Molecular Biologist's Guide to Proteomics; doi: 10.1128/MMBR.66.1.39-63.2002.

4.2.2.3. Protein Function

About one-third of the sequences in organisms whose genomes have been sequenced have no known function, according to one study. The field of structural genomics, whose ultimate goal is to obtain 3-D structures for every protein in a proteome, will benefit from the full identification of every protein in a genome. This is required because many proteins can only have their functions deduced from their three-dimensional structure.

4.2.2.4. Protein Modifications

The characterization of posttranslational protein modifications is poised to become a crucial application of proteomics. Numerous intracellular and extracellular cues have been shown to cause posttranslational modifications in proteins. For instance, protein phosphorylation is a crucial signaling mechanism, and oncogenesis can arise from the dysregulation of protein kinases or phosphatases. A proteomics method allows for the simultaneous analysis of modifications in a large number of proteins that are expressed by a single cell (Fleischmann., Adms et al., 1995).

4.2.2.5. Protein Localization and Compartmentalization

Protein localization is one of the most significant regulatory mechanisms currently understood. It is well known that protein mislocalization significantly affects cellular function (e.g. cystic fibrosis). The goal of proteomics is to pinpoint each protein's subcellular location. By using this data, a three-dimensional protein map of the cell can be made, which will reveal new details about how proteins are regulated.

4.2.2.6. Protein-Protein Interactions

The comprehension of protein-protein interactions is fundamental to biology. Protein complexes facilitate signal transduction, which controls cell division, programmed cell death, and the choice to continue through the cell cycle. The goal of proteomics is to create a comprehensive three-dimensional map of all protein interactions within a cell. With regard to the bacteria Helicobacter pylori, a recent advancement has been made. Using the yeast two-hybrid method to detect protein interactions, 1,200 connections were identified between *H. pylori* proteins covering 46.6% of the genome. A comprehensive two-hybrid analysis has also been performed on all the proteins from the yeast *S. cerevisiae*.

4.2.3. Types of Proteomics

Proteomics has three main types: expression proteomics, functional proteomics, and structural proteomics.

4.2.3.1. Expression Proteomics

Expression proteomics is a novel approach that studies the quantitative and qualitative expression of proteins. It aims to specify the difference in protein expression between two conditions such as patients and controls. In addition, it can identify disease-specific proteins and new proteins in signal transduction. Expression proteomics experiments are usually used to study the patterns of protein expression in different cells. For example, a tumor tissue sample is compared to a normal tissue sample to identify differences in the levels of proteins.

- **Importance of Expression Proteomics:** Expression proteomics is crucial for understanding how the genome is translated into the proteome, providing insights into the regulation of gene expression, protein abundance, and cellular dynamics. Changes in protein expression levels are often associated with cellular responses to environmental stimuli, developmental processes, disease states, and drug treatments. Therefore, characterizing protein expression profiles is fundamental for elucidating biological mechanisms and identifying potential biomarkers and drug targets.

- **Principles of Expression Proteomics:** Expression proteomics encompasses a variety of experimental and computational techniques aimed at quantifying the abundance of proteins within a biological sample. The key principles underlying expression proteomics include protein extraction, separation, quantification, and data analysis. By integrating these principles, expression proteomics enables the systematic profiling of protein expression patterns under different physiological and pathological conditions.
- **Experimental Techniques in Expression Proteomics:**
- Gel-Based Techniques:
  - **Two-Dimensional Gel Electrophoresis (2DE):** 2DE is a classical technique for separating proteins based on their isoelectric point (pI) and molecular weight. In 2DE, proteins are first separated by their pI using isoelectric focusing (IEF) and then by their molecular weight using SDS-PAGE. After separation, proteins can be visualized by staining or labeled for quantification using fluorescent or radioactive probes. 2DE is a powerful tool for comparing protein expression profiles between different samples and identifying differentially expressed proteins.
- Mass Spectrometry-Based Techniques:
  - **Shotgun Proteomics:** Shotgun proteomics, also known as bottom-up proteomics, involves digesting proteins into peptides followed by liquid chromatography-tandem mass spectrometry (LC-MS/MS) analysis. In shotgun proteomics, proteins are first extracted from the sample, digested into peptides using proteolytic enzymes (e.g., trypsin), and then separated by liquid chromatography before being analyzed by tandem mass spectrometry. The resulting mass spectra are matched to protein databases to identify and quantify the proteins present in the sample.
  - **Selected Reaction Monitoring (SRM):** SRM is a targeted mass spectrometry technique used for quantifying specific proteins of interest with high sensitivity and specificity. In SRM, predefined precursor ions and their corresponding fragment ions are selectively monitored using a triple quadrupole mass spectrometer. By measuring the abundance of specific

peptide fragments, SRM enables precise quantification of target proteins in complex biological samples.

- Quantitative Labeling Techniques:
  - **Stable Isotope Labeling:** Stable isotope labeling methods, such as SILAC (Stable Isotope Labeling by Amino acids in Cell culture) and iTRAQ (Isobaric Tags for Relative and Absolute Quantitation), enable quantitative comparison of protein expression levels between different samples. In SILAC, cells are cultured in media containing isotopically labeled amino acids, leading to the incorporation of the labels into newly synthesized proteins. In iTRAQ, peptides from different samples are labeled with isobaric tags, allowing multiplexed quantification of protein expression levels.

Protein Microarrays:

- **Reverse Phase Protein Arrays (RPPAs):** RPPAs are high-throughput platforms for quantifying the expression levels of multiple proteins in parallel using immobilized antibodies. In RPPAs, protein lysates from different samples are printed onto a solid support, such as a glass slide, and probed with specific antibodies against the proteins of interest. By measuring the signal intensity of bound antibodies, RPPAs enable quantitative analysis of protein expression levels across multiple samples simultaneously.

- **Applications of Expression Proteomics:**
- **Biomarker Discovery:** Expression proteomics is widely used for discovering biomarkers associated with various diseases, including cancer, cardiovascular disorders, and neurological conditions. By comparing protein expression profiles between diseased and healthy samples, expression proteomics can identify proteins that are differentially expressed in disease states, providing potential biomarkers for diagnosis, prognosis, and therapeutic monitoring.
- **Drug Target Identification:** Expression proteomics plays a crucial role in identifying potential drug targets by profiling protein expression levels in response to drug treatments. By characterizing the proteomic changes induced by drugs, expression proteomics can reveal key signaling pathways, molecular targets, and mechanisms of drug action, facilitating the development of targeted therapies for various diseases.
- **Functional Proteomics:** Expression proteomics provides valuable insights into the functional roles of proteins within cellular networks and pathways. By analyzing protein expression profiles under different experimental conditions, expression proteomics can elucidate the regulatory mechanisms governing protein expression, localization, and interaction networks, contributing to our understanding of cellular processes and systems biology.

- **Personalized Medicine:** Expression proteomics has the potential to advance personalized medicine by enabling the stratification of patients based on their protein expression profiles. By profiling protein expression levels in patient samples, expression proteomics can identify molecular subtypes, predict treatment responses, and guide personalized therapeutic strategies tailored to individual patients, improving clinical outcomes and patient care.

4.2.3.2. Structural Proteomics

Structural proteomics aims to determine the three-dimensional structures of proteins and their complexes on a genome-wide scale. It employs various experimental and computational techniques to elucidate protein structures, providing insights into their functions and interactions. By characterizing the structural landscape of the proteome, structural proteomics contributes to our understanding of cellular processes at the molecular level.

4.2.3.2.1. Experimental Techniques in Structural Proteomics:

- **X-ray Crystallography:** It is a widely used technique for determining protein structures at atomic resolution. It involves crystallizing the protein of interest and analyzing the diffraction pattern produced by X-rays passing through the crystal. By interpreting the diffraction data, the spatial arrangement of atoms within the protein can be deduced, yielding detailed structural information.
- **Nuclear Magnetic Resonance (NMR) Spectroscopy:** NMR spectroscopy enables the determination of protein structures in solution. It relies on the interaction between the magnetic properties of atomic nuclei and a strong magnetic field. By analyzing the signals emitted by nuclei in response to radiofrequency pulses, NMR provides information about the distances and angles between atoms, facilitating the reconstruction of protein structures.
- **Cryo-Electron Microscopy (Cryo-EM):** Cryo-EM has emerged as a powerful technique for studying the structures of large macromolecular complexes. It involves imaging frozen-hydrated samples with an electron microscope at cryogenic temperatures. Recent advances in hardware and software have significantly enhanced the resolution and applicability of cryo-EM, enabling the visualization of protein complexes at near-atomic resolution.
- **Homology Modeling:** Homology modeling, or comparative modeling, is a computational technique used to predict protein structures based on the known structures of homologous proteins. It relies on the assumption that proteins with similar sequences share similar structures. By aligning the sequence of the target protein with that of a template structure, homology modeling can generate a three-dimensional model of the target protein, providing valuable structural insights.
- **Mass Spectrometry:** It is a versatile technique used in structural proteomics for analyzing protein composition, interactions, and modifications. It involves ionizing proteins and measuring the mass-to-charge ratio of the resulting ions.

By analyzing the mass spectra of proteins, information about their primary sequence, post-translational modifications, and binding partners can be obtained, complementing structural data from other techniques.

4.2.3.2.2. Applications of Structural Proteomics

- **Drug Discovery and Development:** Structural proteomics plays a crucial role in drug discovery by facilitating the identification of druggable targets and the rational design of therapeutic agents. By elucidating the structures of target proteins and their interactions with ligands, structural proteomics aids in the development of small molecule inhibitors, biologics, and other therapeutic modalities.
- **Understanding Disease Mechanisms:** Structural proteomics contributes to our understanding of disease mechanisms by elucidating the structural basis of protein dysfunction in various pathological conditions. By characterizing disease-associated mutations and their effects on protein structure and function, structural proteomics provides insights into the molecular basis of diseases such as cancer, neurodegenerative disorders, and infectious diseases.
- **Functional Annotation of Genomes:** Structural proteomics facilitates the functional annotation of genomes by providing structural information for proteins with unknown functions. By predicting protein structures and inferring their functions based on structural similarities with known proteins, structural proteomics helps assign putative functions to uncharacterized genes, enabling a more comprehensive understanding of cellular processes.
- **Protein Engineering and Design:** Structural proteomics serves as a valuable tool for protein engineering and design by providing insights into the structure-function relationships of proteins. By analyzing protein structures and their interactions with ligands, substrates, and other molecules, structural proteomics aids in the rational design of proteins with desired properties, such as improved stability, specificity, or catalytic activity.

4.2.3.3. Functional Proteomics

Functional proteomics is crucial for bridging the gap between genome sequences and biological functions, as proteins are the primary effectors of cellular processes encoded by the genome. By systematically characterizing the functions and interactions of proteins, functional proteomics provides insights into the molecular mechanisms underlying physiological processes, disease states, and responses to environmental stimuli or therapeutic interventions. Additionally, functional proteomics facilitates the identification of potential drug targets, biomarkers, and diagnostic tools for various diseases.

- **Principles of Functional Proteomics:** Functional proteomics encompasses a variety of experimental and computational techniques aimed at elucidating the functions, interactions, and activities of proteins within biological systems. The key principles underlying functional proteomics include protein purification, interaction analysis, post-translational modification (PTM) profiling, and functional annotation. By integrating these principles, functional proteomics enables the systematic exploration of protein functions and regulatory mechanisms in a holistic manner.

- Experimental Techniques in Functional Proteomics

Protein-Protein Interaction (PPI) Analysis:

- **Yeast Two-Hybrid (Y2H):** Y2H is a widely used technique for detecting protein-protein interactions *in vivo*. In Y2H, a bait protein of interest is expressed as a fusion with a DNA-binding domain in yeast cells, while a prey protein library is expressed as fusions with an activation domain. Protein-protein interactions are detected by the reconstitution of a functional transcription factor when the bait and prey proteins interact, leading to the activation of reporter genes.
- **Co-immunoprecipitation (Co-IP):** Co-IP is a biochemical technique used to validate protein-protein interactions detected by other methods or to investigate specific protein complexes. In Co-IP, a target protein of interest is immunoprecipitated from cell lysates using a specific antibody, along with its interacting partners. The protein complexes are then eluted and analyzed by Western blotting or mass spectrometry to identify the interacting proteins.
- **Affinity Purification-Mass Spectrometry (AP-MS):** AP-MS is a high-throughput method for identifying protein complexes and their interacting partners. In AP-MS, a bait protein of interest is expressed as a fusion with an affinity tag (e.g., FLAG, GFP) and immobilized on a solid support, such as beads. The bait protein and its interacting partners are then co-purified from cell lysates using the affinity tag and analyzed by mass spectrometry to identify the interacting proteins.

Post-Translational Modification (PTM) Analysis:

- **Phosphoproteomics:** It is the study of protein phosphorylation, a reversible PTM involved in the regulation of various cellular processes, including signaling, transcription, and cell cycle control. Phosphoproteomics techniques involve enriching phosphopeptides from complex protein mixtures using methods such as immobilized metal affinity chromatography (IMAC) or titanium dioxide ($TiO2$) chromatography, followed by mass spectrometry analysis to identify and quantify phosphorylation sites.
- **Glycoproteomics:** It is the study of protein glycosylation, a common PTM involved in protein folding, stability, and cell-cell interactions. Glycoproteomics techniques involve enzymatic or chemical enrichment of glycopeptides from protein samples, followed by mass spectrometry analysis to identify and characterize glycosylation sites and glycan structures.

Functional Annotation and Pathway Analysis:

- **Gene Ontology (GO) Analysis:** GO analysis is a computational method for annotating proteins based on their molecular functions, biological processes, and cellular components. GO terms are organized in a hierarchical structure, allowing the functional annotation of proteins at different levels of specificity.

GO analysis enables the systematic characterization of protein functions and the identification of enriched functional categories within a set of proteins.

- **Pathway Analysis:** Pathway analysis is a computational method for identifying biological pathways and networks enriched with proteins of interest. Pathway databases such as KEGG, Reactome, and WikiPathways contain curated information about molecular pathways and their components. Pathway analysis tools enable the integration of protein expression data with pathway information, facilitating the identification of dysregulated pathways and the elucidation of biological mechanisms underlying cellular processes and disease states.

4.2.3.3.1. Applications of Functional Proteomics

- **Elucidating Cellular Signaling Pathways:** Functional proteomics is instrumental in deciphering cellular signaling pathways involved in various physiological processes, including cell proliferation, differentiation, apoptosis, and metabolism. By identifying protein-protein interactions, PTMs, and pathway components, functional proteomics provides insights into the molecular mechanisms underlying signal transduction and regulatory networks.
- **Characterizing Disease Mechanisms:** Functional proteomics contributes to our understanding of disease mechanisms by elucidating the functional roles of proteins in disease processes. By comparing protein expression profiles, PPI networks, and PTM patterns between healthy and diseased samples, functional proteomics can identify dysregulated pathways and key proteins associated with diseases such as cancer, neurodegenerative disorders, and autoimmune diseases.
- **Identifying Drug Targets and Biomarkers:** Functional proteomics plays a crucial role in identifying potential drug targets and biomarkers for various diseases. By characterizing protein-protein interactions, PTMs, and expression profiles in response to drug treatments, functional proteomics can identify key proteins involved in drug action, resistance, and toxicity. Additionally, functional proteomics can discover biomarkers associated with disease progression, treatment response, and patient prognosis, facilitating personalized medicine approaches.
- **Systems Biology and Network Modeling:** Functional proteomics contributes to systems biology approaches aimed at understanding biological systems as integrated networks of molecular interactions. By integrating protein expression data with other omics datasets (e.g., genomics, transcriptomics, metabolomics), functional proteomics enables the reconstruction of cellular networks and predictive modeling of biological processes. Systems biology approaches facilitate the identification of emergent properties, network motifs, and feedback loops underlying complex biological phenomena.

4.2.4. Technology of Proteomics

Proteomics has grown significantly as a result of advancements in protein technologies. There weren't many proteomics tools available when 2-DE was first introduced 26 years ago. Since then, advances have been made in both old and new technologies, spanning from protein separation to protein identification. It is also evident, though, that technological constraints still make many forms of proteomics impractical. To fully realize the promise of proteomics, these issues must be resolved and new technologies created. Generally, a proteomics experiment (like protein expression profiling) falls into one of three categories: (i) protein isolation and separation from a cell line, tissue, or organism; (ii) obtaining protein structure data for protein identification and characterization; or (iii) database utilization.

4.2.4.1. Separation and Isolation of Proteins

Complex protein mixtures will inevitably be encountered in proteomics, by definition. Thus, to visualize, identify, and characterize the proteins, a way must be found to separate these protein mixtures into their constituent parts. The most common technique for isolating and separating proteins is polyacrylamide gel electrophoresis. In contrast to the advances in molecular biology that eventually made it possible to sequence the human genome, certain areas of protein science have not advanced much over time. Technology for protein separation is one of them. Protein electrophoresis has been around for about 32 years, and it's still the best method for resolving complex protein mixtures. This is the point in many applications where the bottleneck appears. This is because 1- or 2-DE is a laborious, slow process that is difficult to automate. Nonetheless, this methodology will continue to be a crucial part of proteomics until something else takes its place.

4.2.4.1.1. One- and Two-Dimensional Gel Electrophoresis

Protein mixtures are best resolved using 1-DE in many proteomics applications. Proteins are sorted in 1-DE according to their molecular mass. Protein solubility is rarely an issue since proteins are solubilized in sodium dodecyl sulfate (SDS). In addition, 1-DE can be used to resolve proteins with molecular masses ranging from 10 to 300 kDa and is easy to execute. Characterization of proteins following protein purification is the most popular use of 1-DE. This is a result of a 1-D gel's low resolving power. 2-DE can be applied in the event of encountering a more complex protein mixture, such as a crude cell lysate. Two unique properties differentiate proteins in 2-DE. They are resolved in the first dimension based on their net charge and in the second dimension based on their molecular mass. When these two methods are combined, the resolution produced is far higher than with 1-DE.

The capacity of 2-DE to resolve proteins that have undergone posttranslational modification is one of its strongest points. This resolution in 2-DE is made possible by the fact that a variety of protein modifications alter the protein's mass and charge. Phosphorylation of proteins is one such instance. 2-DE is frequently used to separate a protein's phosphorylated from nonphosphorylated forms. In this instance, a single phosphoprotein on a 2-D gel will show up as several spots. Furthermore, 2-DE is capable of identifying various protein forms resulting from proteolytic processing or alternative mRNA splicing (Lewis and Hunt, 2000). Protein expression profiling remains the main use case for 2-DE. With this method, it is possible to compare the protein expression of any two samples both quantitatively and qualitatively. Spots can reveal information about differential protein expression when they appear or disappear, and their intensity can reveal quantitative information about the levels of protein expression. Protein expression profiling is applicable to samples derived from tissues, cell lines, organisms as a whole, or body fluids. This method can be used to compare tissues that are healthy and those that are sick, as well as cells that have been exposed to different medications or stimuli. Figure 4.9 provides an illustration of 2-DE in protein profiling.

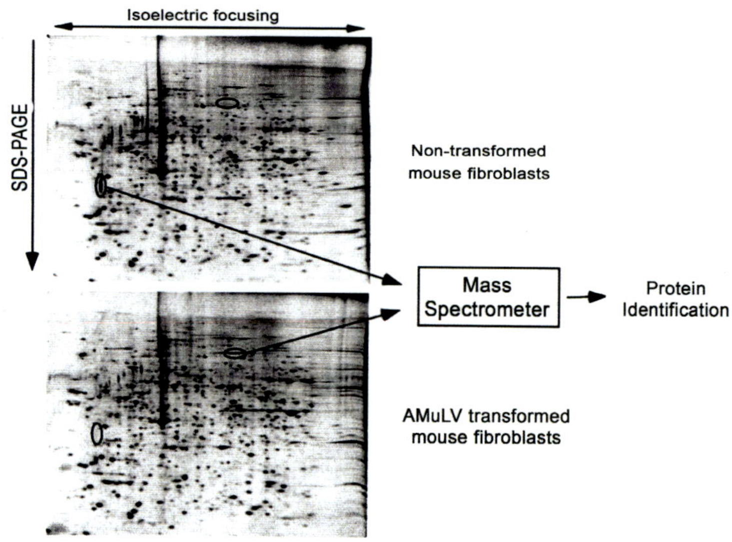

Figure 4.9. Protein expression profiling by 2-DE. Whole-cell lysates from nontransformed and Abelson murine leukemia virus (AMuLV)-transformed mouse fibroblasts were resolved by 2-DE, and proteins were visualized by silver staining. Differentially expressed proteins were excised from the gel and identified by MS.

Source: Paul R. Graves and Timothy A. J. Haystead; Molecular Biologist's Guide to Proteomics; doi: 10.1128/MMBR.66.1.39-63.2002.

A further use for 2-DE is in the field of cell map proteomics. Proteins from cellular organelles, protein complexes, and microorganisms can all be mapped using 2-DE. Proteins in subproteomes that have been produced by purifying a proteome can also be resolved and characterized using this method. A single 2-DE gel can resolve thousands of proteins, making it an effective tool for protein cataloging. Numerous 2-DE databases have been created and are accessible via the internet.

Over time, 2-DE has undergone a number of developments. The addition of immobilized pH gradients, which significantly increased 2-DE's reproducibility, was one of the largest advancements. The sensitivity of protein detection has increased with the use of fluorescent dyes, and more proteins can be resolved by specific pH gradients. 2-D gels can now be run in the minigel format, and 2-DE processing speed has increased. Furthermore, attempts have been made to automate 2-DE. The 2-DE process, from gel running to image analysis and spot picking, has been automated by Hochstrasser's group. The analysis of intricate 2-D gel images has been made easier by the use of computers. This is a crucial component of 2-DE since, to prevent artifacts, spot detection and annotation must be done very accurately. A molecular scanner designed to capture 2-DE images was created recently. Melanie is one example of software that compares computer images of 2-D gels and makes it easier to identify and measure the differences in protein spots between samples. Minden and associates have made an intriguing development in 2-DE recently. This technique, known as difference gel electrophoresis (DIGE), makes use of fluorescent tagging with two distinct dyes for two protein samples. The tagged proteins are run on the same 2-D gel, and pattern differences are identified by superimposing two images that were created using postrun fluorescence imaging of the gel. The dyes are amine-reactive and are designed to ensure that proteins common to both samples have the same relative mobility regardless of the dye used to tag them. This technique circumvents the need to compare several 2-D gels. In their original paper, DIGE was used to detect differences between exogenous proteins in two *D. melanogaster* embryo extracts at nanogram levels. Moreover, an inducible protein from *Escherichia coli* was detected after 15 min of induction. This technology is now commercially available from Amersham/Pharmacia.

But there are still a lot of issues with 2-DE. Even with attempts to automate 2-DE protein analysis, it remains a labor-intensive and time-consuming procedure. Only one sample can be examined per gel in a typical 2-DE experiment, which can take two days. The quantity and kind of proteins that can be resolved also place restrictions on 2-DE. For instance, a single 2-D gel cannot fully resolve the protein mixture that was extracted from a lysate of eukaryotic cells due to its complexity. The first dimension of the gel will not allow many large or hydrophobic proteins to enter, and proteins with extreme acidity or basicity (proteins with pIs below pH 3 and above pH 10) are not well represented. Various solubilization conditions and pH gradients can help solve some of these issues. When a total-cell lysate is analyzed, 2-DE is also limited in its ability to identify low-copy proteins. Low-copy protein detection can be challenging in a crude cell extract because the most abundant proteins may predominate on the gel. No proteins classified as low-copy proteins were found to be visible by 2-DE in the analysis of yeast proteins. However, more than half of yeast's 6,000 genes are thought to encode low-copy proteins. It is estimated that the dynamic range of protein expression in mammalian cells is approximately 7–9 orders of magnitude. Merely adding more protein to the gel won't solve the issue because it will reduce resolution and increase protein comigration. Due to these drawbacks, the analysis of protein complexes or subproteomes rather than entire proteomes is likely to be the most common use of 2-DE in the future.

4.2.4.2. Alternatives to Electrophoresis

Protein gel electrophoresis can be avoided in several ways thanks to the limitations of 2-DE. One method involves turning a whole protein mixture into peptides (usually by digesting it with trypsin), then purifying the peptides before using MS. to analyze them. Several techniques have been developed for peptide purification, such as liquid chromatography, capillary electrophoresis, and a combination of methods like reverse-phase (RP) chromatography and cation-exchange chromatography or multidimensional protein identification. These techniques have the benefit of allowing for the representation of a higher number of proteins in the mixture because a 2-D gel is not required. The drawback is that deconvoluting the acquired data may take a very long time and a lot of processing power. Additionally, the analysis of insignificant proteins may require a significant investment of time and energy. Isotope-coded affinity tags are one of the most intriguing methods to surface as a substitute for protein electrophoresis (ICAT). With this technique, quantitative protein profiling between various samples is possible without the need for electrophoresis.

4.2.5. Acquisition of Protein Structure Information

4.2.5.1. Edman Sequencing

N-terminal amino acid sequences were obtained by microsequencing using Edman chemistry, which was one of the first techniques for protein identification. Edman chemistry hasn't changed much since it was first used, but sequencing technology advancements have made Edman sequencing more sensitive and user-friendly. Edman sequencing is becoming less common in proteomics, but it's still a very helpful tool for a number of reasons. First off, a sizable portion of researchers still employ Edman sequencing since it was a sequencing technique that was used before MS. When identifying low-copy proteins, Edman sequencing is a feasible substitute for mass spectrometry in cases where mass spectrometers are either unavailable or in high demand. Lastly, Edman sequencing is used to find a protein's true start by obtaining its N-terminal sequence, if at all possible.

Edman first described the N-terminal sequencing of proteins in 1949. The most common application of Edman sequencing nowadays is the identification of proteins following their membrane transfer. Edman sequencing became a more useful sequencing technique for the identification of proteins separated by SDS-polyacrylamide gel electrophoresis with the development of membranes compatible with sequencing chemicals. N-terminal modification of proteins is a major issue that has hindered the effectiveness of Edman sequencing in the past. Precious samples were frequently lost in unsuccessful sequencing attempts because it is challenging to determine whether a protein is N-terminally blocked before it is sequenced. We created a brand-new method called mixed-peptide sequencing to solve this issue. In mixed-peptide sequencing, a protein is split into peptides by skatole or cyanogen bromide (CNBr) cleavage and the resulting peptides are simultaneously sequenced in an Edman sequencer.

In a nutshell, polyacrylamide gel electrophoresis (1-D or 2-D) is used to separate a complex protein mixture, and electroblotting is used to transfer the proteins to an inert membrane (Figure 4.10). The target proteins are visible on the membrane surface, removed, and chemically broken up into multiple large peptide fragments at methionine (by CNBr) or tryptophan (by skatole). Three to five peptide fragments are produced on average, which is in line with how frequently tryptophan and methionine occur in most proteins. Without any additional handling, the membrane piece is put straight into an automated Edman sequencer. The mixed-sequence data are fed into the FASTF or TFASTF algorithms, which sort and match the data against protein (FASTF) and DNA (TFASTF) databases to clearly identify the protein, after six to 12 automated Edman cycles (four to eight hours). The FASTF and TFASTF programs were written in collaboration with William Pearson. Because minimal sample handling is involved, mixed-peptide sequencing can be a sensitive approach for identifying proteins in polyacrylamide gels at the 0.1- to 1-pmol level. An example of mixed-peptide sequencing is shown in Figure 4.11a. The mixed-sequence approach has the advantage of enabling subsequent searches to be carried out against unannotated or non-species-specific DNA databases as well as annotated protein databases. This is because the T/FASTF algorithms utilize actual amino acid sequences and are therefore able to tolerate errors in the database as well as polymorphisms or conservative substitutions. (Chung et al., 2007)

A recent variation of T/FASTF has been devised for MS (Figure 4.11b).

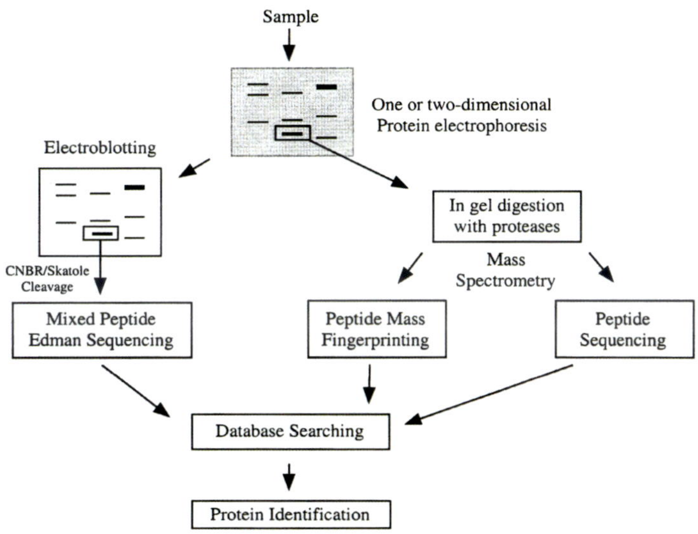

Figure 4.10. Strategies for protein identification. The identification of proteins from a polyacrylamide gel by mixed-peptide sequencing or MS is depicted. For mixed-peptide sequencing, proteins are transferred to a membrane and cleaved with CNB or skatole, and the resulting peptides are sequenced simultaneously by Edman degradation. For MS, proteins are in-gel digested with proteases and the resulting peptides are mass fingerprinted or sequenced. Information from all these methods is used to search nucleotide and protein databases for protein identification.

Source: Paul R. Graves and Timothy A. J. Haystead; Molecular Biologist's Guide to Proteomics; doi: 10.1128/MMBR.66.1.39-63.2002.

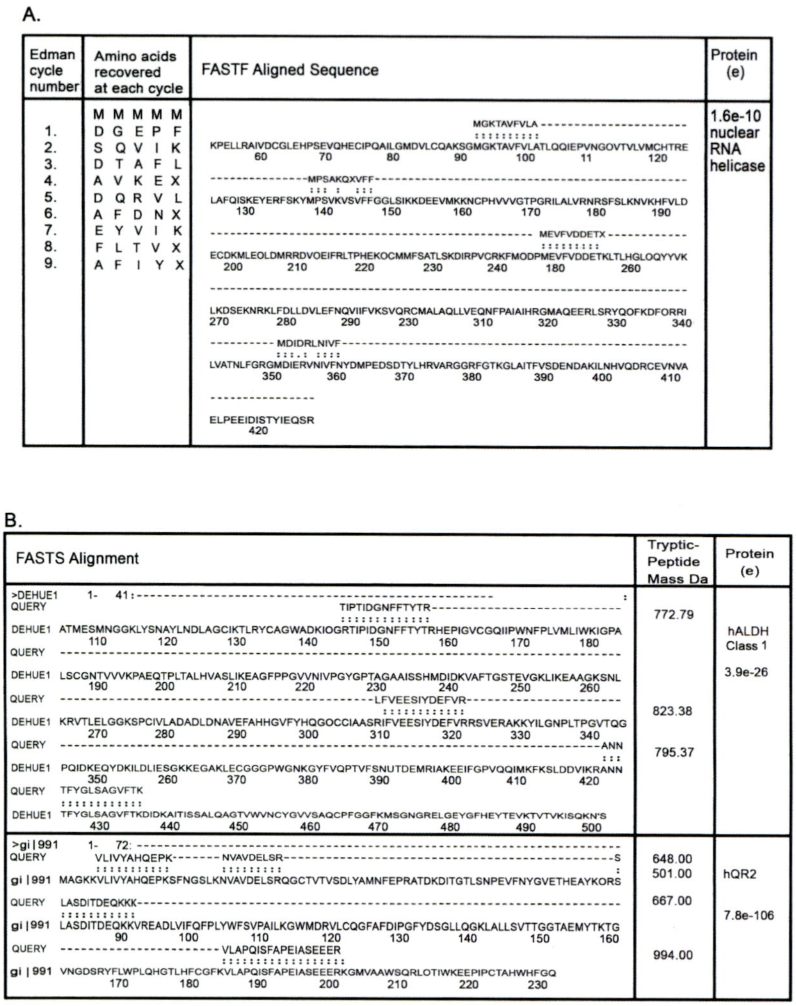

Figure 4.11. The FASTF and FASTS search programs.(A) Example of a FASTF search where the amino acid sequence is obtained by Edman sequencing of a mixture of peptides. The information is then deconvoluted by a computer algorithm, and the results are given an expectation score (e). (B) With the FASTS program, a similar type of search is conducted except that peptide sequences obtained from MS are used.

Source: Paul R. Graves and Timothy A. J. Haystead; Molecular Biologist's Guide to Proteomics; doi: 10.1128/MMBR.66.1.39-63.2002.

4.2.5.2. Mass Spectrometry

Protein structural data, such as peptide masses or amino acid sequences, can be obtained thanks to mass spectrometry. By utilizing nucleotide and protein databases, this information can be utilized to identify the protein (Figure 4.10). The kind and location of protein alterations can also be ascertained using this method. Three steps make up the extraction of protein information using mass spectrometry: sample preparation, sample ionization, and mass analysis.

4.2.5.2.1. Sample Preparation

In most proteomics, a 1- or 2-D polyacrylamide gel is used to separate a protein from a mixture. It is difficult to extract whole proteins from gels; however, many of the constituent peptides can be extracted from the gel if a protein is "in-gel" digested with a protease. The challenge is to extract the protein or its constituent peptides from the gel, purify the sample, and analyze it using MSdot. It was possible to develop a technique for in-gel protein digestion, which is now widely used for both 1- and 2-D gels. When it comes to sample recovery, in-gel digestion outperforms other widely used techniques like electroblotting. Furthermore, breaking down a protein into its individual peptides reveals more information than the protein as a whole. The peptides recovered after in-gel digestion must be purified to eliminate gel contaminants for a variety of applications. Common impurities from electrophoresis such as salts, buffers, and detergents can interfere with MS. In addition, peptide samples often require concentration before being analyzed by MS. One method of peptide purification commonly employed for this purpose is reverse-phase chromatography, which is available in a variety of formats. Peptides can be purified with ZipTips (Millipore) or Poros R2 perfusion material (PerSeptive Biosystems, Framingham, Mass.) or by high-pressure liquid chromatography (HPLC).

4.2.5.2.2. Sample Ionization

Molecules in biological samples need to be charged and dry in order for MS to analyze them. They are changed into desolvated ions to achieve this. Electrospray ionization (ESI) and matrix-assisted laser desorption/ionization (MALDI) are the two most widely used techniques for this. Protons are added or removed to transform peptides into ions in both processes. "Soft" ionization techniques like ESI and MALDI enable ion formation without significantly compromising sample integrity. This is significant because it makes it possible to gather precise mass data about proteins and peptides in their natural forms.

- **Electrospray Ionization:** In ESI, a liquid sample enters the mass spectrometer through a microcapillary tube. A potential difference between the capillary and the mass spectrometer's inlet causes a fine mist of charged droplets to form. Desolvated ions are formed as a result of the droplet sizes decreasing as the solvent evaporates. The advent of nanospray ionization brought about a major advancement in ESI technology. The microcapillary tube in nanospray ionization has a spraying orifice of 1 to 2 μm and can have flow rates as low as 5 to 10 nl/min. Because nanospray ionization can achieve low flow rates, less sample is used and more time is available for analysis. There are various methods for delivering the sample to the mass spectrometer in ESI. Filling individual microcapillary tubes with samples is the simplest approach. Cross-contamination is prevented because each sample is placed in a fresh microcapillary tube. Peptides in ESI need to be purified in some way following in-gel digestion, and this can be done right in the microcapillary tubes. Both the manual loading of microcapillary tubes and the purification process have the disadvantage of being laborious and slow. Alternatively, liquid chromatography (LC) systems that automatically purify and deliver the sample to the mass spectrometer have been linked to electrospray sources. Reverse-phase microcapillary LC (RP-µLC), reverse-phase LC (RP-LC), and LC are a few examples of this technique.

- **Matrix-Assisted Laser Desorption/Ionization:** In MALDI, the sample is combined with matrix molecules and exposed to laser radiation. Molecular ion formation is aided by the laser. Usually, the matrix consists of a small molecule that absorbs energy, like α-cyano-4-hydroxycinnamic acid or 2,5-dihydroxybenzoic acid. Crystals form as a result of the analyte and matrix being spotted on a metal plate and then allowed to evaporate. After the plate—which has a 96-well format option—is inserted into the mass spectrometer, the laser is programmed to automatically target particular regions of the plate. The entire procedure, including data collection and analysis, can be automated because a robot can apply samples. This is the greatest benefit all together. The fact that samples from in-gel digestion can frequently be used immediately without further purification is another benefit of MALDI over ESI.

4.2.5.2.3. Mass Analysis

Mass analysis comes after peptides or proteins are transformed into molecular ions. The molecular ions in a mass spectrometer are resolved based on their mass and charge in a vacuum by the mass analyzers in the apparatus.

- **Quadrupole Mass Analyzers:** The quadrupole mass analyzer is among the most widely used mass analyzers. Ions travel through an electric field produced by a quadrupole, a configuration of four parallel metal rods. All ions can be transmitted by a quadrupole, or it can function as a mass filter to permit the transmission of ions with a specific mass-to-charge (m/z) ratio. Information about a peptide's amino acid sequence can be obtained by combining multiple quadrupoles. The reader is directed to several outstanding reviews for a more thorough examination of the fundamentals of a quadrupole mass analyzer's operation.
- **Time of Flight(TOF):** TOF device is among the most basic types of mass analyzers. It calculates an ion's m/z ratio by timing how long it takes for it to travel the length of a flight tube. Ions are reflected back through the flight tube to a detector by an ion mirror located at the end of some TOF mass analyzers. The ion mirror extends the flight tube's length in this manner. Additionally, the ion mirror adjusts for minute energy variations between ions. These two elements work together to boost mass resolution.
- **Ion Trap:** A 3-D electric field is used by ion-trap mass analyzers to trap molecular ions. The primary benefit of an ion trap mass analyzer over a quadrupole mass analyzer is the latter's inability to discard ions before analysis, whereas the former allows ions to be "stored" and subsequently selectively ejected from the ion trap, thereby enhancing sensitivity. (Chung et al., 2007)

CASE STUDY

ADVANCING PRECISION MEDICINE THROUGH GENOMICS AND PROTEOMICS

- **Background:** XYZ Biomedical Institute is at the forefront of precision medicine research, aiming to tailor medical treatments based on individual genetic and protein profiles. They are conducting a study focused on understanding the genetic and proteomic basis of a rare neurological disorder known as NeuroX.
- **Problem Statement:** NeuroX is a complex disorder with varying symptoms and severity among affected individuals. Traditional diagnostic and treatment approaches have been limited due to the heterogeneous nature of the disease. XYZ Biomedical Institute seeks to uncover the genetic mutations and protein biomarkers associated with NeuroX to develop targeted therapies and diagnostic tools.
- **Solution:** To address these challenges, XYZ Biomedical Institute integrates genomics and proteomics technologies to comprehensively analyze the molecular basis of NeuroX. This approach involves sequencing the genomes of affected individuals to identify genetic variations and using advanced proteomics techniques to profile protein expression patterns in affected tissues.

Implementation

Genomics Analysis:

- **DNA Sequencing:** Researchers sequence the genomes of a cohort of NeuroX patients using next-generation sequencing (NGS) technologies. This allows them to identify genetic mutations, single nucleotide polymorphisms (SNPs), and structural variations that may contribute to the development and progression of NeuroX.
- **Bioinformatics Analysis:** Bioinformatics tools are employed to analyze the sequencing data, annotate genetic variants, and prioritize candidate genes implicated in NeuroX pathogenesis. This includes pathway analysis to understand the molecular mechanisms underlying the disorder.

Proteomics Analysis:

- **Protein Profiling:** Using mass spectrometry-based proteomics, researchers analyze protein expression profiles in cerebrospinal fluid (CSF) and brain tissues from NeuroX patients and healthy controls. This identifies dysregulated proteins and potential biomarkers associated with disease pathology.
- **Data Integration:** Genomic and proteomic data sets are integrated to

correlate genetic mutations with altered protein expression levels. This multi-omics approach helps identify key molecular pathways disrupted in NeuroX and pinpoint potential therapeutic targets.

Validation and Clinical Translation:

- **Experimental Validation:** Candidate genetic mutations and protein biomarkers identified through genomic and proteomic analyses are experimentally validated using functional assays and animal models.
- **Clinical Application:** Validated findings are translated into clinical applications, including the development of diagnostic tests for early detection of NeuroX and personalized treatment strategies targeting specific genetic and protein targets.

Results: Through the integration of genomics and proteomics technologies, XYZ Biomedical Institute identifies several genetic mutations and protein biomarkers associated with NeuroX. They uncover novel insights into the molecular mechanisms underlying the disorder, paving the way for the development of targeted therapies and diagnostic tools.

Conclusion: This case study highlights the transformative impact of genomics and proteomics in advancing precision medicine for complex disorders like NeuroX. By leveraging high-throughput sequencing and advanced proteomics techniques, researchers can unravel the intricate interplay between genetic variations and protein dysregulation, leading to personalized approaches in diagnosis and treatment. XYZ Biomedical Institute's pioneering research exemplifies how interdisciplinary approaches in genomics and proteomics can drive innovation in healthcare and improve patient outcomes.

CLASS ACTIVITY

EXPLORING GENOMICS AND PROTEOMICS

- **Objective:** To understand the fundamental concepts of genomics and proteomics, their techniques, applications, and their importance in modern biology.
- **Materials Needed:** Computers with internet access, Access to scientific databases (NCBI, UniProt), Worksheets, Projector and screen for presentations.
- Assessment: Group Discussion
  - Divide students into small groups;
  - Assign each group a specific topic to discuss:

    Group 1: Importance of genomics in personalized medicine.

    Group 2: Role of proteomics in disease diagnosis.

    Group 3: Techniques used in genomics.

    Group 4: Techniques used in proteomics.

Each group will prepare a short presentation.

SUMMARY

- The study of nucleic acids began with the discovery of DNA, progressed to the study of genes and small fragments, and has now exploded in the field of genomics. Genomics is the study of entire genomes, including the complete set of genes, their nucleotide sequence and organization, and their interactions within a species and with other species.
- Transcription at the whole genome level is assessed as a function of clinical conditions, as a consequence of mutations, in response to natural or toxic agents, in different cells or tissues, or at different times during biological processes, such as cell division or the development of an organism.
- A DNA microarray (DNA chip, or gene chip) experiment consists of hybridizing a nucleic acid sample (target) derived from the messenger RNAs (mRNAs) of a cell or tissue to single-stranded DNA sequences (probes) that are bound in an ordered arrangement to a solid platform.
- Microarray analysis is also used to identify genes that are coexpressed under different conditions or over a period of time, to determine which gene products function in a given pathway. A number of computational strategies are available that organize the data into related groups (clusters).
- Proteomics is the study of the interactions, function, composition, and structures of proteins and their cellular activities. Proteomics provides a better understanding of the structure and function of the organism than genomics.
- Protein interaction maps can reveal novel functions for proteins, place proteins with unknown functions among those with known cellular roles, and elucidate new biological pathways and cellular machines.
- Expression proteomics is a novel approach that studies the quantitative and qualitative expression of proteins. It aims to specify the difference in protein expression between two conditions such as patients and controls. In addition, it can identify disease-specific proteins and new proteins in signal transduction. Expression proteomics experiments are usually used to study the patterns of protein expression in different cells.
- Structural proteomics aims to determine the three-dimensional structures of proteins and their complexes on a genome-wide scale. It employs various experimental and computational techniques to elucidate protein structures, providing insights into their functions and interactions.

REVIEW QUESTIONS

1. What is functional genomics, and how does it differ from structural genomics?
2. What is the proteome, and why is it considered dynamic?
3. Describe the process of protein identification using mass spectrometry.
4. Why are there so many more proteins than genes in humans?

5. What are some high-throughput techniques commonly used in functional genomics, and how do they facilitate large-scale gene function analysis?
6. What are the objectives of protein–protein interaction studies?
7. What types of molecules can be captured by a functional protein microarray?

REFERENCES

1. Anderson, N. G., & Anderson, N. L. (1996). Twenty years of two-dimensional electrophoresis: Past, present and future. *Electrophoresis, 17*(3), 443–453. https://doi.org/10.1002/elps.1150170302.
2. Bertone, P., & Snyder, M. (2005). Advances in functional protein microarray technology. *FEBS Journal, 272*(21), 5400–5411. https://doi.org/10.1111/j.1742-4658.2005.04942.x.
3. Brasch, M. A., Hartley, J. L., & Vidal, M. (2004). ORFeome cloning and systems biology: Standardized mass production of the parts from the parts list. *Genome Research, 14*(10a), 2001–2009. https://doi.org/10.1101/gr.2634704.
4. Buyse, M., Loi, S., van't Veer, L., Viale, G., Delorenzi, M., Glas, A. M., d'Assignies, M. S., Bergh, J., Lidereau, R., Ellis, P., Harris, A., Bogaerts, J., Therasse, P., Floore, M., Amakrane, F., Piette, F., Rutgers, E., Sotiriou, C., Cardoso, F., & Piccart, M. J. (2006). Validation and clinical utility of a 70-gene prognostic signature for women with node-negative breast cancer. *Journal of the National Cancer Institute, 98*(17), 1183–1192. https://doi.org/10.1093/jnci/djj320.
5. Chung, C. H., Levy, S., Chaurand, P., & Carbone, D. P. (2007). Genomics and proteomics: Emerging technologies in clinical cancer research. *Critical Reviews in Oncology/Hematology, 61*(1), 1–25. https://doi.org/10.1016/j.critrevonc.2006.09.001.
6. Droit, A., Poirier, G. G., & Hunter, J. M. (2005). Experimental and bioinformatic approaches for interrogating protein-protein interactions to determine protein function. *Journal of Molecular Endocrinology, 34*(2), 263–280. https://doi.org/10.1677/jme.1.01770.
7. Emanuelsson, O., Nagalakshmi, U., Zheng, D., Rozowsky, J. S., Urban, A. E., Du, J., Lian, Z., Stolc, V., Weissman, S., Snyder, M., & Gerstein, M. B. (2007). Assessing the performance of different high-density tiling microarray strategies for mapping transcribed regions of the human genome. *Genome Research, 17*(6), 886–897. https://doi.org/10.1101/gr.5775807.
8. Fleischmann, R. D., Adams, M. D., White, O., Clayton, R. A., Kirkness, E. F., Kerlavage, A. R., Bult, C. J., Tomb, J. F., Dougherty, B. A., Merrick, J. M., et al. (1995). Whole-genome random sequencing and assembly of *Haemophilus influenzae* Rd. *Science, 269*(5222), 496–512. https://doi.org/10.1126/science.7542800.
9. Glas, A. M., Floore, A., Delahaye, L. J., Witteveen, A. T., Pover, R. C., Bakx, N., Lahti-Domenici, J. S., Bruinsma, T. J., Warmoes, M. O., Bernards, R., Wessels, L. F., & Van't Veer, L. J. (2006). Converting a breast cancer microarray signature into a high-throughput diagnostic test. *BMC Genomics,* 7, 278. https://doi.org/10.1186/1471-2164-7-278.

10. Handelsman, J. (2004). Metagenomics: Application of genomics to uncultured microorganisms. *Microbiology and Molecular Biology Reviews, 68*(4), 669–685. https://doi.org/10.1128/MMBR.68.4.669-685.2004.
11. Hoheisel, J. D. (2006). Microarray technology: Beyond transcript profiling and genotype analysis. *Nature Reviews Genetics, 7*(3), 200–210. https://doi.org/10.1038/nrg1801.
12. Lewis, T. S., Hunt, J. B., Aveline, L. D., Jonscher, K. R., Louie, D. F., Yeh, J. M., Nahreini, T. S., Resing, K. A., & Ahn, N. G. (2000). Identification of novel MAP kinase pathway signaling targets by functional proteomics and mass spectrometry. *Molecular Cell, 6*(5), 1343–1354. https://doi.org/10.1016/S1097-2765(00)00145-8.
13. Morris, J. S., Clark, B. N., & Gutstein, H. B. (2008). Pinnacle: A fast, automatic and accurate method for detecting and quantifying protein spots in 2-dimensional gel electrophoresis data. *Bioinformatics, 24*(4), 529–536. https://doi.org/10.1093/bioinformatics/btm628.
14. Mulder, N., & Apweiler, R. (2007). InterPro and InterProScan: Tools for protein sequence classification and comparison. *Methods in Molecular Biology, 396*, 59–70. https://doi.org/10.1385/1-59745-234-3:059.
15. Noguchi, H., Park, J., & Takagi, T. (2006). MetaGene: Prokaryotic gene finding from environmental genome shotgun sequences. *Nucleic Acids Research, 34*(12), 5623–5630. https://doi.org/10.1093/nar/gkl706.
16. Parrish, J. R., Gulyas, K. D., & Finley, R. L. Jr. (2006). Yeast two-hybrid contributions to interactome mapping. *Current Opinion in Biotechnology, 17*(4), 387–393. https://doi.org/10.1016/j.copbio.2006.07.005.
17. Rapp, B. A., & Wheeler, D. L. (2005). Bioinformatics resources from the National Center for Biotechnology Information: An integrated foundation for discovery. *Journal of the American Society for Information Science and Technology, 56*(6), 538–550. https://doi.org/10.1002/asi.20118.
18. Riesenfeld, C. S., Schloss, P. D., & Handelsman, J. (2004). Metagenomics: Genomic analysis of microbial communities. *Annual Review of Genetics, 38*, 525–552. https://doi.org/10.1146/annurev.genet.38.072902.091216.
19. Sanchez-Carbayo, M. (2006). Antibody arrays: Technical considerations and clinical applications in cancer. *Clinical Chemistry, 52*(10), 1651–1659. https://doi.org/10.1373/clinchem.2006.072184.

CHAPTER 5

Recombinant DNA and Genetic Analysis

LEARNING OBJECTIVES

After studying this chapter, you will be able to:

- Explain the construction of gene libraries;
- Discuss cloning vectors.

KEY TERMS FROM THIS CHAPTER

cDNA libraries	DNA molecules
DNA polymerase	DNA technology
Electrophoresis	Gene cloning
Gene library	Genetic analysis
Microarray analysis	Protein synthesis

INTRODUCTION

Significant advancements in bioinformatics analysis, sequencing technologies, and microarray technology are now starting to offer meaningful insights into the growth and upkeep of cells and tissues. In fact, fields of study like transcriptomics, metabolomics, and systems biology are now well-established and enable the simultaneous analysis of enormous numbers of samples. The primary engine behind biological research and analysis today is this kind of extensive parallel analysis. Molecular biology and genetic analysis, on the other hand, owe their roots to approaches that were developed several decades prior. The discovery of restriction endonucleases in the early 1970s served as a major catalyst for the development of molecular biology analysis. This discovery allowed for more effective DNA analysis as well as the ability to cut different DNA molecules apart for later recombinant DNA fragment creation. A new era in the manipulation, analysis, and utilization of biological molecules was ushered in by the creation of DNA molecules. Numerous discoveries and insights into the structure, function, and regulation of genes have been made possible by this procedure, known as gene cloning. The techniques for creating gene libraries have been improved upon and developed ever since they were first used. Gene cloning techniques are still necessary in many situations, even though techniques like microarray analysis and the polymerase chain reaction (PCR) have made gene analysis quicker and easier. The following provides an account of the process of gene cloning and other methods based on recombinant DNA technology.

5.1. CONSTRUCTING GENE LIBRARIES

An organism's various DNA sequences that have been cloned into a vector to facilitate purification, storage, and analysis are gathered into a gene library. A gene library is a collection of cloned DNA that is arranged so that any given piece of the original DNA will most likely be found in the collection.

5.1.1. Digesting Genomic DNA Molecules

Restriction endonucleases are enzymes that can be used to specifically fragment genomic DNA after it has been isolated and purified. Due to their specificity for specific DNA sequences, these enzymes are essential to molecular cloning. It is significant to remember that when an enzyme is used to break down a DNA molecule, all copies of that molecule from that particular organism will yield the same set of fragments. When treated with the same enzyme, DNA from different organisms will typically yield different sets of fragments. It is feasible to consistently split an organism's genome into many tiny pieces, each roughly the size of a single gene, by breaking down its complex genomic DNA. Certain enzymes produce blunt or flush ends by cutting the DNA straight through. At either end of the digested DNA, short single-stranded projections are created by other restriction enzymes performing staggered single-strand cuts. These ends are referred to as cohesive or sticky ends because they are complementary, identical, and will base-pair with one another. Furthermore, the phosphate groups are consistently retained in the DNA's 5′ end projection.

More than 600 enzymes that can identify more than 200 distinct restriction sites have been described. Several factors influence the choice of enzyme to use. For instance, if each of the four bases has a random sequence, the recognition sequence of six bp will occur, on average, every 4096 (46) bases. This indicates that after genomic DNA is digested using EcoR1, which can identify the sequence 5'-GAATTC-3,' fragments with an average size of slightly more than 4 kb will be produced. Much longer fragments are produced by enzymes possessing 8 bp recognition sequences. As a result, enzymes that break down very large genomes—like the human genome—usually result in lengthy DNA fragments. As a result, fewer of those fragments need to be cloned and then examined, which simplifies the steps that follow (Table 5.1).

Table 5.1. *Numbers of Clones Required for Representation of DNA in a Genome Library*

		No. of clones required	
Species	**Genome Size (kb)**	**17kb fragments**	**35kb fragments**
Bacteria (E. coli)	4000	700	340
Yeast	20000	3500	1700
Fruit fly	165000	29000	14500
Man	3000000	535000	258250
Maize	15000000	2700000	1350000

Source: Keith Wilson and John Walker; Principles and Techniques of Biochemistry and Molecular Biology; ISBN 978-0-521-51635-8.

5.1.2. Ligating DNA Molecules

Any other DNA fragments treated with the same restriction enzyme can be joined to the DNA products to form sticky ends that result from restriction digestion. Because of this, base-pairing between sticky ends will cause the two sets of fragments to anneal together when they are combined, even though their original DNA sources are different. Naturally, a process known as reannealing will also involve the pairing of fragments that originated from the same initial DNA molecules. Because the few bases in the sticky ends of all these pairings have weak hydrogen bonds, they are all ephemeral. However, by using an enzyme called DNA ligase, a process known as ligation, these pairings can be stabilized.

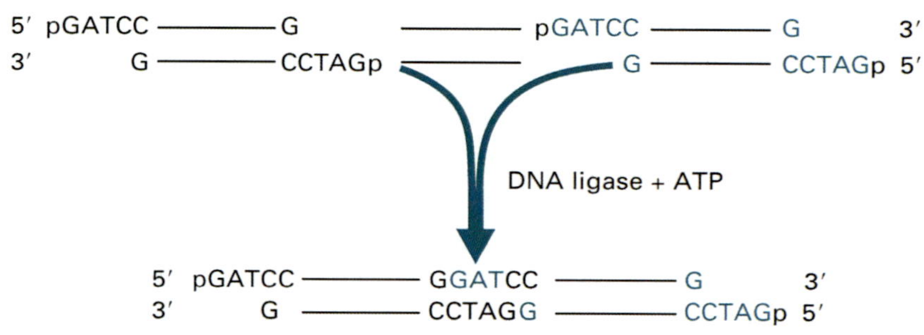

Figure 5.1. *Ligation molecules with cohesive ends.*

Source: Keith Wilson And John Walker; Principles and Techniques of Biochemistry and Molecular Biology; ISBN 978-0-521-51635-8.

Known as T4 DNA ligase, this enzyme is typically isolated from bacteriophage T4 and forms a covalent bond between the 3' hydroxyl of the adjacent strand and the 5' phosphate at the end of one strand (Figure 5.1). To decrease the kinetic energy of molecules and lessen the likelihood of base-paired sticky ends splitting before they have been stabilized by ligation, the ATP-dependent reaction is frequently conducted at 10°C.

Long reaction times are nevertheless required to make up for DNA ligase's decreased activity in the cold. DNA molecules can also be joined at their blunt ends, though this process is far less effective than ligations with sticky ends (Brooker, 2005; Brown, 2006).

Recombinant molecules created by ligation of sticky ends can be cleaved again at the "joins" using the same restriction enzyme that was used to create the fragments in the first place because ligation reconstructs the site of cleavage. Digested DNA from an organism must be joined or ligated with a specific DNA carrier molecule known as a vector to be replicated. As a result, each DNA fragment is ligated into the vector DNA molecule, enabling the continuous replication of the entire recombined DNA within microbial cells (Figure 5.2). This method of cloning a DNA fragment allows for enough material to be used for additional manipulation or in-depth analysis. As a result, every piece of DNA that is taken out of an organism and digested by a restriction enzyme will produce clones. A gene library is a collection of clones like this one.

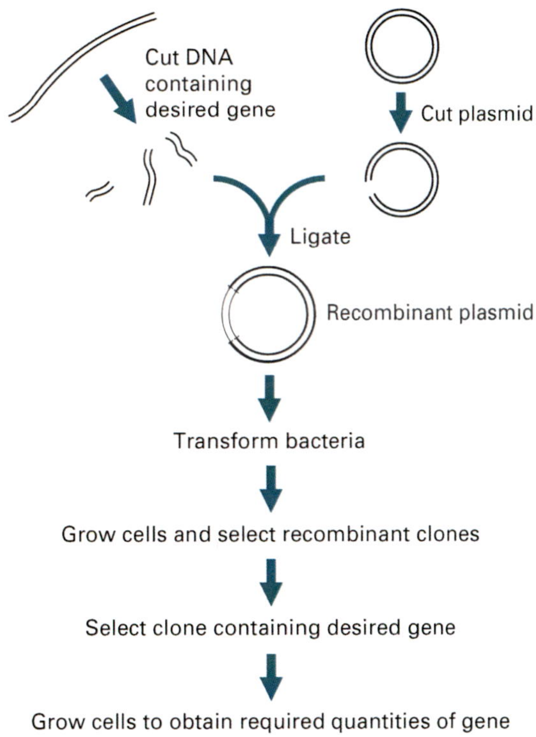

Figure 5.2. *Outline of gene cloning.*

Source: Keith Wilson And John Walker; Principles and Techniques of Biochemistry and Molecular Biology; ISBN 978-0-521-51635-8.

5.1.3. Aspects of Gene Libraries

Gene libraries can be broadly classified into two types: genomic libraries, which comprise an organism's entire chromosomal DNA, and cDNA libraries, which solely represent the mRNA from a specific cell or tissue at a given time (Figure 5.3). The selection of a specific gene library type is contingent upon several factors, the primary one being the ultimate use of any DNA fragment obtained from the library. Genomic libraries must be used if the ultimate goal is to comprehend the regulation of protein production for a specific gene or the analysis of its architecture. cDNA libraries are more suitable, though, if the objective is the synthesis of fresh or altered proteins or the identification of tissue-specific expression and timing patterns. Therefore, the nucleic acid starting material is the primary factor to be taken into account when building genomic or cDNA libraries. Since an organism's genome is fixed, almost any type of cell can have its chromosomal DNA isolated to create genomic libraries. On the other hand, cDNA libraries solely depict the mRNA generated from a particular cell type during a given period of time. Therefore, when building cDNA libraries, it is crucial to carefully consider the type of cell or tissue from which the mRNA is to be derived (Brooker, 2005; Brown, 2006).

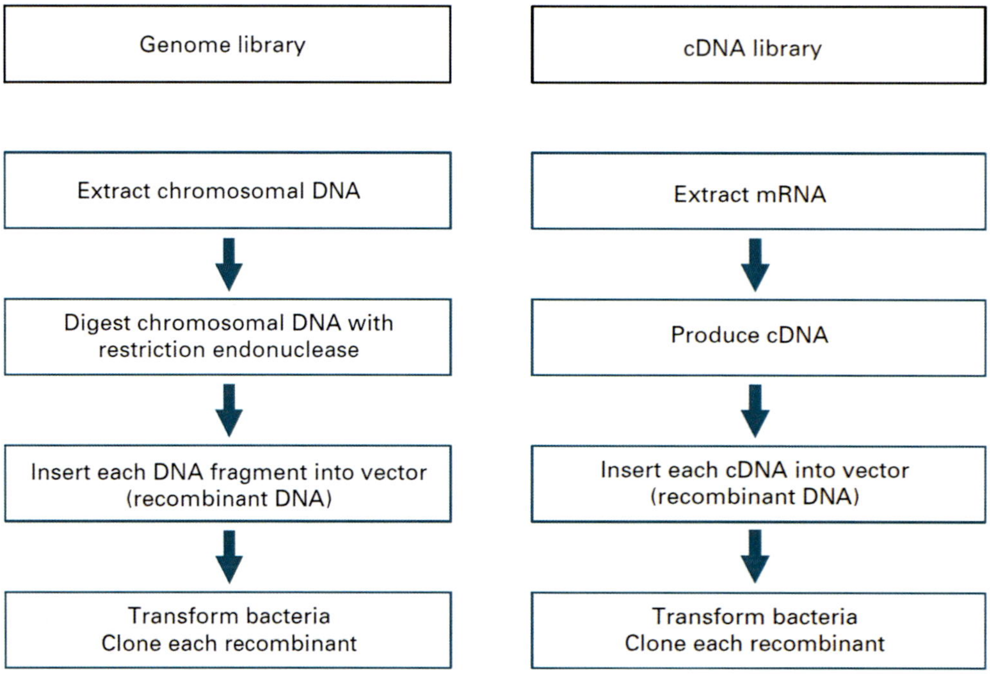

Figure 5.3. *Comparison of the general steps involved in the construction of genomic and complementary DNA (cDNA) libraries.*

Source: Keith Wilson and John Walker; Principles and Techniques of Biochemistry and Molecular Biology; ISBN 978-0-521-51635-8.

Many cloning vectors are available; these include viruses that infect bacteria or bacterial plasmids, which are examples of naturally occurring molecules. Whether a cDNA library or a genomic library is built determines which vector should be used.

5.1.4. Genomic DNA Libraries

The process of creating genomic libraries involves separating the entire chromosomal DNA from a cell and using restriction endonucleases to break it up into pieces with the appropriate average length. Using an enzyme that detects tetranucleotide sequences, partial restriction digestion can be used to accomplish this. If the enzyme is allowed to cleave only a few of its potential restriction sites before the reaction is stopped, each DNA molecule will be cut into relatively large fragments instead of the large number of very short fragments that would result from complete digestion with such an enzyme. The relative concentrations of DNA and restriction enzyme, as well as the circumstances and length of incubation, will determine the average fragment size (Figure 5.4). DNA fragments can also be physically sheared, though to make the fragments flush-ended, the ends may need to be repaired. Klenow polymerase, a modified DNA polymerase, can be used to accomplish this. This is made by subtilizing DNA polymerase and cleaving it, resulting in a large enzyme fragment that functions as a 5' to 3' polymerase but lacks 5' to 3' exonuclease activity. By using the proper dNTPs, this will fill in any recessed 3' ends on the sheared DNA.

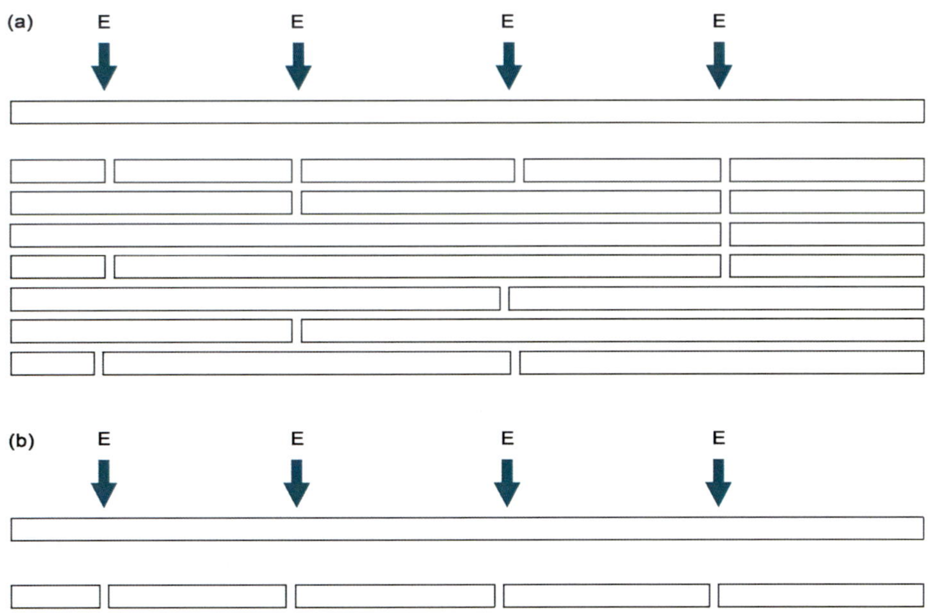

Figure 5.4. *Comparison of (a) partial and (b) complete digestion of DNA molecules at restriction enzyme sites (E).*

Source: Keith Wilson and John Walker; Principles and Techniques of Biochemistry and Molecular Biology; ISBN 978-0-521-51635-8.

After using a vector to ligate the mixture of DNA fragments, the mixture is cloned. There is a very high likelihood that any specific DNA fragment, like a gene, will be present in at least one of the clones if enough of them are made. For prokaryotic libraries, fragments around 10 kb in length are required; however, for mammalian libraries, the length must be increased to roughly 40 kb to maintain a manageable number of clones. The number of clones required in a gene library to provide a probability of obtaining a specific DNA sequence can be computed. This formula is:

$$N = \frac{\ln(1-P)}{\ln(1-f)}$$

Where; N is the probability of recombinants, P is the probability and f is the fraction of the genome in one insert. Thus, for the *E. coli* DNA chromosome of 5 10^6 bp and with an insert size of 20 kb the number of clones needed (N) would be 1 10^3 with a probability of 0.99.

5.1.5. cDNA Libraries

A cell can produce thousands of different proteins at any given time, each of which has a corresponding mRNA molecule. Clones of each individual mRNA must be created to identify any one of those mRNA molecules. cDNA libraries are those that contain the mRNA from a specific cell or tissue. Cloning cannot directly use mRNA because it is too unstable. Nonetheless, it is feasible to create complementary DNA molecules (cDNAs) for every mRNA in the chosen tissue. It is possible to clone the cDNA after inserting it into a vector. The RNA-containing retroviruses that contain the enzyme reverse transcriptase are used to produce cDNA or complementary DNA.

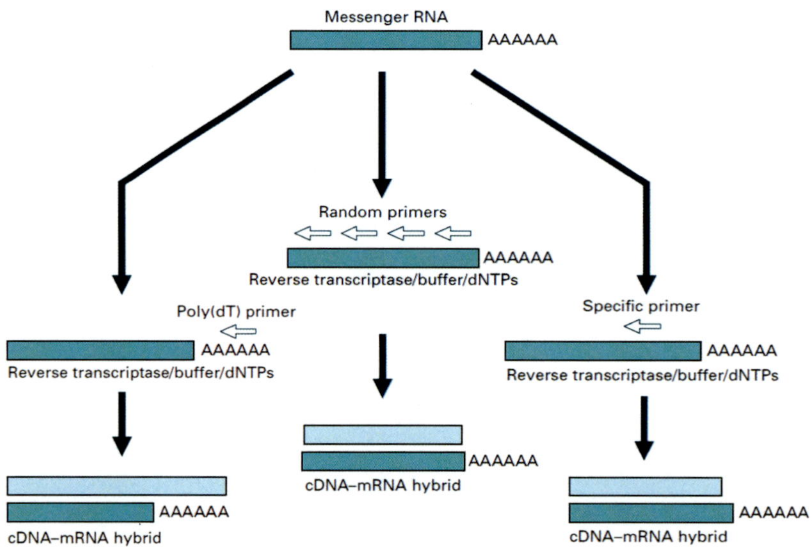

Figure 5.5. *Strategies for producing first-strand cDNA from mRNA.*

Source: Keith Wilson and John Walker; Principles and Techniques of Biochemistry and Molecular Biology; ISBN 978-0-521-51635-8.

Reverse transcriptase is an RNA-dependent DNA polymerase that uses a combination of the four dNTPs to create a first-strand DNA complementary to an mRNA template. As with all polymerase enzymes, the presence of a brief oligonucleotide primer is also necessary (Figure 5.5). It is possible to use a complementary oligo(dT) primer with eukaryotic mRNA that has a poly(A) tail. An alternative is to employ random hexamers, which anneal to the complex's mRNAs at random. These primers offer a free 3' hydroxyl group, which reverse transcriptase uses as its initial starting point. A prerequisite for creating first-strand cDNA is high-quality, undamaged mRNA, no matter how it is prepared. The standard procedure for determining the integrity of RNA is gel electrophoresis. Alternatively, a portion of the extract can be employed in a cell-free translation system, which will control the synthesis of the proteins that the mRNA molecules in the sample represent if intact mRNA is present.

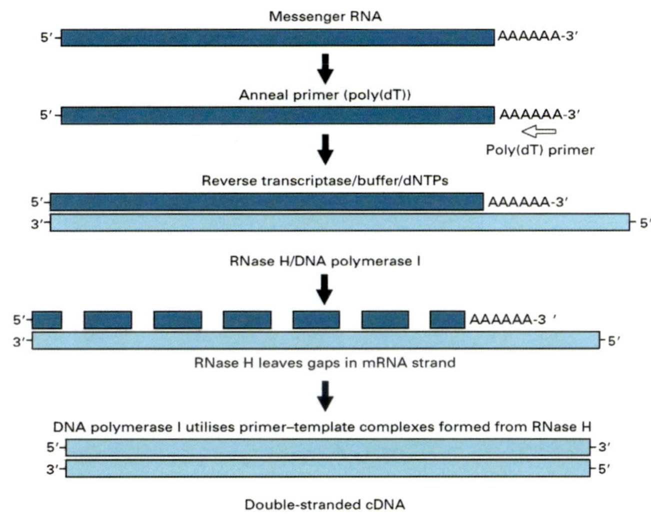

Figure 5.6. *Second-strand cDNA synthesis using the RNase H method.*

Source: Keith Wilson and John Walker; Principles and Techniques of Biochemistry and Molecular Biology; ISBN 978-0-521-51635-8.

Using terminal transferase and dCTP, a poly(dC) tail is added to the 3' end of the first DNA strand after it has been synthesized. Incidentally, this will also add a poly(dC) tail to the mRNA's poly(A) strand. After that, the RNA strand is eliminated by alkaline hydrolysis, leaving single-stranded DNA that, like mRNA, can be utilized to control the synthesis of a complementary DNA strand. DNA polymerase I's Klenow fragment catalyzes the poly(dC) tail and base pairs it with an oligo(dG) primer for second-strand synthesis. Double-stranded DNA is the end result, with one strand complementary to the mRNA. RNase H is used in cDNA synthesis as an additional technique. Here, reverse transcriptase is used to create first-strand cDNA in the same manner as before, but the resulting mRNA-CDNA hybrid is kept. The RNA strand is then nicked using low concentrations of RNase H. When a second strand of cDNA is added to the RNA, DNA polymerase uses the exposed 3' hydroxyl groups in the nicks to act as a primer (Figure 5.6).

5.1.6. Treatment of Blunt cDNA Ends

Since ligation of blunt-ended DNA fragments is less effective than ligation of sticky ends, extra steps are taken with cDNA molecules before ligation using cloning vectors. One method is to incorporate nucleic acid linkers—small double-stranded molecules with a single internal site for a restriction end nuclease—into the cDNA. For many of the most widely used restriction enzymes, there are a variety of commercially available linkers with internal restriction sites. Linkers are blunt-end ligated to the cDNA, but the ligation process is reasonably successful because they are added in large excess of the cDNA. The linkers are then digested using the proper restriction enzyme, supplying the sticky ends needed for a successful ligation to a vector that has also been digested by the same enzyme. Adding adaptors instead of linkers, which are identical except for the preformed sticky ends and the lack of restriction digestion after ligation, may simplify this process (Figure 5.7).

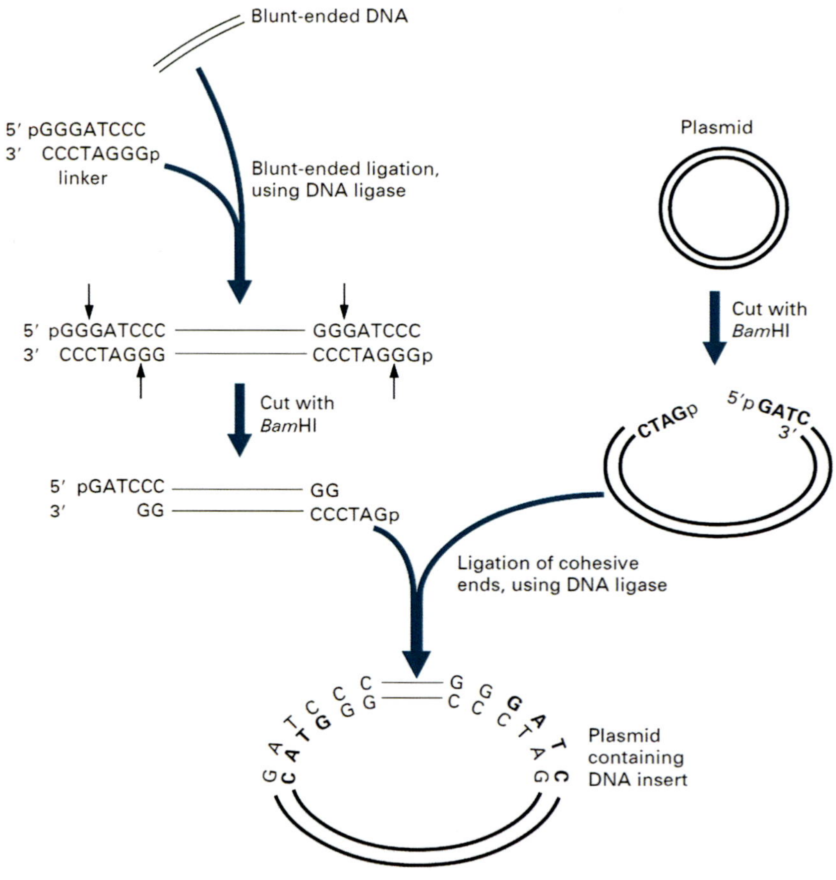

Figure 5.7. *Use of linkers. In this example, blunt-ended DNA is inserted into a specific restriction site on a plasmid, after ligation to a linker containing the same restriction site.*

Source: Keith Wilson And John Walker; Principles and Techniques of Biochemistry and Molecular Biology; ISBN 978-0-521-51635-8.

5.1.7. Enrichment Methods for RNA

Often, the goal is to isolate the mRNA that is transcribed from a target gene in a specific cell or tissue that generates a large quantity of the protein. Therefore, a significant portion of the total mRNA will code for the protein if the cell or tissue produces one of the cell's major proteins. An illustration of this would be the pancreatic B cells, which have high concentrations of pro-insulin mRNA. In these situations, employing antibodies against the ribosomal proteins allows one to precipitate polysomes that are actively translating the mRNA can then be separated from the precipitated ribosomes. The necessary mRNA typically makes up a very small portion of the total cellular mRNA. In such circumstances, sucrose density gradient centrifugation can be used to separate total mRNA by size. Next, an *in vitro* translation system is employed to control the synthesis of proteins utilizing each fraction.

5.1.8. Subtractive Hybridization

Genes are frequently differentially activated at a specific stage of cellular growth or transcribed at a specific cell type, frequently at very low levels. By using subtractive hybridization, those mRNA transcripts can be separated. The mRNA species specific to each cell type or tissue are typically left for analysis after the mRNA species shared by the various cell types are eliminated (Figure 5.8). To do this, the mRNA from the so-called subtractor cells can be isolated, and a first-strand cDNA can be created. Following the degradation of the original mRNA from the subtractor cells, the mRNA from the target cells is separated and combined with the cDNA. Every complementary mRNA-CDNA molecule shared by the two cell types will hybridize, releasing unbound mRNA that can be separated and examined in more detail. Using PCR, a quicker method for examining gene expression differences has been created.

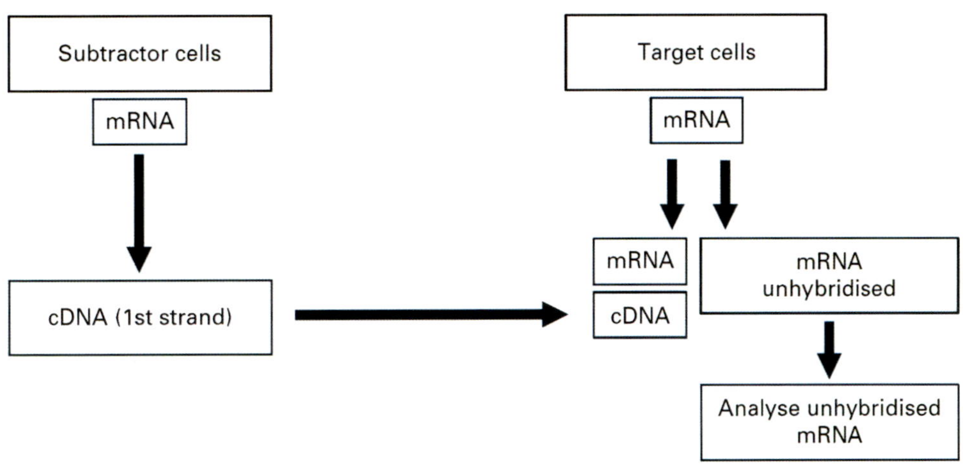

Figure 5.8. *Scheme of analyzing specific mRNA molecules by subtractive hybridization.*

Source: Keith Wilson And John Walker; Principles and Techniques of Biochemistry and Molecular Biology; ISBN 978-0-521-51635-8.

5.1.9. Cloning PCR Products

Although cloning has been largely superseded by PCR as a technique for producing large amounts of a desired DNA fragment, cloning of PCR-amplified DNA is still necessary in some situations. For example, inserting the DNA fragment into the proper plasmid or phage cloning vector is the best way to accomplish some techniques, like *in vitro* protein synthesis. Cloning techniques for PCR closely resemble the cloning of DNA fragments obtained through traditional DNA manipulation. Either cohesive-ended or blunt-ended cloning are the methods that can be used to accomplish this. Taq DNA polymerase and 7th DNA polymerase are examples of thermostable DNA polymerases that produce PCR products with a 3' overhanging A residue. The PCR product can be cloned into dT vectors, a process known as dA:dT cloning. This takes advantage of the fact that vectors prepared with T residue overhangs can be successfully ligated to the terminal additions of A residues, allowing for efficient ligation of the PCR product (Figure 5.9). As in other ligation reactions, DNA ligase catalyzes the reaction.

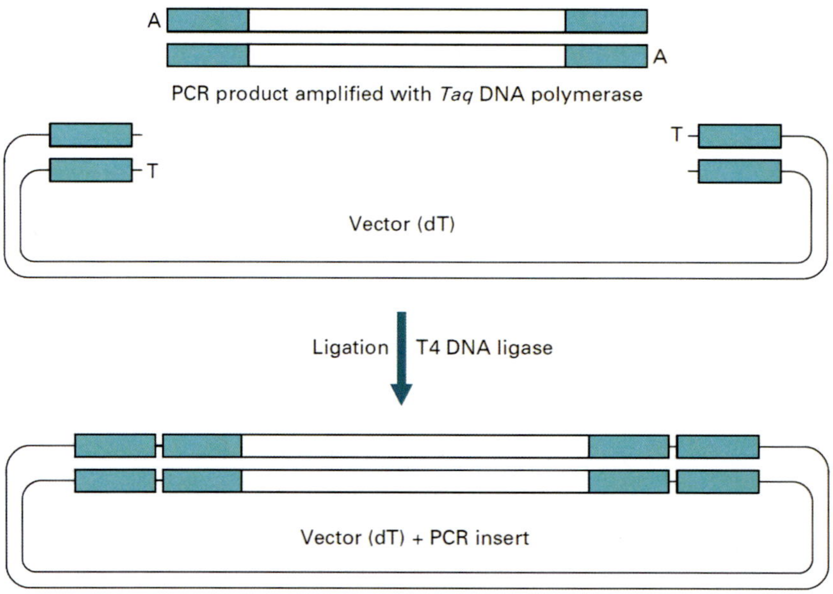

Figure 5.9. *Cloning of PCR products using dA: dT cloning.*

Source: Keith Wilson And John Walker; Principles and Techniques of Biochemistry and Molecular Biology; ISBN 978-0-521-51635-8.

Cohesive-ended cloning using PCR products is another option. In this instance, restriction endonuclease sites are integrated into the design of oligonucleotide primers. The restriction site is typically located at the 5' end of the primer because the complementarity of the primers must be absolute at the 3' end. This needs to be carefully designed because the absence of extra nucleotides at the 5' end that are not involved in recognition reduces the efficiency of digestion with a specific restriction endonuclease. The digestion and ligation reactions in this instance are identical to those carried out for conventional reactions (Brooker, 2005; Brown, 2006).

5.2. CLONING VECTORS

Each DNA molecule must be integrated into a cloning vector in order for it to be cloned. These are DNA segments that the vector can replicate to sustain and multiply them in a host organism. A bacterium like *E. coli* is an example of a typical host organism. That multiplies and grows quickly. Any vector that has an origin of replication in *E. coli* will multiply effectively, along with any added DNA. Therefore, any DNA that has been cloned into a vector will permit both the amplification of the foreign DNA fragment that has been inserted and any further analysis that may be carried out. Although there are some significant distinctions between the two methods, the cloning process is similar to the PCR in this sense. Cloning allows one to create an infinite number of copies of any specific DNA fragment in addition to storing one copy (Figure 5.10).

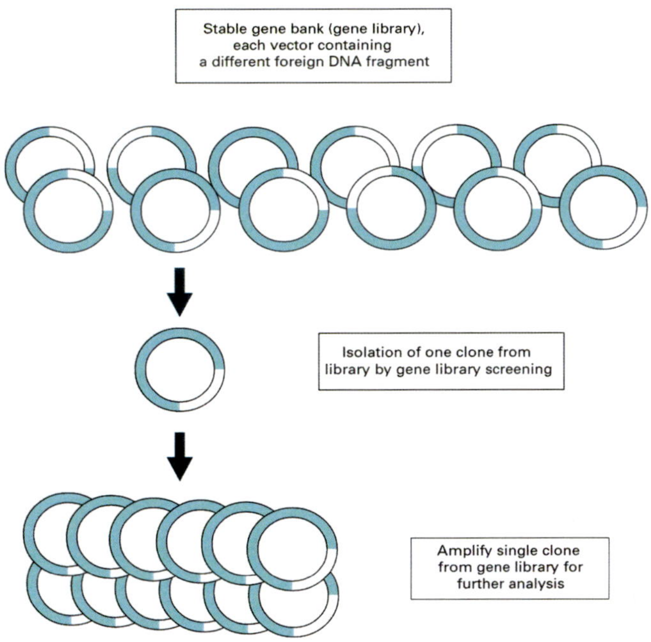

Figure 5.10. *Production of multiple copies of a single clone from a stable gene bank or library.*

Source: Keith Wilson And John Walker; Principles and Techniques of Biochemistry and Molecular Biology; ISBN 978-0-521-51635-8.

The complexity, selectivity, ease of manipulation, and insert capacity—the amount of DNA sequence that a vector can hold—all differ amongst cloning vectors. Generally speaking, vectors have been created from naturally occurring molecules like bacteriophages, bacterial plasmids, or combinations of their constituent parts like cosmids. There are different vector types to choose from when building gene libraries, and these choices are typically based on the vector's maximum size for a foreign DNA insert and how easy the manipulations are to complete (Table 5.2). Large insert capacities mean that vectors with these advantages are typically harder to work with, but there are a lot more variables to take into account, as the treatment of vector systems that follows will show.

Table 5.2. *Comparison of Vectors Generally Available for Cloning DNA Fragments*

Vector	Host cell	Vector structure	Insert range (kb)
M13	E. coli	Circular virus	1-4
Plasmid	E. coli	Circular plasmid	1-5
Phage λ	E. coli	Liner virus	2-25
Cosmids	E. coli	Circular plasmid	35-45
BACs	E. coli	Circular plasmid	50-300
YACs	S. cerevisiae	Linear chromosome	100-2000

Source: Keith Wilson And John Walker; Principles and Techniques of Biochemistry and Molecular Biology; ISBN 978-0-521-51635-8.

5.2.1. Plasmids

A plasmid is an extra-chromosomal DNA element found in many bacteria. It is a small, circular molecule that is covalently closed and contains genes for conjugation, antibiotic resistance, and the metabolism of "unusual" substrates. Certain bacteria, like *E. coli*, have a high rate of plasmid replication. Making them great candidates to act as vectors. Several naturally occurring plasmids were synthetically altered and created as cloning vectors at the beginning of the 1970s through a difficult sequence of digestion and ligation reactions. One of the most famous plasmids, known as pBR322 after its creators Bolivar and Rodriguez (pBR), was extensively used and, as shown below (Figure 5.11), exemplifies the ideal characteristics of a cloning vector.

- Due to its smaller size compared to a natural plasmid, the plasmid is more resilient to shearing damage and enhances the efficiency of bacterial uptake, a process known as transformation.
- Plasmid replication within the host cell is guaranteed by bacterial DNA replication origins. Certain replication origins exhibit strict replication regulation, whereby replication rounds are triggered concurrently with cell division. The majority of plasmids, such as pBR322, have a relaxed origin of replication, meaning that they replicate much more frequently than chromosomal replication because of this loose connection between their activity and cell division. Plasmid molecules will therefore be produced in large quantities per cell.
- Two genes that encode antibiotic resistance have been identified. One of these enables the selection of plasmid-containing cells: only plasmid-containing cells will proliferate into colonies when plated on a medium containing the proper antibiotic. It is possible to identify plasmids with inserted DNA using the other resistance gene.

- The circular plasmid can be opened or linearized using the single recognition sites for several restriction enzymes located at different locations throughout the plasmid. A DNA fragment can be inserted and the circle closed by linearizing a plasmid. The diversity of sites not only facilitates the selection of a restriction enzyme appropriate for the vector and the foreign DNA to be inserted, but some of the sites are located within genes that confer antibiotic resistance, making it possible to identify the presence of an insert by measuring the reduction in antibiotic resistance. We call this process "insertional inactivation."

A helpful selection technique for locating recombinant vectors with inserts is insertional inactivation. An isolated and purified chromosomal DNA fragment, for instance, would be the result of BamH1 digestion. Following a single-site digestion of the plasmid pBR322 with BamH1, the restriction enzyme would be rendered inactive by deproteinizing both samples. In the presence of T4 DNA ligase, ligation between the plasmid and digested DNA fragments is possible because BamH1 cleaves to produce sticky ends. The ligation process will yield a plasmid with a single DNA fragment as the insert as well as undesirable byproducts like plasmid that has circularized without an insert, plasmid dimers, fragments joined together, and plasmid with an insert made up of multiple fragments. The majority of these undesirable molecules can be removed in later stages. Agarose gel electrophoresis is typically used to identify the products of these types of reactions.

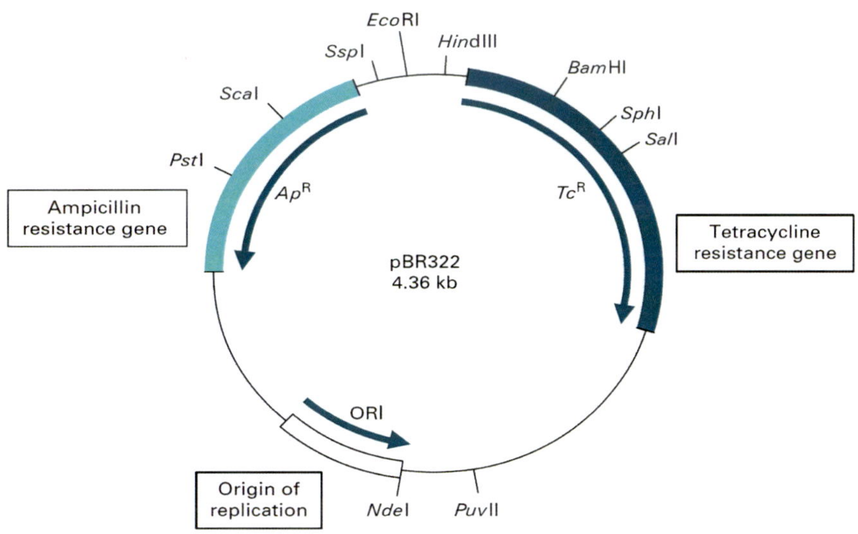

Figure 5.11. *Map and important features of pBR322.*

Source: Keith Wilson And John Walker; Principles and Techniques of Biochemistry and Molecular Biology; ISBN 978-0-521-51635-8.

It is now necessary to transform *E. coli* using the ligated DNA. Although bacteria do not normally take up DNA from their surroundings, they can be made to do so by treating them beforehand with Ca^{2+} at 4°C. In this case, the bacteria are referred to as competent because DNA added to their suspension will be taken up during a brief temperature increase known as heat shock. Long, linear molecules are not absorbed by the bacteria, while small, circular molecules are most effectively taken up.

The cells are plated onto medium containing the antibiotic, e.g. ampicillin, after a brief incubation to allow expression of the antibiotic resistance genes. The plasmid-containing cells are required to generate colonies on these plates because they carry the ampicillin resistance gene. At this point, it is not feasible to differentiate colonies that have plasmids with inserts from those that only have circularized plasmids. To accomplish this, the colonies are replica plated onto plates with tetracycline in their medium using a sterile velvet pad. The presence of an insert will inactivate the tetracycline resistance gene because the BamHI site is located within it; however, in plasmids that have only recircularized, the gene will remain intact (Figure 5.12). Plasmids containing inserts are therefore required in colonies that grow on ampicillin but not on tetracycline. Replicate plating results in an identical pattern of colonies on both sets of plates, making it simple to identify insert-containing colonies and remove them from the ampicillin plate for additional growth. This demonstrates how crucial a second gene is for a vector's ability to resist antibiotics.

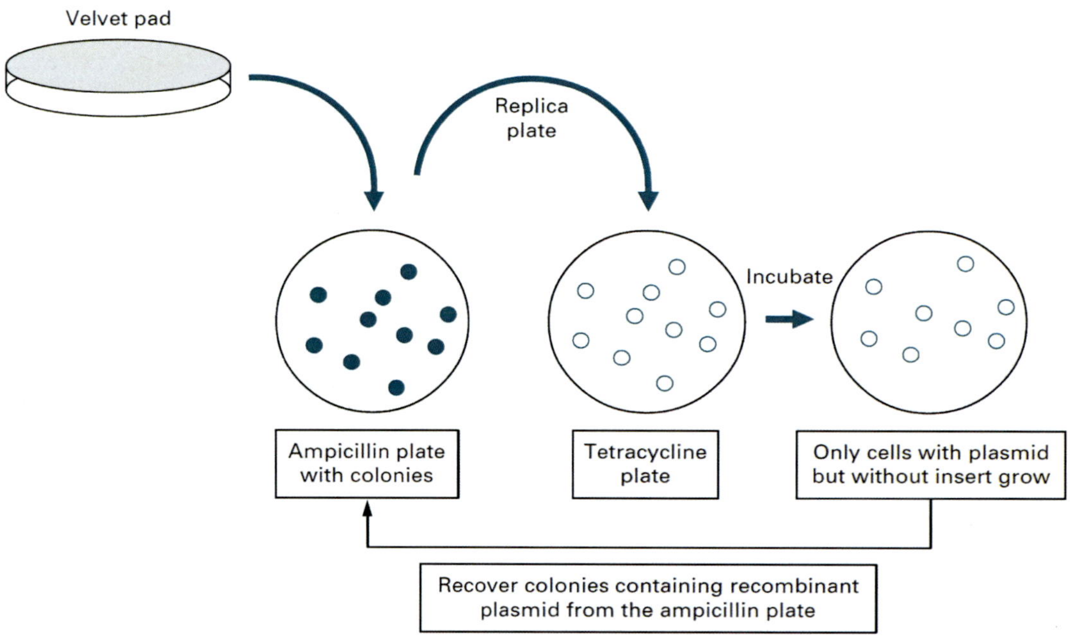

Figure 5.12. *Replica plating to detect recombinant plasmids. A sterile velvet pad is pressed onto the surface of an agar plate, picking up some cells from each colony growing on that plate.*

Source: Keith Wilson And John Walker; Principles and Techniques of Biochemistry and Molecular Biology; ISBN 978-0-521-51635-8.

Reciprocalized plasmid reduces the yield of recombinant plasmid-containing inserts, even though it can be selected against. Recircularization will not occur if the digested plasmid is treated with the enzyme alkaline phosphatase before ligation, as this enzyme eliminates the 5' phosphate groups required for ligation. The 5' phosphate of the insert and the 3' hydroxyl of the plasmid can still form links, thus the only products that will form are recombinant plasmids and chains of linked DNA fragments. One ligated strand of the recombinant DNA does not matter because bacteria transformed with these molecules will repair the nick (Brooker, 2005; Brown, 2006).

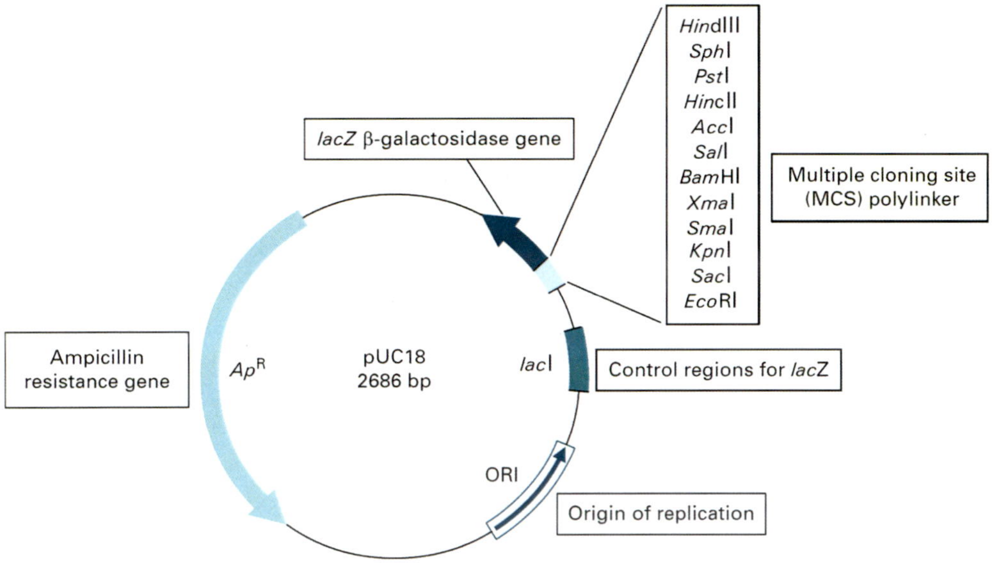

Figure 5.13. *Map and important features of pUC18.*

Source: Keith Wilson And John Walker; Principles and Techniques of Biochemistry and Molecular Biology; ISBN 978-0-521-51635-8.

The development of a set of plasmids known as pUC (made at the University of California) has improved the useful characteristics of pBR322 (Figure 5.13). Tetracycline has a gene for antibiotic resistance, and *E. coli*. has an origin of replication. Furthermore, the most well-liked restriction sites are grouped together in an area known as the multiple cloning site (MCS). Furthermore, the MCS codes for a segment of the polypeptide β-galactosidase and is a part of a separate gene. After the host cell *E. coli*, has been transformed by the pUC plasmid. The inducer IPTG (isopropyl-β-$_{D}$-thiogalactopyranoside) can be added to turn on the gene. The production of the enzyme b-galactosidase is triggered by its presence.

The functional enzyme can hydrolyze X-gal (5-bromo-4-chloro-3-indolylb-galactopyranoside), a colorless substance, into 5,5'-dibromo-4,4'-dichloro indigo, a blue insoluble material (Figure 5.14). Nevertheless, in the event that a foreign DNA fragment disrupts the gene, an inoperable enzyme arises that is unable to hydrolyze X-gal. Therefore, in the presence of X-gal, a recombinant pUC plasmid can be easily identified as it is white or colorless, whereas an intact non-recombinant pUC plasmid will be blue because its gene is intact and fully functional. This sophisticated method, known as "blue/white selection," has been incorporated into several later vector systems and enables rapid recombinant identification. However, other than the vector's status, this selection process and the insertional inactivation of antibiotic resistance genes do not reveal anything about the nature of the DNA insert. Hybridization to gene probes is necessary to filter gene libraries for the desired insert.

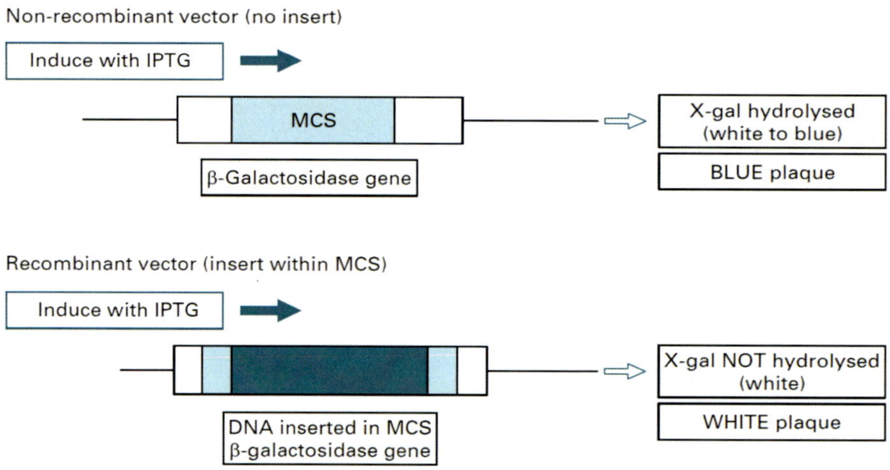

Figure 5.14. *Principle of blue/white selection for the detection of recombinant vectors.*

Source: Keith Wilson And John Walker; Principles and Techniques of Biochemistry and Molecular Biology; ISBN 978-0-521-51635-8.

5.2.2. Virus-based Vectors

Any cloning vector's ability to accept or insert a certain amount of DNA before it loses viability is a helpful feature. Bacteriophages, or bacterial viruses, have evolved as vectors to spread larger DNA fragments in bacterial cells because inserts larger than 5 kb cause plasmid size to increase to the point where bacterial cell transformation efficiency sharply declines. Cloning vectors originating from λ bacteriophage are frequently employed due to their roughly 16-fold boost in cloning efficacy when compared to the most proficient plasmid cloning vectors.

Phage λ is a linear double-stranded phage approximately 49 kb in length (Figure 5.15). It infects *E. coli* with great efficiency by injecting its DNA through the cell membrane. In the wild-type phage λ the DNA follows one of two possible modes of replication. First, there are two possible outcomes for the DNA integration into the *E. coli* chromosome, where it waits to be excised until a signal causes it to do so. We call this the "lysogenic life cycle." It could also have a lytic life cycle, in which new functional phages are assembled, phage head and tail proteins are synthesized quickly, and DNA is replicated upon entry into the cell. The next step involves lysing the cell membrane to release the phage and infect additional *E. coli* cells in the vicinity. The 12 bp sequences known as cos (cohesive) sites are located at the extremities of the phage λ. They enable the circularization of the phage DNA and resemble restriction sites despite their asymmetry. In this way, the phage can be replicated very effectively, producing concatemers of multiple phage genomes that are cleaved at the cos sites and inserted into freshly formed phage protein heads.

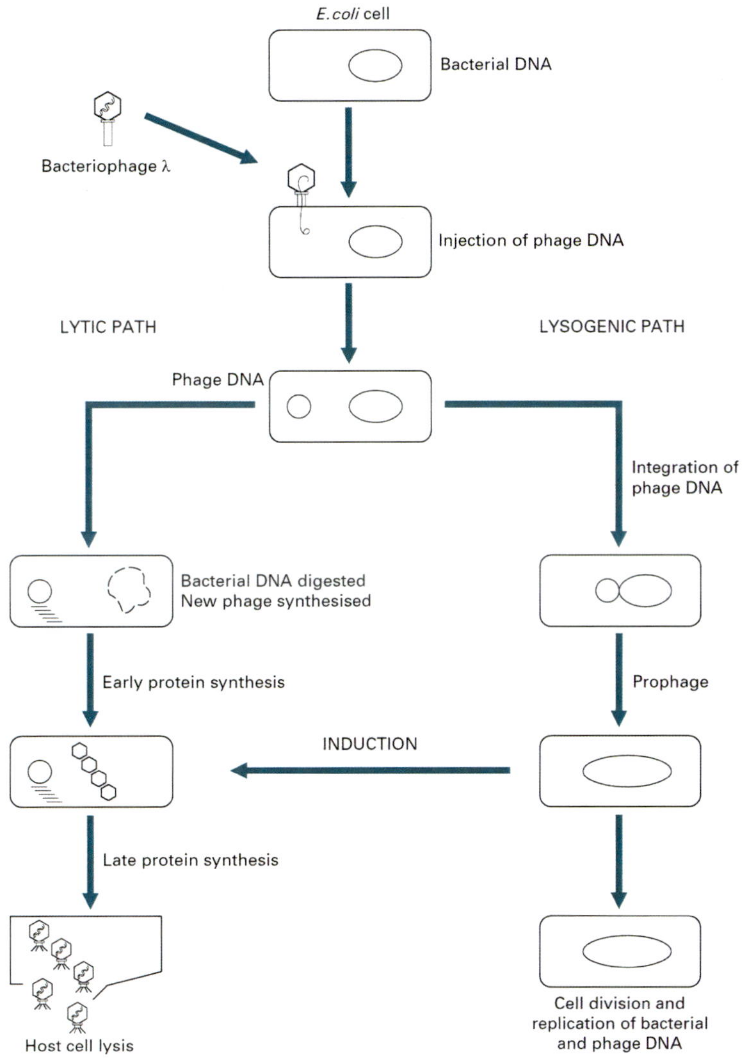

Figure 5.15. *The lysogenic and lytic cycles of bacteriophage λ.*

Source: Keith Wilson And John Walker; Principles and Techniques of Biochemistry and Molecular Biology; ISBN 978-0-521-51635-8.

Phage λ has been extensively employed in the synthesis of gene libraries primarily due to its rapid entry into the *E. coli* cell as well as the possibility of stable integration of larger DNA fragments. The foreign DNA insert is substituted for a large portion of the non-essential λ DNA, which codes for the lysogenic life cycle, to clone long DNA fragments up to approximately 25 kb. Next, in a procedure known as "*in vitro* packaging," the recombinant phage is put together into pre-formed viral protein particles. Plated-out bacterial cells on agar medium are infected using these recently formed phages (Figure 5.16).

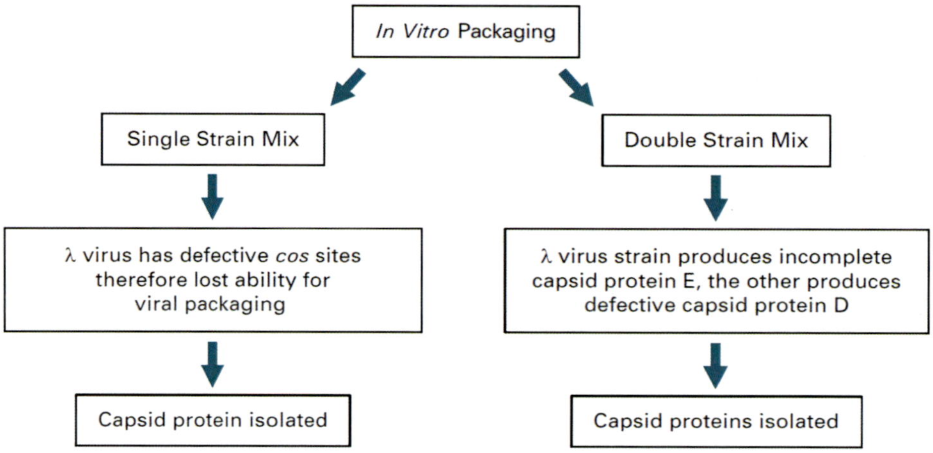

Figure 5.16. *Two strategies for producing in vitro packaging extracts for bacteriophage λ.*

Source: Keith Wilson And John Walker; Principles and Techniques of Biochemistry and Molecular Biology; ISBN 978-0-521-51635-8.

The recombinant viral DNA replicates once it is within the host cells. Since all the genes required for regular lytic growth are still present in the phage DNA, the virus multiplies by infecting neighboring cells and lysing new cells, creating plaques of lysed cells on a lawn of bacterial cells. Viral DNA, including cloned foreign DNA, can be extracted from the viruses in these plaques and subjected to additional analysis techniques such as agarose gel electrophoresis and restriction mapping.

Generally speaking, two different kinds of λ phage vectors have been created: λ replacement vectors and λ insertion vectors (Figure 5.17). Since the foreign DNA is only inserted into a section of the phage genome with the proper restriction sites—common examples are λgt10 and λcharon16A—the λ insertion vectors accept less DNA than the replacement type. Using a replacement vector, a stuffer fragment—a central DNA region not required for lytic growth—is eliminated by double digestion using enzymes like EcoRI and BamHI. The left and right arms are the two remaining DNA fragments. To create a functional recombinant λ phage, foreign DNA is inserted between the arms to replace the central stuffer fragment. The vectors λEMBL and λZap are the most prominent instances of λ replacement vectors.

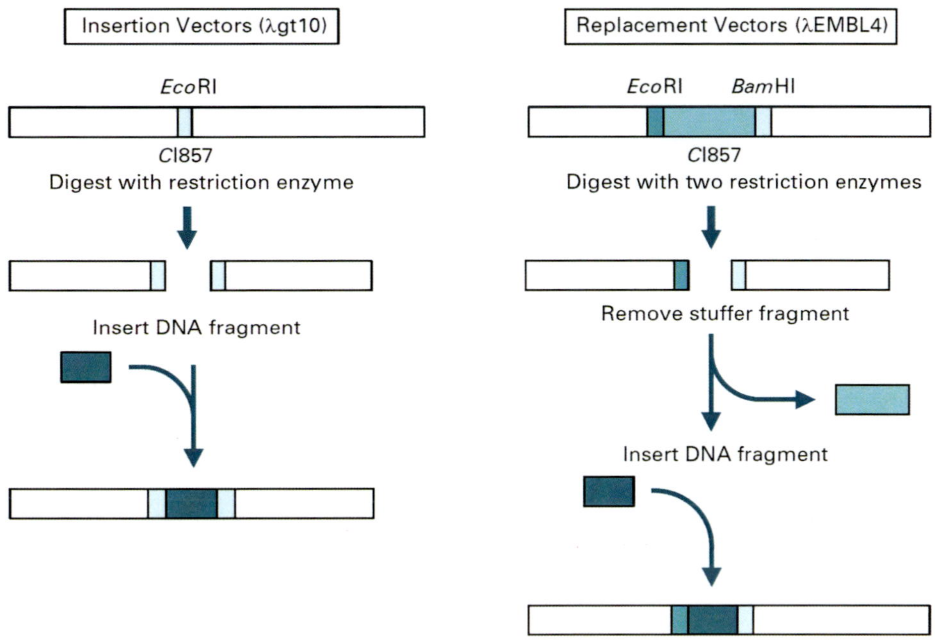

Figure 5.17. *General schemes used for cloning in λ insertion and λ replacement vectors. CI857 is a temperature-sensitive mutation that promotes lysis at 42°C after incubation at 37°C.*

Source: Keith Wilson And John Walker; Principles and Techniques of Biochemistry and Molecular Biology; ISBN 978-0-521-51635-8.

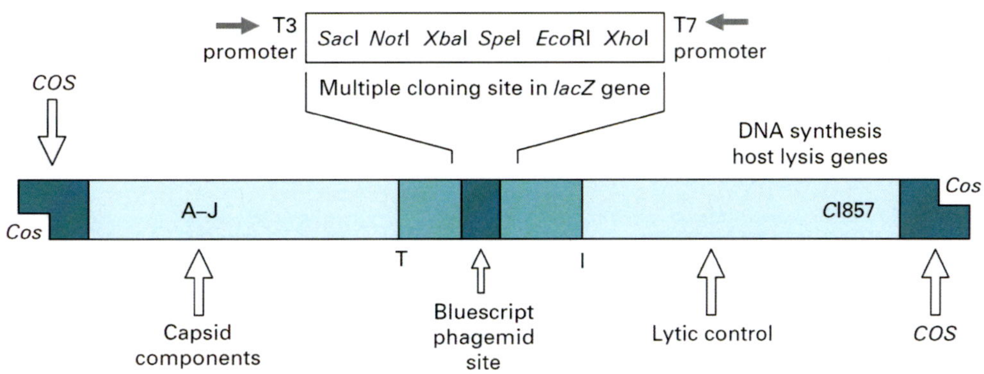

Figure 5.18. *General map of λZap cloning vector, indicating important areas of the vector. The multiple cloning site is based on the lacZ gene, providing blue/white selection based on the β-galactosidase gene. In between the initiator (I) site and terminator (T) site lie sequences encoding the phagemid blue script.*

Source: Keith Wilson And John Walker; Principles and Techniques of Biochemistry and Molecular Biology; ISBN 978-0-521-51635-8.

A multiple cloning site (MCS) made of distinct cloning sites clustered together is called λZap, a cloning vector that is commercially produced (Figure 5.18). Additionally, the MCS is situated in a lacZ region, which offers an insertional inactivation-based blue/white screening system. Using this vector, foreign cloned DNA can also be expressed.

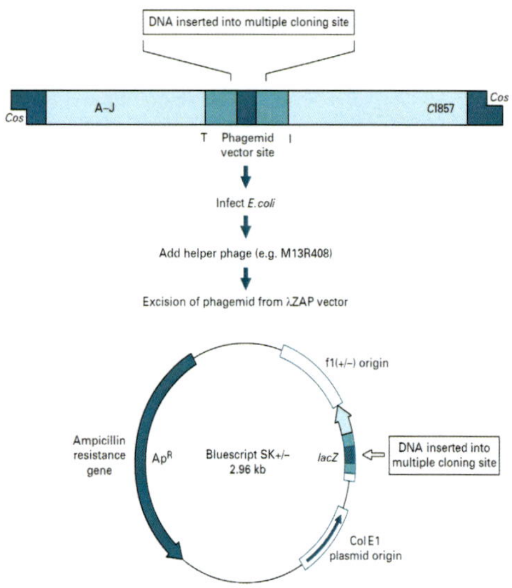

Figure 5.19. *Single-stranded DNA rescue of phagemid from λZap. The single-stranded phagemid pBluescript SK may be excised from λZap by addition of helper phage. This provides the necessary proteins and factors for transcription between the I and T sites in the parent phage to produce the phagemid with the DNA cloned into the parent vector.*

Source: Keith Wilson And John Walker; Principles and Techniques of Biochemistry and Molecular Biology; ISBN 978-0-521-51635-8.

Some λ vectors have this very helpful feature, which allows you to screen for protein products instead of the DNA that was inserted into the vector. Consequently, antibody probes targeting the target protein are used for this screening. This cloning vector's additional capabilities, such as its capacity to generate RNA transcripts known as cRNA or riboprobes, are advantageous. This is made possible by the presence of two RNA polymerase enzyme promoters in the vector, T7 and T3, which flank the MCS.

The ability of λZap to automatically remove a small 2.9 kb colony-producing vector known as a phagemid, pBluescript SK, *in vivo* is one of its most helpful features. This method, which is also known as single-stranded DNA rescue, is the outcome of a procedure known as super-infection in which helper phages are added to the cells and allowed to grow for an extra four hours or so (Figure 5.19).

The helper phage displaces a strand within the λZap which contains the foreign DNA insert. This is circularized and packaged as a filamentous phage similar to M13. The packaged phagemid is secreted from the *E. coli* cell and may be recovered from the supernatant. Thus, the λZap vector allows a number of diverse manipulations to be undertaken without the necessity of recloning or subcloning foreign DNA fragments. The process of subcloning is sometimes necessary when the manipulation of a gene fragment cloned in a general purpose vector needs to be inserted into a more specialized vector for the application of techniques such as *in vitro* mutagenesis or protein production (Brooker, 2005; Brown, 2006).

5.2.3. M13 and Phagemid-based Vectors

Single-stranded bacteriophage vectors, like M13, and vectors called phagemids—which combine the characteristics of a phage and a plasmid—have been widely utilized. The genome of filamentous coliphage M13 is made of single-stranded circular DNA. After contracting *E. coli*, the DNA first replicates as a double-stranded molecule before creating single-stranded virions that are used to infect additional bacterial cells (lytic growth). These vectors' characteristics make them perfect for procedures like *in vitro* mutagenesis and chain termination sequencing, which both need single-stranded DNA.

M13 or phagemids such as pBluescript SK infect *E. coli* harboring a male-specific structure termed the F-pilus. They enter the cell by adsorption to this structure and once inside the phage DNA is converted to a double-stranded replicative form or RF DNA. Replication then proceeds rapidly until some 100 RF molecules are produced within the *E. coli* cell. At the bacterial periplasm, DNA synthesis then shifts to producing single strands, which are assembled and packaged into capsids. The major coat protein, gene VIII protein, which has about 2,800 copies overall, encapsulates the bacteriophage DNA. At one end of the particle, there are three to six copies of the gene III protein. The rate at which the *E. coli* grows is slowed down by the bacteriophage's extrusion through the bacterial periplasm cell lysis as opposed to host cell lysis, and it manifests as a clearing region on a bacterial lawn. In a single cell division, about 1,000 packaged phage particles could be released into the medium.

The coliphage vectors have several other characteristics that make them desirable as cloning vectors in addition to their ability to produce single-stranded DNA. DNA ligation, restriction digestion, and mapping are among the common DNA manipulations that can be carried out on bacteriophage DNA because it is replicated as a double-stranded RF DNA intermediate. Lysing infected *E. coli* is how RF DNA is made using the same techniques as for plasmid isolation to isolate the supercoiled circular phage DNA from *E. coli* cells. One can use reagents like polyethylene glycol to precipitate intact single-stranded DNA packaged in the phage protein coat found in the supernatant, and then purify the DNA using phenol/chloroform. Therefore, the bacteriophage can produce DNA in the manner of a virus at times and act as a plasmid in others. Nowadays, a family of vectors derived from M13 is commonly referred to as M13mp8/9, mp18/19, etc. each of which has several very practical features. All have a synthetic MCS that is situated inside the lacZ gene without causing the gene's reading frame to be disrupted. This makes it possible to select people effectively using the blue/white screening method. The number of restriction sites increased asymmetrically as the vector series was developed. Because there are more restriction sites in the MCS of M13mp8, mp12, mp18, and sister vectors with the same MCS but the opposite orientation (M13mp9, mp13, and mp19, respectively), the vector is more useful because there is a wider selection of restriction enzymes available. Nevertheless, the instability and spontaneous loss of inserts larger than 6 kb is a common issue with M13.

Phagemids replicate similarly to M13 and are very similar to each other. A portion of another phage, known as f1, that contained elements for its morphogenesis and a phage origin of replication was inserted into a pUC8 plasmid to create pEMBL, one of the first phagemid vectors. Single-stranded DNA can be produced after the f1 origin is activated by a super-infection with a helper phage. Similar to M13, the phage is assembled into a phage coat that is extruded through the periplasm and secreted into the culture medium. The phagemid replicates as a pUC-type plasmid in the absence of superinfection, and the isolated double-stranded DNA is in the replicative form (RF). This makes it possible to carry out additional operations like restriction digestion, ligation, and mapping analysis. In addition to being a phagemid, the pBluescript SK vector can also be used as a cloning vector and handled similarly to a plasmid. Similar to M13, it can be employed in site-directed mutagenesis and nucleotide sequencing. Additionally, it can generate RNA transcripts that can be utilized to make labeled cRNA probes or riboprobes (Brooker, 2005; Brown, 2006).

5.2.4. Cosmid-Based Vectors

In particular, the replication process of phage λ DNA holds significance for the creation of cosmids, which are larger insert cloning vectors. These are a crucial component of many genome mapping projects and are particularly helpful for the analysis of extremely complex genomes.

Phage λ has an insert capacity that goes up to about 21 kb. The reason for this is that essential genes must be present, and the maximum distance between cos sites is 52 kb. As a result, cosmid vectors have been created that combine the cos sites from phage λ with the fundamental components of a plasmid, including the replication origin of the plasmid, a drug resistance gene, and multiple distinct restriction sites for cloning DNA. Upon linearizing a cosmid preparation via restriction digestion and ligating it to DNA for cloning, alternating cosmid vector and insert concatamers will be generated. Therefore, the only prerequisite for a DNA length to be packaged into viral heads is that it must have cos sites that are the appropriate distance apart from one another (in actuality, this spacing can vary from 37 to 52 kb). If phage head precursors, tails, and packaging proteins are available, such DNA can be packaged *in vitro*. Given the tiny size of the cosmid, inserts up to 40 kb in length will be packaged the easiest. After entering the cell, the DNA circularizes through its cos sites and begins to function just like a plasmid.

5.2.5. Large Insert Capacity Vectors

When looking for the foreign DNA of interest, fewer clones need to be screened when using vectors that accept larger DNA fragments than phage λ or cosmids. They are widely used in the human genome mapping project and have had a significant impact on the mapping of the genomes of other organisms like the mouse. Large insert capacity vectors based on human artificial chromosomes, bacterial artificial chromosomes (BACs), mammalian artificial chromosomes (MACs), and the virus P1 (PACs) and P1 artificial chromosomes can now be produced thanks to recent advancements. However, vectors based on yeast artificial chromosomes (YACs) may be the most important development to date.

5.2.6. Yeast Artificial Chromosome (YAC) Vectors

The linear molecules known as yeast artificial chromosomes (YACs) are made up of a replication origin known as an ARS element (autonomous replicating sequence), a telomere, and a centromere. Restriction enzymes are used to break down the YAC at the BamHI sites that separate the telomere sequences and the SUP4 site, which is a suppressor tRNA gene marker. This results in the production of two arms, which are then ligated to create a functional YAC construct. YACs are replicated in yeast cells, but to leave a spheroplast, the yeast must have its outer cell wall removed. They must be embedded in a solid matrix, like agar, and are osmotically unstable. Only correctly constructed YACs with corresponding selectable markers are replicated in the yeast strains after the yeast cells are transformed. Repeated DNA fragments can occasionally be challenging to clone in bacterial-based vectors, but they can occasionally be successfully cloned in YAC systems. YAC-based vectors' primary benefit is their capacity to clone extremely large DNA fragments. YAC vectors have been utilized for the stable maintenance and replication of foreign DNA fragments up to 2000 kb, making them the primary vector of choice for various genome mapping and sequencing projects. (Brooker, 2005; Brown, 2006)

CASE STUDY

RECOMBINANT DNA AND GENETIC ANALYSIS IN AGRICULTURAL BIOTECHNOLOGY

Background: ABC Biotech is a leading agricultural biotechnology company specializing in developing genetically modified (GM) crops to enhance crop yield and resistance to diseases. They have recently embarked on a project to develop a new strain of maize that can thrive in drought-prone regions while maintaining high nutritional value.

Problem Statement: ABC Biotech faces the challenge of integrating multiple beneficial traits into a single maize variety without compromising its agronomic performance. Traditional breeding methods are time-consuming and may not yield the desired combinations of traits within a reasonable timeframe.

Solution: To overcome these challenges, ABC Biotech turns to recombinant DNA technology and genetic analysis. By leveraging these advanced biotechnological tools, they aim to precisely manipulate the maize genome to incorporate drought tolerance genes from a desert plant, pest resistance genes from a wild maize variety, and enhanced nutritional content genes from other plant species.

Implementation

- **Isolation of Genes:** Researchers at ABC Biotech identify and isolate specific genes responsible for drought tolerance, pest resistance, and enhanced nutritional content from various donor organisms.
- **Recombinant DNA Technology:** Using molecular biology techniques, they insert these beneficial genes into the maize genome at precise locations. This involves creating recombinant DNA molecules by cutting and splicing DNA using restriction enzymes and ligases.
- **Genetic Analysis:** After integrating the genes, the modified maize plants undergo extensive genetic analysis. This includes PCR (Polymerase Chain Reaction) assays to confirm the presence of the inserted genes, DNA sequencing to verify the integrity of the modified genome, and bioinformatics analysis to predict the expression patterns of the introduced genes.
- **Field Trials:** Modified maize plants showing promising genetic profiles are subjected to field trials. These trials assess their performance under varying environmental conditions, including drought stress and pest infestation. Data on yield, nutritional content, and agronomic traits are collected and analyzed.

- **Regulatory Approval:** Upon successful field trials and comprehensive safety assessments, ABC Biotech submits its findings to regulatory authorities for approval of the GM maize variety. This involves demonstrating the safety of the modified crop for human consumption and the environment.
- **Results:** Through the application of recombinant DNA technology and genetic analysis, ABC Biotech successfully develops a new strain of maize that exhibits enhanced drought tolerance, resistance to pests, and improved nutritional content. The GM maize variety proves to be a viable solution for farmers in drought-prone regions, offering increased yield potential and nutritional benefits compared to conventional varieties.

Conclusion: This case study illustrates how recombinant DNA technology and genetic analysis can be effectively utilized in agricultural biotechnology to address complex challenges such as enhancing crop resilience and nutritional value. By integrating advanced genetic techniques with rigorous scientific research and regulatory compliance, companies like ABC Biotech pave the way for sustainable agriculture and food security in a changing climate.

SUMMARY

- A gene library is a collection of different DNA sequences from an organism, which has been cloned into a vector for ease of purification, storage and analysis. Gene library is a collection of cloned DNA designed so that there is a high probability of finding any particular piece of the source DNA in the collection.
- Genomic libraries are constructed by isolating the complete chromosomal DNA from a cell, and then digesting it into fragments of the desired average length with restriction endonucleases. This can be achieved by partial restriction digestion using an enzyme that recognizes tetranucleotide sequences.
- Ligation of blunt-ended DNA fragments is not as efficient as ligation of sticky ends, therefore with cDNA molecules additional procedures are undertaken before ligation with cloning vectors.
- Frequently an attempt is made to isolate the mRNA transcribed from a desired gene within a particular cell or tissue that produces the protein in high amounts. Thus, if the cell or tissue produces a major protein of the cell a large fraction of the total mRNA will code for the protein.
- The vectors used for cloning vary in their complexity, their ease of manipulation, their selection and the amount of DNA sequence they can accommodate (the insert capacity). Vectors have in general been developed from naturally occurring molecules such as bacterial plasmids, bacteriophages or combinations of the elements that make them up, such as cosmids.
- The ligated DNA must now be used to transform *E. coli*. Bacteria do not normally take up DNA from their surroundings, but can be induced to do so by prior treatment with Ca2+ at 4°C; they are then termed competent since DNA added to the suspension of competent cells will be taken up during a brief increase in temperature termed heat shock.
- M13 or phagemids such as pBluescript SK infect *E. coli* harboring a male-specific structure termed the F-pilus. They enter the cell by adsorption to this structure and once inside the phage DNA is converted to a double-stranded replicative form or RF DNA.
- Phagemids are very similar to M13 and replicate in a similar fashion. One of the first phagemid vectors, pEMBL, was constructed by inserting a fragment of another phage termed f1 containing a phage origin of replication and elements for its morphogenesis into a pUC8 plasmid.
- Yeast artificial chromosomes (YACs) are linear molecules composed of a centromere, telomere and a replication origin termed an ARS element (autonomous replicating sequence). The YAC is digested with restriction enzymes at the SUP4 site (a suppressor tRNA gene marker) and BamHI sites separating the telomere sequences.

REVIEW QUESTIONS

1. Describe the process of digesting genomic DNA molecules with restriction enzymes. What role do these enzymes play in preparing genomic DNA for library construction?
2. Explain the concept of a genomic DNA library. What types of information can be obtained from a genomic DNA library?
3. Compare and contrast cDNA libraries and genomic DNA libraries. When might one type of library be preferred over the other for specific research purposes?
4. What are enrichment methods for RNA, and why are they important in molecular biology research? Provide examples of common RNA enrichment techniques.
5. Discuss the characteristics and advantages of cloning vectors. How do they facilitate the insertion and replication of foreign DNA fragments in host cells?
6. Explain the structure and function of virus-based vectors in molecular cloning. What are the advantages and potential limitations of using viral vectors?
7. What are cosmid-based vectors, and how do they differ from other types of cloning vectors? Give examples of applications where cosmid vectors are particularly useful.

REFERENCES

1. Augen, J. (2005). *Bioinformatics in the Post-Genomic Era*. Addison-Wesley.
2. Brooker, R. J. (2005). *Genetics: Analysis and Principles* (2nd ed.). McGraw-Hill.
3. Brown, T. A. (2006). *Gene Cloning and DNA Analysis*. Wiley-Blackwell.
4. Cross, D. M., & Bayliss, M. K. (2000). A commentary on the use of hepatocytes in drug metabolism studies during drug discovery. *Drug Metabolism Reviews, 32*(2), 219–240. https://doi.org/10.1081/DMR-100101844.
5. Eichler, H. G., Pinatti, F., Flamion, B., Leufkens, H., & Breckenridge, A. (2008). Balancing early market access to new drugs with the need for benefit/risk data: A mounting dilemma. *Nature Reviews Drug Discovery, 7*(9), 818–826. https://doi.org/10.1038/nrd2664.
6. Ekins, S., Mestres, J., & Testa, B. (2007). *In silico* pharmacology for drug discovery: Methods for virtual ligand screening and profiling. *British Journal of Pharmacology, 152*(1), 9–20. https://doi.org/10.1038/sj.bjp.0707328.
7. Gomez-Hens, A., & Aguilar-Caballos, M. P. (2007). Modern analytical approaches to high-throughput drug discovery. *Trends in Analytical Chemistry, 26*(2), 171–180. https://doi.org/10.1016/j.trac.2006.12.007.
8. Imming, P., Sinning, C., & Meyer, A. (2006). Drugs, their targets and the nature and number of drug targets. *Nature Reviews Drug Discovery, 5*(12), 821–832. https://doi.org/10.1038/nrd2134.
9. Joy, T., & Hegele, R. A. (2008). Is raising HDL a futile strategy for atheroprotection? *Nature Reviews Drug Discovery, 7*(2), 143–154. https://doi.org/10.1038/nrd2461.

10. Karlberg, J. P. E. (2008). Trends in disease focus of drug development. *Nature Reviews Drug Discovery, 7*(8), 639–640. https://doi.org/10.1038/nrd2658.
11. Primrose, S. B., & Twyman, R. M. (2006). *Principles of Gene Manipulation and Genomics* (7th ed.). Wiley-Blackwell.
12. Rader, D. J., & Daugherty, A. (2008). Translating molecular discoveries into new therapies for atherosclerosis. *Nature, 451*(7181), 904–912. https://doi.org/10.1038/nature06668.
13. Rognan, D. (2007). Chemogenomic approaches to rational drug design. *British Journal of Pharmacology, 152*(1), 38–52. https://doi.org/10.1038/sj.bjp.0707336.
14. Rose, A. D. (2008). Pharmacogenetics in drug discovery and development: A translational perspective. *Nature Reviews Drug Discovery, 7*(10), 807–817. https://doi.org/10.1038/nrd2636.
15. Starkuviene, V., & Pepperkok, R. (2007). The potential of high-content high-throughput microscopy in drug discovery. *British Journal of Pharmacology, 152*(1), 62–71. https://doi.org/10.1038/sj.bjp.0707314.
16. Strachan, T., & Read, A. P. (2004). *Human Molecular Genetics* (3rd ed.). Bios.
17. Walker, J. M., & Rapley, R. (2008). *Molecular Bio Methods Handbook* (2nd ed.). Humana Press.
18. Wang, C. Y., Liu, P. Y., & Liao, J. K. (2008). Pleiotropic effects of statin therapy: Molecular mechanisms and clinical results. *Trends in Molecular Medicine, 14*(1), 32–44. https://doi.org/10.1016/j.molmed.2007.11.003.
19. Watson, J. D., Caudy, A. A., Myers, R. M., & Witkowski, J. A. (2007). *Recombinant DNA: Genes and Genomes*. W. H. Freeman.

CHAPTER 6

Chemical Synthesis, Amplification, and Sequencing of DNA

LEARNING OBJECTIVES

After studying this chapter, you will be able to:

- Define chemical synthesis of DNA;
- Explain the polymerase chain reaction;
- Describe the DNA-sequencing techniques.

KEY TERMS FROM THIS CHAPTER

Annealing	Cloning	Denaturation
DNA polymerase	DNA template	Electropherogram
Extension	Gel electrophoresis	High-throughput sequencing
Nucleotide	Plasmid	Polymerase chain reaction
Primer	Restriction enzyme	Sanger sequencing
Ligation	Next-generation sequencing (NGS)	

INTRODUCTION

Research is significantly impacted by technological advancements in all fields of science. Novel experiments are generated by new protocols and laboratory techniques that were once challenging to execute become considerably simpler. Numerous technological advancements, many of which are now commonplace and available to both large and small research facilities, are the foundation of molecular biotechnology. For instance, chemical synthesis of a DNA molecule, amplification of DNA via polymerase chain reaction (PCR), and acquisition of its nucleotide sequence are now considered standard procedures. Basic research on the composition of DNA and the process of DNA replication served as the foundation for each of these protocols. Furthermore, the isolation, characterization, and expression of cloned genes depend on these experimental techniques (Bermingham and Luettich, 2003; Garibyan and Avashia, 2013).

6.1. CHEMICAL SYNTHESIS OF DNA

The techniques of molecular cloning and DNA characterization have greatly benefited from the rapid, easy, and affordable synthesis of a DNA strand with a desired nucleotide sequence. Single-stranded DNA oligonucleotides that are chemically synthesized can be used to sequence DNA, amplify particular DNA sequences, introduce mutations into cloned genes, screen gene libraries, assemble entire genes, and aid in the process of gene cloning.

Gene machines, also known as DNA synthesizers, are devices that automate the chemical reactions required for DNA synthesis, thereby turning the process of producing single-stranded oligonucleotides (≤50 nucleotides) into a largely routine process. DNA synthesizers are typically made up of a series of pumps and valves that are programmed to introduce specific nucleotides and the reagents needed to couple each subsequent nucleotide to the growing chain in the proper order. Each incoming nucleotide is coupled to the 5′ hydroxyl terminus of the growing chain during the chemical process, which differs from the biological direction of DNA synthesis. The length of each reaction and the washing stages are computer-controlled, and each reaction is conducted one after the other in a single reaction column.

6.1.1. The Phosphoramidite Method

The preferred method for chemical DNA synthesis at the moment is the phosphoramidite method. To avoid unwanted side reactions during chain growth, the amino groups of the bases adenine, guanine, and cytosine are derivatized before being introduced into the reaction column by the addition of benzoyl, isobutyryl, and benzoyl groups, respectively. Since thymine doesn't have an amino group, it is not treated. Synthesis in solid phase, i.e., the growing DNA strand to a solid support. This allows all of the reactions to take place in a single reaction vessel, the reagents from one step to be easily removed before the reagents for the next are added, and the reagents to be used excessively in an effort to complete the reactions.

DNA is chemically synthesized through a series of steps (Figure 6.1). The spacer molecule is covalently attached to an inert support, which is frequently a controlled-pore glass (CPG) bead (a glass bead with uniformly sized pores), by the initial nucleoside (base and sugar only), which will be the 3′-terminal nucleotide of the synthesized strand (Figure 6.2). To stop the 5′ hydroxyl group from reacting in an unspecific way before the addition of the second nucleotide, a dimethoxytrityl (DMT) group is joined to the 5′ terminus of the first nucleoside. Every nucleotide that joins the growing chain is accompanied by a diisopropylamine group attached to a 3′ phosphite group that is shielded by a β-cyanoethyl (CH_2CH_2CN) group and a 5′ DMT protective group (Figure 6.3). We refer to this group of molecules as a phosphoramidite.

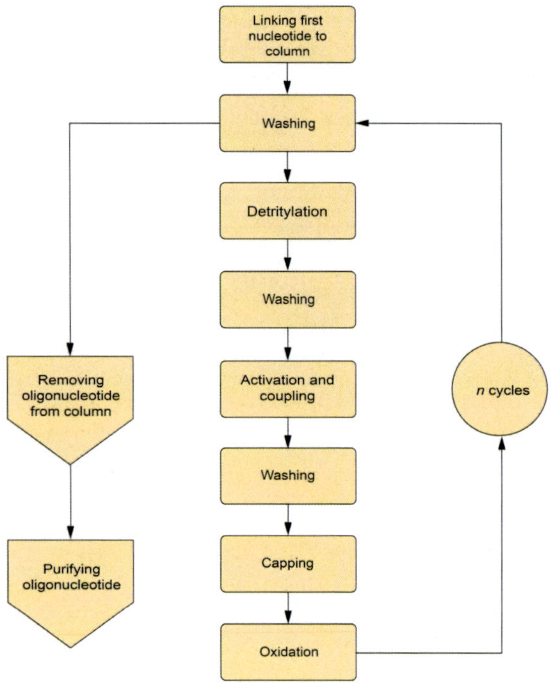

Figure 6.1. *Flowchart for the chemical synthesis of DNA oligonucleotides. After n coupling reactions (cycles), a single-stranded piece of DNA with n + 1 nucleotides is produced.*

Source: Bernard R. Glick, Jack J. Pasternak, and Cheryl L. Patten.; Molecular biotechnology: principles and applications of recombinant DNA; ISBN 978-1-55581-498-4.

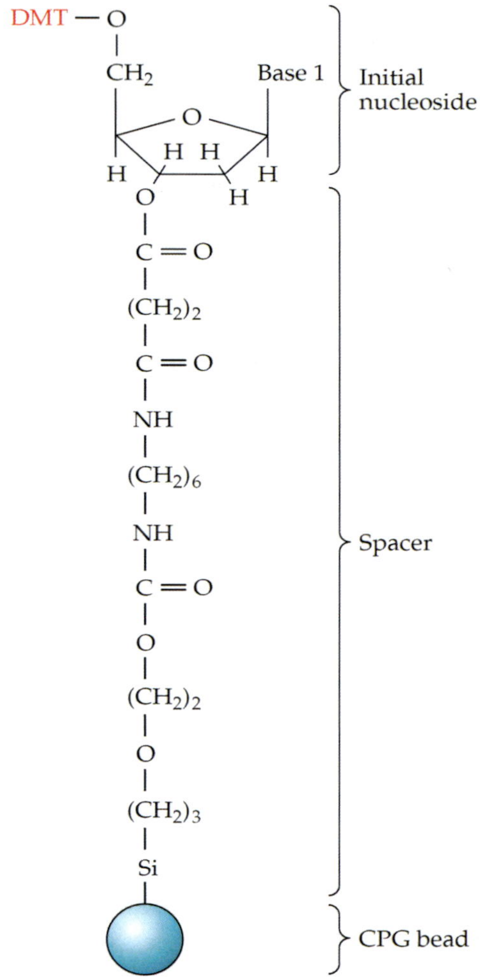

Figure 6.2. *Starting complex for the chemical synthesis of a DNA strand. The initial nucleoside has a protective DMT group attached to the 5′ hydroxyl group of the deoxyribose moiety and a spacer molecule attached to the hydroxyl group of the 3′ carbon of the deoxyribose. The spacer unit is attached to a solid support, which is usually a CPG bead.*

Source: Bernard R. Glick, Jack J. Pasternak, and Cheryl L. Patten.; Molecular Biotechnology: Principles and Applications of Recombinant DNA; ISBN 978-1-55581-498-4.

The cycle starts when the first nucleoside binds to the CPG beads. Initially, an anhydrous reagent such as acetonitrile is used to thoroughly wash the reaction column to eliminate any possible nucleophiles and water. To remove the acetonitrile, argon is flushed through the column. The attached nucleoside is then treated with trichloroacetic acid (TCA) to remove the 5′ DMT group, resulting in the production of a reactive 5′ hydroxyl group (Figure 6.4).

Following this detritylation step, acetonitrile is used to remove the TCA from the reaction column, and argon is used to remove the acetonitrile. For the activation and coupling steps, the machine is set up to introduce phosphoramidite, the next recommended base, and tetrazole simultaneously. The phosphoramidite's 3′ phosphite forms a covalent bond with the initial nucleoside's 5′ hydroxyl group when the tetrazole activates it (Figure 6.5). By flushing the column with argon, unincorporated phosphoramidite and tetrazole are eliminated.

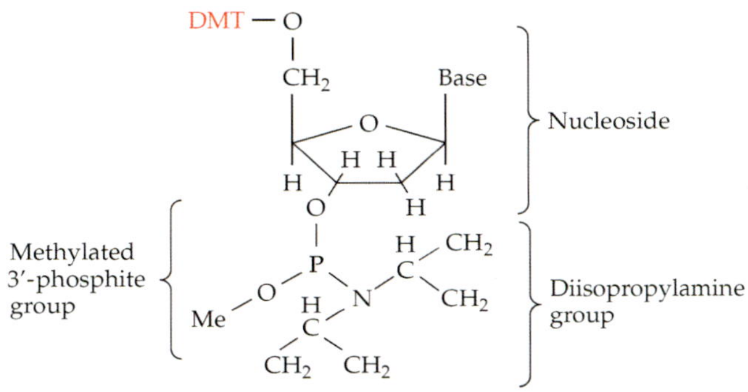

Figure 6.3. *Structure of a phosphoramidite. Phosphoramidites are available for each of the four bases (A, C, G, and T) that are used for the chemical synthesis of a DNA strand. A diisopropylamine group is attached to the 3′ phosphite group of the nucleoside. A β-cyanoethyl group protects the 3′ phosphite group, and a DMT group is bound to the 5′ hydroxyl group of the deoxyribose sugar.*

Source: Bernard R. Glick, Jack J. Pasternak, and Cheryl L. Patten.; Molecular biotechnology: principles and applications of recombinant DNA; ISBN 978-1-55581-498-4.

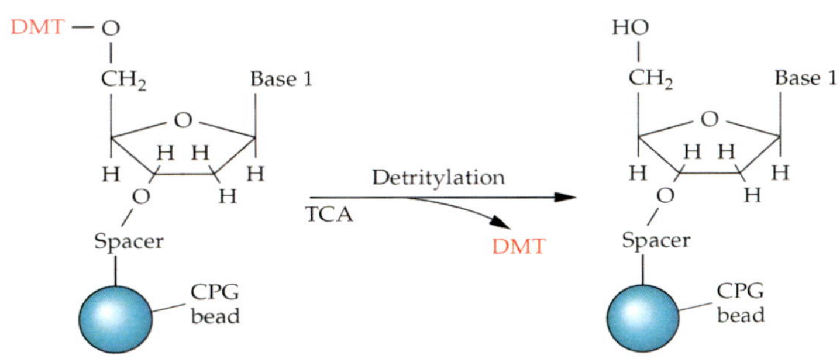

Figure 6.4. *Detritylation. The 5′ DMT group is removed by treatment with TCA. In this example, the detritylation of the first nucleoside is depicted.*

Source: Bernard R. Glick, Jack J. Pasternak, and Cheryl L. Patten.; Molecular biotechnology: principles and applications of recombinant DNA; ISBN 978-1-55581-498-4.

The unlinked residues need to be kept from linking to the next nucleotide in the subsequent cycle since not all of the support-bound nucleosides are linked to a phosphoramidite during the initial coupling reaction. To acetylate the unreacted 5′ hydroxyl groups, acetic anhydride and dimethylaminopyridine are added (Figure 6.6). The length and nucleotide sequence of the growing chains will vary after a number of cycles if this capping step is skipped.

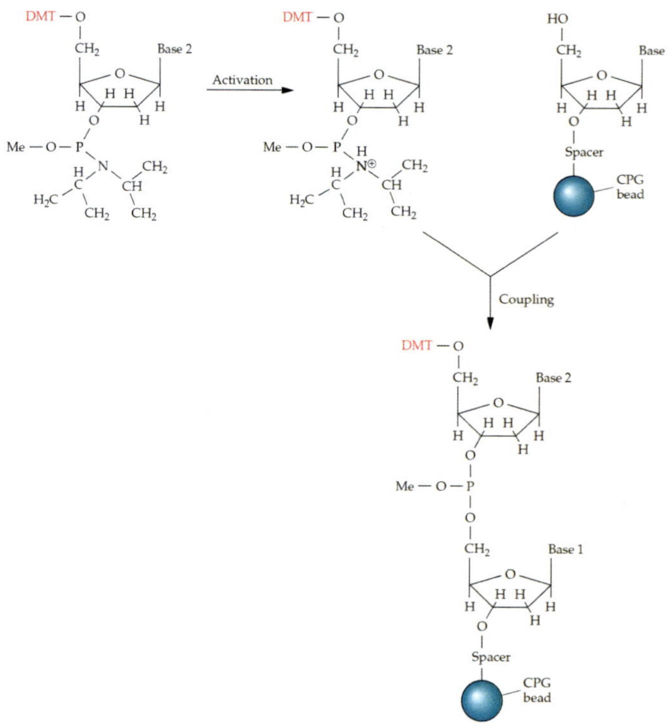

Figure 6.5. *Activation and coupling. The activation of a phosphoramidite enables its 3′ phosphite group to attach to the 5′ hydroxyl group of the bound detritylated nucleoside.*

Source: Bernard R. Glick, Jack J. Pasternak, and Cheryl L. Patten.; Molecular biotechnology: principles and applications of recombinant DNA; ISBN 978-1-55581-498-4.

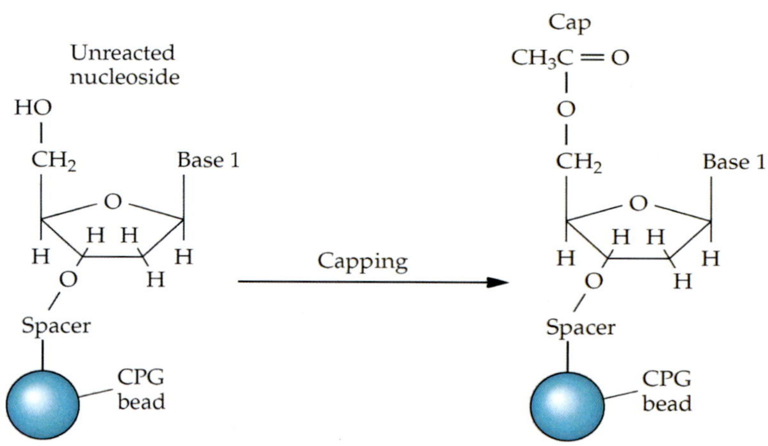

Figure 6.6. *Capping. The available 5′ hydroxyl groups of unreacted detritylated nucleosides are acetylated to prevent them from participating in the coupling reaction of the next cycle.*

Source: Bernard R. Glick, Jack J. Pasternak, and Cheryl L. Patten.; Molecular biotechnology: principles and applications of recombinant DNA; ISBN 978-1-55581-498-4.

The connection between the nucleotides at this point in the process is a phosphite triester bond, which is brittle and prone to breaking in the presence of an acid or a base. To create the more stable pentavalent phosphate triester, the phosphite triester is oxidized using an iodine mixture (Figure 6.7). The cycle of detritylation, phosphoramidite activation, coupling, capping, and oxidation is repeated following this oxidation step and a subsequent wash of the reaction column (Figure 6.1). This cycle is carried out with every additional phosphoramidite until the growing chain contains the final programmed residue. The freshly synthesized DNA strands are bound to the CPG beads after the last cycle is finished. Every phosphate triester has a β-cyanoethyl group, every guanine, cytosine, and adenine has an amino-protecting group, and the last nucleotide's 5′ terminus has a DMT group.

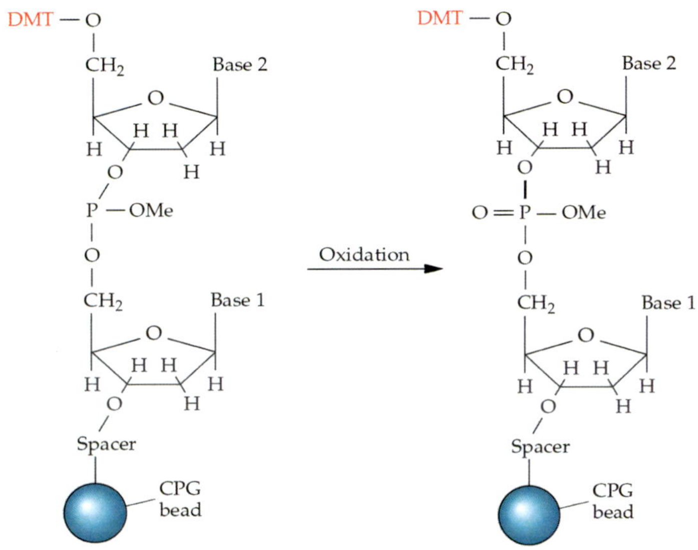

Figure 6.7. *Oxidation. The phosphite triester internucleotide linkage is oxidized to the pentavalent phosphate triester. This reaction stabilizes the phosphodiester bond and makes it less susceptible to cleavage under either acidic or basic conditions.*

Source: Bernard R. Glick, Jack J. Pasternak, and Cheryl L. Patten.; Molecular biotechnology: principles and applications of recombinant DNA; ISBN 978-1-55581-498-4.

In the reaction column, the β-cyanoethyl groups are eliminated through a chemical process. After that, the spacer molecule is broken apart by the DNA strands, leaving a 3′ hydroxyl terminus. After the DNA is eluted from the reaction column, it undergoes detritylation as the benzoyl and isobutyryl groups are successively removed. The DNA strand's 5′ terminus is phosphorylated chemically or via a T4 polynucleotide kinase reaction. After detritylation, phosphorylation can also be done while the oligonucleotide is still attached to the support.

To obtain a reasonable total yield of an oligonucleotide, each step's coupling efficiency needs to be higher than 98 percent. Tritium groups that have been released are monitored spectrophotometrically to ascertain the coupling efficiency of each cycle. Since the first base is bound to the spacer and is not involved in a coupling step, for instance, if the efficiency is 99% at each cycle during the production of a 20-unit oligonucleotide (20-mer), which involves 19 coupling reactions, then 83% (i.e., $0.99^{19} \times 100$) of the product will be 20 nucleotides long. About 55% of the final product will contain all 60 nucleotides if a 60-mer is synthesized with 99% efficiency at each cycle. The yield of full-length oligonucleotides decreases with the number of cycles required, with an average coupling efficiency that is continuously less than 98 % (Table 6.1). For each step, the majority of commercial DNA synthesizers have an average coupling efficiency of 99.5%. Reverse-phase high-pressure liquid chromatography or gel electrophoresis may be required to purify the final product, though, depending on the length and stringency of an oligonucleotide's intended use. These techniques distinguish the shorter "failure" sequences from the longer target oligonucleotides.

Table 6.1. *Overall Yields of Chemically Synthesized Oligonucleotides with Different Coupling Efficiencies*

Coupling efficiency (%)	Overall yield of oligonucleotide (%)				
	20-mer	**40-mer**	**60-mer**	**80-mer**	**100-mer**
90	14	1.6	0.2	0.02	0.003
95	38	14	4.8	1.7	0.6
98	68	46	30	20	14
99	83	68	55	45	37
99.5	91	82	74	67	61

Source: Bernard R. Glick, Jack J. Pasternak, and Cheryl L. Patten.; Molecular biotechnology: principles and applications of recombinant DNA; ISBN 978-1-55581-498-4.

6.1.2. Uses of Synthesized Oligonucleotides

Chemically produced oligonucleotides (less than 100 Mers) serve a variety of purposes. By inferring the codons from a protein's amino acid sequence, single-stranded hybridization oligonucleotide probes (20–40 Mers) can be created, which can then be used to screen a genomic library for the gene (Figure 6.8A). A single haphazard synthetic probe might not contain enough complementary bases (matches) to significantly hybridize with a heterologous sequence since codon redundancy, particularly at the third position, obscures the actual codons encoding a conserved amino acid sequence. Because of this, screening a genomic library typically involves using a set of mixed probes. Creating a mixed probe sample is a simple process. In summary, during the chemical synthesis of DNA, a mixture of different bases is added to the reaction rather than a specific phosphoramidite for a given nucleotide site. For instance, four distinct probes are created when equal concentrations of four different bases are added for a single nucleotide position. This will result in the synthesis of 16 (4^2) distinct probes for two sites (Figure 6.8B); for n sites, 4n different oligonucleotides are available. Furthermore, by altering the ratios of bases in the reaction mixture for particular sites, the frequencies of different probes in the mixture can be distorted. As a result, a set of mixed probes is more likely to contain sequences that are highly complementary to a heterologous gene than a single probe. Primers, also known as single-stranded oligonucleotides (~17–24mers), are also necessary for PCR and DNA sequencing. During the synthesis process, extra sequences are occasionally added to the PCR primers to produce molecules with restriction endonuclease sites for cloning or sequences that contain regulatory elements for the amplified DNA's transcription and translation after it is inserted into a vector.

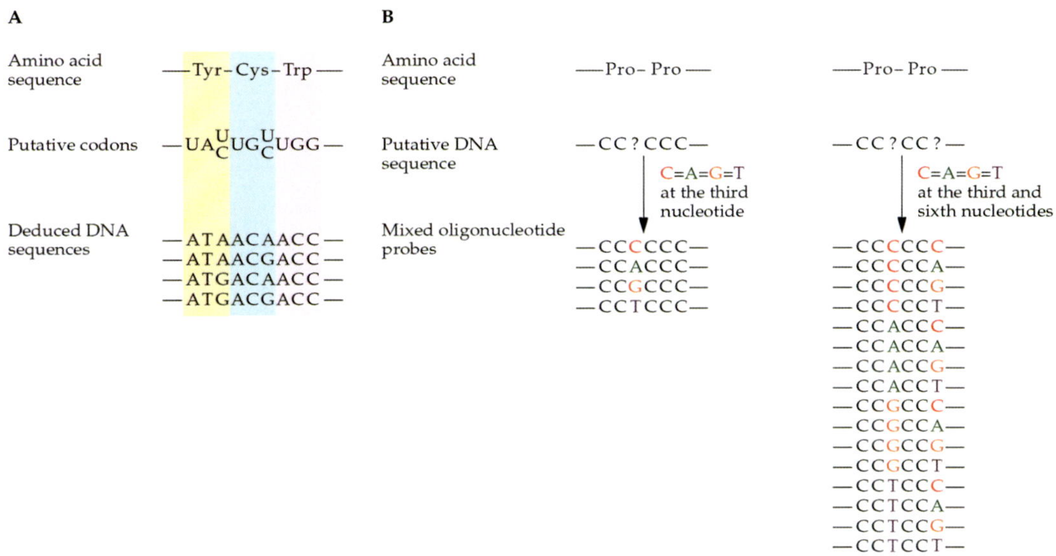

Figure 6.8. *(A) All possible DNA sequences deduced from a protein sequence. For simplicity, only 3 amino acids, 2 of which have twofold degeneracy and 1 that has a single codon, are shown. (B) Formation of sets of mixed probes. A question mark denotes a site where equal concentrations of A, C, G, and T are supplied during chemical synthesis. The remaining sites have the same deoxynucleotide (C). Four and 16 different oligonucleotides are produced when there is a 25% probability that any 1 of the 4 nucleotides will be incorporated at one or two sites, respectively.*

Source: Bernard R. Glick, Jack J. Pasternak, and Cheryl L. Patten.; Molecular biotechnology: principles and applications of recombinant DNA; ISBN 978-1-55581-498-4.

Oligonucleotides are very helpful in the process of cloning DNA molecules and in the creation of distinct cloning sites. For instance, linkers are brief, blunt-ended, double-stranded DNA molecules that have a restriction endonuclease site and self-complementary (palindromic) strands (Figure 6.9A and B). The double-stranded linker that results from annealing complementary single-stranded oligonucleotides is usually blunt-end ligated to unclonable DNA molecules due to the lack of an appropriate restriction endonuclease site. In actuality, the linker molecules are added one after the other and to the ends of the source DNA during the ligation reaction. Following treatment with the linker sequence-corresponding restriction endonuclease, the big DNA fragments, i.e., source DNA with restriction endonuclease extensions are isolated from linker DNA residues and cloned into a cloning vector's corresponding site (Figure 6.10).

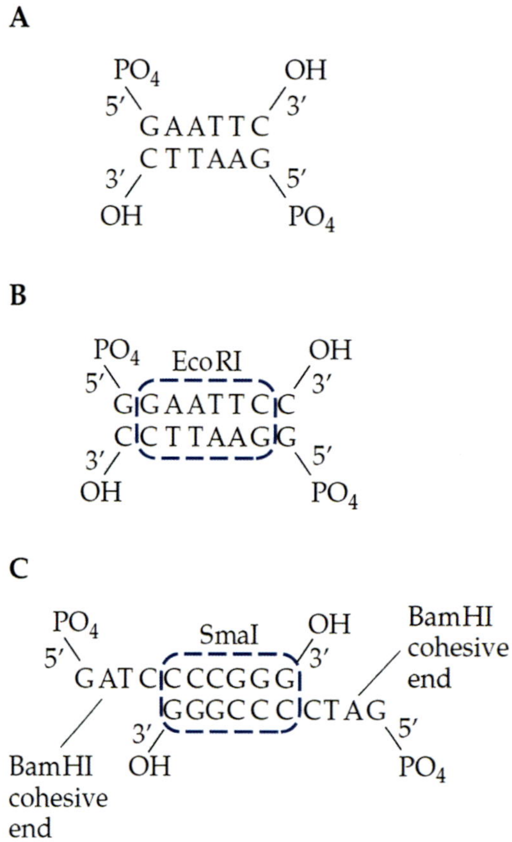

Figure 6.9. *Typical linker and adaptor sequences.*

Note: (A) A 6-mer EcoRI linker; (B) an 8-mer EcoRI linker; (C) a BamHI– SmaI adaptor. Both the BamHI cohesive ends and the SmaI recognition sequence are indicated.

Source: Bernard R. Glick, Jack J. Pasternak, and Cheryl L. Patten.; Molecular biotechnology: principles and applications of recombinant DNA; ISBN 978-1-55581-498-4.

Linker variations called adaptors are frequently employed in vectors to produce unique cloning sites. These are brief molecules of double-stranded DNA with one or more restriction endonuclease sites. They can have two extended ends or one blunt end and one extended end. To create a novel SmaI site, for instance, the BamHI–SmaI adaptor (Figure 6.9C) is inserted into a distinct BamHI site of a cloning vector. The intended cloning DNA fragment can be blunt-ended ligated into the SmaI site, and the insert can be extracted by treating the construct with BamHI following selection (Figure 6.11). The cloned DNA fragment must not contain the same restriction endonuclease sites as the linker sequence in order for this process to work; otherwise, the insert will be cut during the restriction endonuclease treatment.

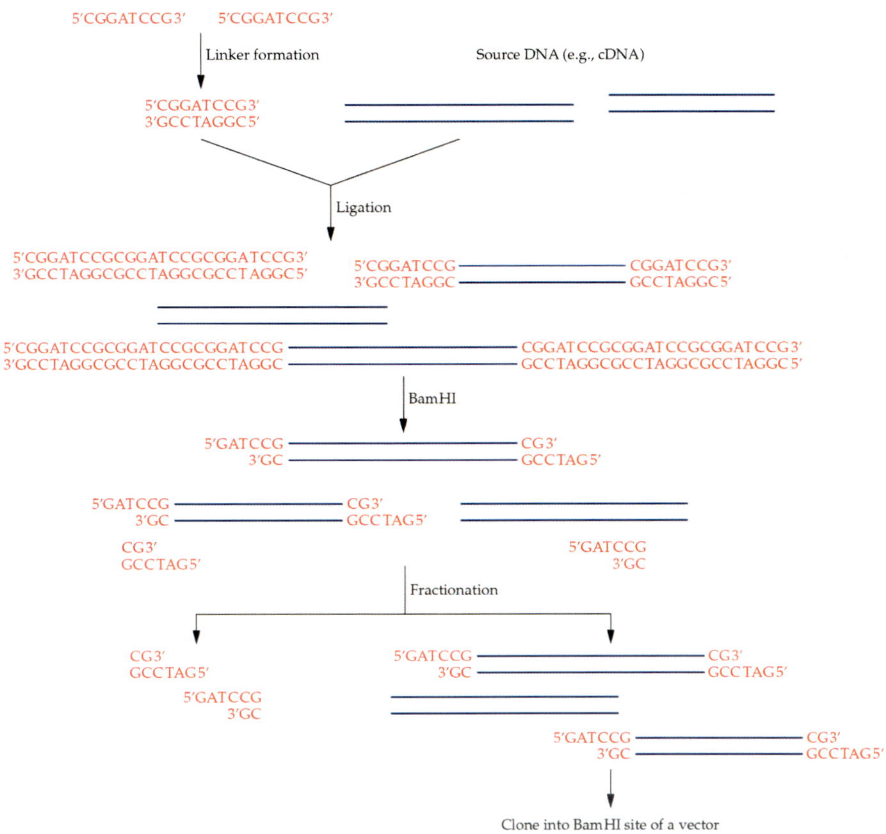

Figure 6.10. *Cloning with a linker. A palindromic oligonucleotide (red) with a BamHI recognition site base pairs to itself to form a BamHI linker. During the ligation reaction, linkers are successively attached to the source DNA and to each other. The ligation mixture is treated with BamHI. The large DNA molecules are purified and cloned into the BamHI site of a cloning vector. DNA molecules without linkers are not cloned.*

Source: Bernard R. Glick, Jack J. Pasternak, and Cheryl L. Patten.; Molecular biotechnology: principles and applications of recombinant DNA; ISBN 978-1-55581-498-4.

The essential elements needed to put genes together are oligonucleotides. Synthetic genes have many uses, such as producing proteins on a large scale, evaluating the functionality of proteins after altering particular codons, and generating nucleotide sequences that encode proteins with unique characteristics. Technically speaking, producing short genes (60–80 base pairs [bp]) is simple and can be achieved by annealing the complementary strands after they have been synthesized. However, since each cycle of chemical DNA synthesis never has a 100% coupling efficiency, unique strategies need to be developed for the synthesis of longer genes (>300 bp). For instance, the fraction of full-length single DNA strands remaining after the final cycle is a pitiful 0.007% if a gene has 1,000 base pairs and the average coupling efficiency is 99.5 percent. Synthetic (double-stranded) genes are created by modularly assembling oligonucleotides with a length of roughly 60 nucleotides to get around this issue.

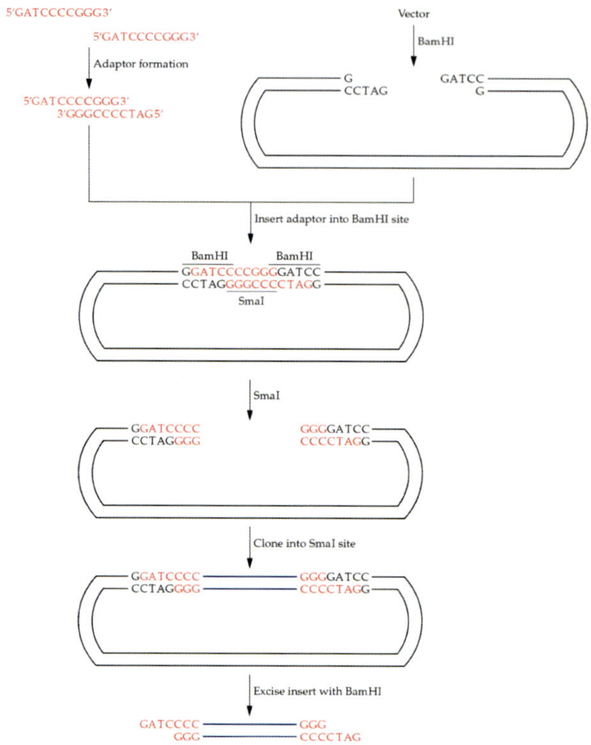

Figure 6.11. *Creating a restriction endonuclease site in a vector with an adaptor. After self-hybridization, an adaptor molecule (red) with two BamHI 5′ extensions and a SmaI site (BamHI–SmaI adaptor) is formed. The BamHI–SmaI adaptor is inserted into a unique BamHI site of a vector to create a unique SmaI site. DNA (blue) is cloned into the SmaI site by blunt-end ligation. Although the SmaI site is destroyed by the insertion of a DNA molecule, the insert can be retrieved by cutting the vector with BamHI.*

Source: Bernard R. Glick, Jack J. Pasternak, and Cheryl L. Patten.; Molecular biotechnology: principles and applications of recombinant DNA; ISBN 978-1-55581-498-4.

A set of overlapping, complementary oligonucleotides, each typically ranging in length from 20 to 60 nucleotides, must first be created to construct a synthetic gene. A set of oligonucleotides with complementary 3′- and 5′-terminal extensions make up each internal section of the gene. These extensions are intended to base pair precisely with another oligonucleotide that also has complementary terminal extensions (Figure 6.12). A gene's two ends are composed of oligonucleotides that are aligned to produce blunt ends. Therefore, using T4 DNA ligase to seal the nicks along the two strands' backbones is the last step needed to finish the process once the gene has been assembled. Synthetic genes can be designed with restriction endonuclease sites at their ends to aid in insertion into a cloning vector, in addition to the protein coding sequence. If required, they can also contain additional sequences that contain signals for the appropriate initiation and termination of transcription and translation. Without changing the amino acid sequence, a gene from one organism can have its codons changed to those that the host cell prefers to optimize translation.

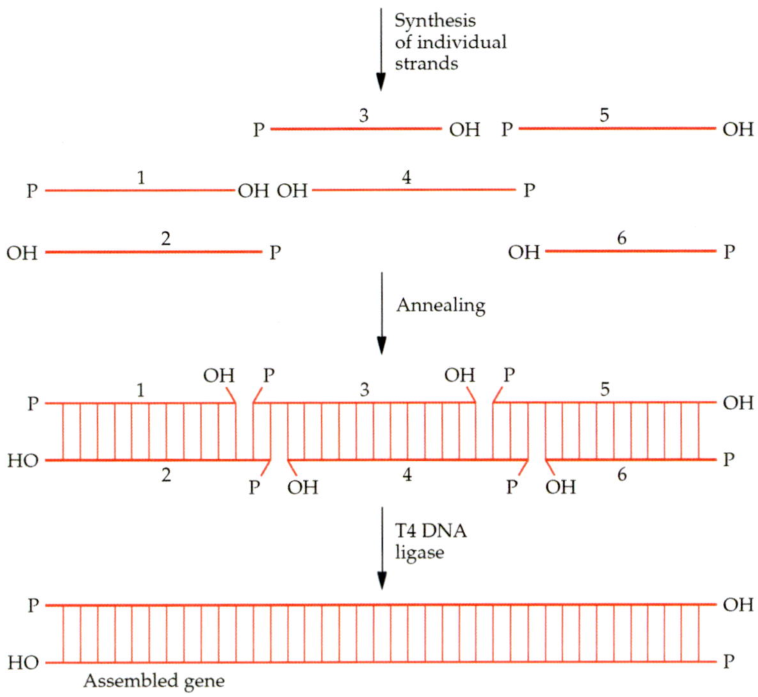

Figure 6.12. *Assembly of a synthetic gene from short oligonucleotides. Individual oligonucleotides (20- to 60-mers) are synthesized chemically. Their sequences are designed so that they form a double-stranded array after annealing. T4 DNA ligase is used to seal the nicks and produce an intact version of the gene.*

Source: Bernard R. Glick, Jack J. Pasternak, and Cheryl L. Patten.; Molecular biotechnology: principles and applications of recombinant DNA; ISBN 978-1-55581-498-4.

Making a specific set of overlapping oligonucleotides with a length of about 60 nucleotides and overlaps of roughly 20 bases is another method for preparing a full-size gene (Figure 6.13). The base-paired regions are long and stable enough to hold the structure together, even after the 3′ and 5′ extensions are annealed, leaving large gaps in between. Following the combination of all the oligonucleotides, *Escherichia coli* DNA polymerase I is used to enzymatically synthesize new DNA to fill in the gaps. The single-stranded regions serve as templates for this enzyme, which uses the 3′ hydroxyl groups as replication initiation sites. T4 DNA ligase is used to seal the nicks following the completion of the enzymatic synthesis. Smaller portions of larger genes (≥1,000 bp) are first assembled into units (chunks, or synthons) of roughly 500 bases in length. These units are then combined with other units of 500 bases, and these larger kilobase segments are joined together one after the other until the entire sequence is completed. Computer programs (like Gene Design) that facilitate the process of choosing the optimal combination of oligonucleotides and overlaps for gene construction are freely and commercially available online. These programs also let users modify any codon, choose a specific codon usage, and designate restriction endonuclease sites at particular locations. The proper nucleotide sequence is the final and most important requirement for a chemically synthesized gene. As a result, the sequence of each of the 500-base building blocks is determined for larger genes, while small synthetic genes are sequenced directly.

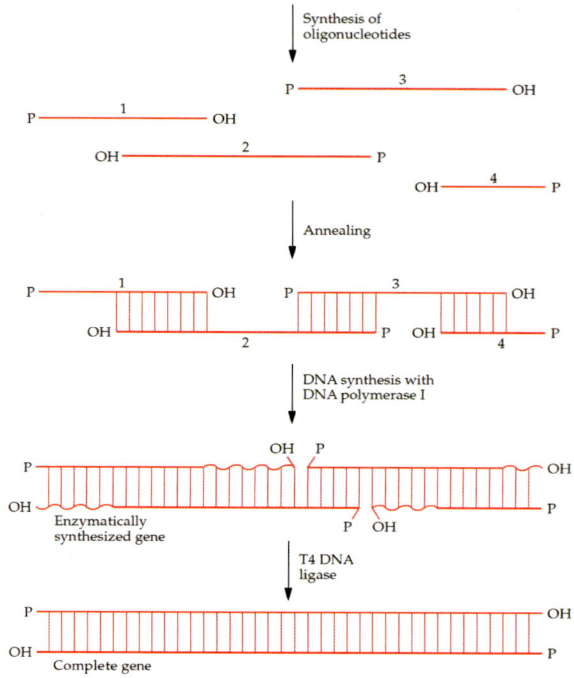

Figure 6.13. *Assembly and in vitro enzymatic DNA synthesis of a gene. Individual oligonucleotides are synthesized chemically. Their sequences are designed to enable them to form a stable molecule, with base-paired regions separated by open regions (gaps). The gaps are filled in by in vitro enzymatic DNA synthesis. The nicks are sealed with T4 DNA ligase.*

Source: Bernard R. Glick, Jack J. Pasternak, and Cheryl L. Patten.; Molecular biotechnology: principles and applications of recombinant DNA; ISBN 978-1-55581-498-4.

6.2. POLYMERASE CHAIN REACTION

In contemporary molecular biology, one of the most widely used techniques is the polymerase chain reaction (PCR). The method was created in 1984 by Nobel laureate Kary Mullis. A target DNA molecule is multiplied extremely precisely *in vitro*, facilitating easy handling and examination using standard molecular biological techniques. Since its invention, PCR has made a substantial contribution to the advancement and modification of biological sciences. In 1988, the first PCR device was released onto the market. The outcome of PCR-based methods is the Human Genome Project. Over the past few decades, many variations of PCR techniques have emerged due to its wide range of applications. (Luettich and Birmingham, 2003).

An oligonucleotide primer is annealed (hybridized) to the target template after the strands of the target DNA molecule (referred to as the template) have been separated (denatured). To add new nucleotides (also known as deoxynucleoside triphosphates, or dNTPs), the DNA polymerase starts with the annealed primers. Based on the template, a nucleotide addition sequence is established. This complete template amplification process, i.e., *in vitro* through cyclical temperature changes that facilitate polymerization, annealing, and separation. The thermophilic microorganisms that thrive between 41 and 122°C, primarily archaea, are the source of the DNA polymerases used in PCR. Double-stranded DNA must melt or separate during PCR, which requires the capacity to tolerate high temperatures.

6.2.1. Principles of PCR

Generally, PCR entails 20–40 iterations of temperature changes, or cycles, with each cycle consisting of three distinct temperature shifts (Figure 6.16).

- Denaturation: 94–96°C;
- Primer annealing (depending on the primer): 45–60°C;
- Primer extension: usually 72°C.

A temperature step known as "hold," where product extension is carried out at (>90) and (~72°C), respectively, frequently precedes and ends the cycling steps. The finished product is stored or subjected to analysis at 4°C. Most PCR techniques can amplify DNA fragments up to about 10 kilobase pairs (kb). On the other hand, some methods can amplify up to 40 kbps.

6.2.2. The Components of PCR

Many of the components, reagents, and setup requirements for a basic PCR are listed below (Figure 6.14).

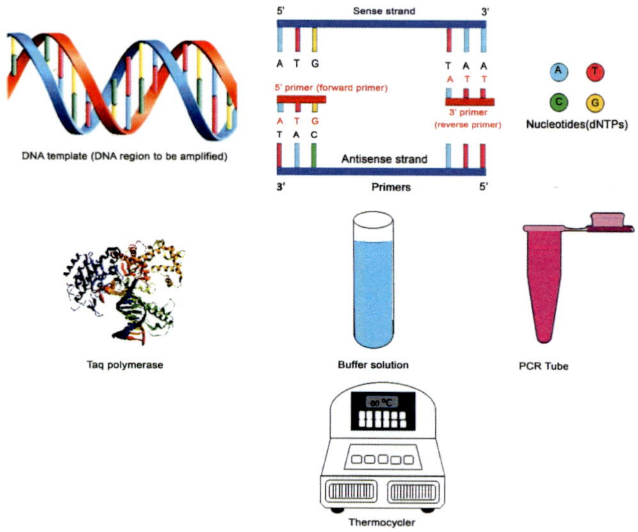

Figure 6.14. *The component of PCR.*

Source: Shaheen Shahzad, Mohammad Afzal, Shomaila Sikandar and Imran Afzal; Polymerase Chain Reaction; DOI: 10.5772/intechopen.81924.

6.2.2.1. DNA Template

Target or template DNA is the name given to the double-stranded DNA molecule that is amplified through PCR. The sequence that new nucleotides are added to during the PCR process is specified by the template. Primers can hybridize, polymerize, and separate DNA strands in this *in vitro* process thanks to cyclically changing temperatures.

If the target sequence is present in the DNA isolate, it can be utilized as a template for PCR. Blood, tissue, forensics specimens, paleontological samples, and laboratory-grown microbes and tissue cells can all provide DNA for polymerase chain reaction (PCR). Regardless of the source, some knowledge of the target DNA sequence is required to design PCR primers. These days, it's very simple to design PCR primers thanks to the abundance of software tools that simply require target sequence information. Primers can be very difficult to design if the sequence information is unknown. One workaround for this issue is to use degenerate primers. (Avashia & Garibyan, 2013).

6.2.2.2. Primers

Single-stranded DNA molecules known as primers are typically produced commercially, i.e., polynucleotides with different lengths. The free 3′ hydroxyl group, also known as the 3′ end, is present in these brief polynucleotide DNA strands. The DNA polymerase needs the free 3′ hydroxyl group on the primer to add new nucleotides during polymerization and create a new complementary strand. Denaturation, or the separation of two complementary DNA strands, is a prerequisite for the binding of DNA primer to the target and is typically accomplished by heating. Two primers are required for PCR to enhance both strands of the template: one primer is called the "forward primer," which is the beginning of the template, and the other is called the "reverse primer," which is for the complementary strand, also known as the antisense strand. Consequently, the 5′ ends of the sense and antisense strands are bound by both primers.

When it comes to accurately identifying the designated target complementary regions, primer length is a significant factor. Lengthening primers increases their specificity—the likelihood that they will match the target. These primers typically have a size of 18–25 nucleotides and are produced commercially.

To guarantee that the right sequence is amplified, primers must anneal or hybridize to the template with good specificity and strength. The primer sequences also influence the precise temperature required for primer annealing, e.g., the higher the annealing temperature, the longer the primer. Therefore, the best primer sequences and concentrations are necessary for PCR to achieve maximum specificity and efficiency. In turn, this is dependent on how primers are made and applied. Inappropriate primers can cause PCR results to completely fail, amplify unwanted DNA segments (nonspecific products), or reduce the yield of specific products. Primers that bind to their target sequences preferentially can be designed and validated to avoid these undesirable results. Primers that have self-complementary sequences will hybridize with one another and create "primer-dimmers." As a result, the PCR efficiency will be lowered since the primers won't bind to their target template. Additionally, hairpin loop structures are formed when complementary sequences are present in a primer.

Primer containing guanine at the 3′ end should be chosen for a strong bond with the template because guanine and cytosine bases (GC) have a stronger bond than adenine and thymine (AT) bases. Nevertheless, runs of three or more C or G bases shouldn't be present in the primers because this could cause mispriming, or nonspecific binding to G- or C-rich sequences in the DNA that aren't the target sequence.

6.2.2.3. DNA Polymerase

The discovery of DNA polymerase in 1955 marked the beginning of PCR technology, which takes advantage of the bacterial enzyme's capacity to transcribe a target DNA strand into a complementary strand. DNA polymerase begins synthesizing new DNA at the template's 3′ end. The primers bind to the 3′ ends of the two template stands, where the DNA polymerase subsequently extends them. Taq DNA polymerase, which was isolated from the thermophilic bacterium Thermus aquaticus, is the most widely used DNA polymerase. At 95°C, Taq polymerase reaches its enzymatic half-life in 40 minutes and adds approximately 1.0 kb of new DNA per minute to the chain. As an alternative, Taq DNA polymerase lacks the 3′–5′ exonuclease activity (proofreading) of Pfu, the DNA polymerase from *Pyrococcus furiosus*, which is another reason why Pfu is also widely used. Through proofreading, Pfu is able to minimize errors in the synthesis of new DNA by removing nucleotides that were added incorrectly during polymerization. KOD DNA polymerase is a recombinant DNA polymerase that is derived from the thermophilic Solfatara bacterium *Thermococcus kodakarensis* KOD1 type strain. Its 3′–5′ exonuclease proofreading activity results in blunt-ended DNA products, and it functions best at 85°C. For small amplicons, KOD DNA polymerase shows excellent fidelity and processivity. However, because of the enzyme's strong 3′–5′ exonuclease activity, the amplification is reduced for amplicons longer than 5 kb. This issue can be resolved by combining the wild-type enzyme with the mutant form, which has reduced 3′–5′ exonuclease activity. This will lead to a more accurate amplification of the amplicons ranging from 5 to 15 kb. *Thermus thermophilus* (Tth) and *Thermus flavus* (Tfl) are two thermophilic species that are additional sources of DNA polymerases used in PCR.

6.2.2.4. Nucleotides

To create new DNA strands, PCR needs four distinct deoxynucleoside triphosphates, or dNTPs: adenine (A), guanine (G), cytosine (C), and thymine (T). A concentration of 200 µM of dNTPs is typically added to the reaction mixture. All four of these dNTPs must have the same concentration in the reaction mixture because the DNA polymerase will misincorporate nucleotides if there is an uneven concentration of even one dNTP.

6.2.2.5. Buffer Solution

The purpose of the PCR buffer solution is to give the DNA polymerase the right environment and substances for maximum activity and stability. The buffers frequently include KCl, Tris-Hcl, and occasionally $MgCl_2$. PCR buffers, which include the compounds listed in Table 6.2, are frequently offered in 10× concentration and occasionally tailored to a particular Taq formulation.

Table 6.2. *Concentrations of PCR Buffers*

Component	Function
100 mM Tris-HCl (pH 8.8. at 25°C)	Maintains reaction pH
500 mM KCl	Stabilizes primer-template annealing
15 mM $MgCl_2$	Cofactor for DNA polymerase
0.8% (v/v) Nonidet P40 (Optional)	Suppresses secondary structure formation

Source: Shaheen Shahzad, Mohammad Afzal, Shomaila Sikandar and Imran Afzal; Polymerase Chain Reaction; DOI: 10.5772/intechopen.81924.

6.2.2.6. Monovalent Cations

When amplifying DNA fragments using PCR, potassium chloride (KCl) is typically used at a final concentration of 50 mM.

6.2.2.7. Divalent Cations

The cofactor magnesium ions are required by the enzyme DNA polymerase. Magnesium or manganese ions are examples of divalent cations. Mg^{2+} is typically employed, but Mn^{2+} can also be used for PCR-mediated DNA mutagenesis because a higher Mn^{2+} concentration raises the error rate during DNA synthesis. (Gelfand D, Innis M, 1999).

6.2.2.8. PCR Tube

The PCR tube is a tiny plastic tube with thin walls used for performing PCR. The tube is made especially to allow for the equilibration of favorable thermal conductivity during thermal cycling.

6.2.2.9. Thermal Cycler

The reaction mixtures are quickly heated and cooled in a thermal cycler, also known as a thermocycler, as they are cycled between the three PCR temperature steps. This temperature ramping is accomplished by reversing the electric current in many contemporary thermocyclers, which makes use of the Peltier effect. To avoid condensation of the reaction mixture during PCR operation, contemporary thermocyclers are additionally equipped with heated lids. This feature was absent from older thermocyclers, and evaporation was stopped by covering the PCR mixture's surface with wax or oil balls.

6.2.3. Procedure

Denaturation, annealing, and extension are the three main steps in a PCR cycle or round, which is repeated 30 or 40 times on a thermocycler (Figure 6.15). The temperature range and length of each cycle step are determined by several factors (Figure 6.16), e.g., the polymerase employed in the synthesis of DNA, the primers' melting temperature (Tm), and the reagent concentration, i.e., divalent ions as well as dNTPs. A primer's length and particular nucleotide sequence determine its melting point. The single-stranded form of DNA makes up half of the molecules at Tm.

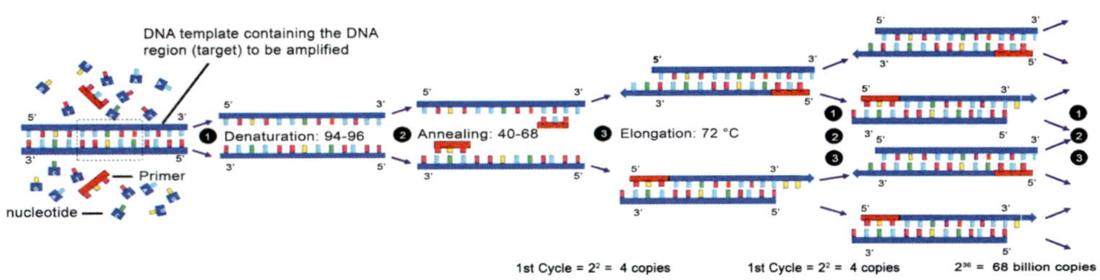

Figure 6.15. *A basic PCR protocol – DNA synthesis cycle.*

Source: Shaheen Shahzad, Mohammad Afzal, Shomaila Sikandar and Imran Afzal; Polymerase Chain Reaction; DOI: 10.5772/intechopen.81924.

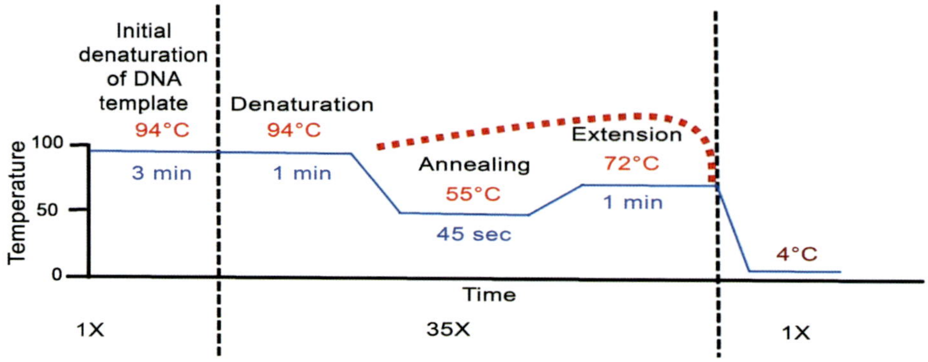

Figure 6.16. *PCR – a simple thermocycling protocol.*

Source: Shaheen Shahzad, Mohammad Afzal, Shomaila Sikandar and Imran Afzal; Polymerase Chain Reaction; DOI: 10.5772/intechopen.81924.

6.2.3.1. Denaturation

Heating the reaction mixture to 94–98°C for 20–30 seconds is the first cycling step. The single-stranded DNA templates are created when the two complementary strands' hydrogen bond is broken by the elevated temperature. Denaturation thus gets the DNA template ready for primer binding.

6.2.3.2. Annealing

The PCR reaction is heated to 50–65°C and held there for 20–60 seconds during the annealing cycling step, which comes after denaturation. Primers and single-stranded templates are encouraged to hybridize as a result. Three to five degrees Celsius below the primer Tm is the optimal annealing temperature. To bind firmly to their target template during annealing, the primers need to be long enough and contain enough GC.

6.2.3.3. Extension

By adding complementary dNTPs to the template bases in a 5′–3′ direction, DNA polymerase synthesizes (polymerizes) new DNA molecules. Depending on the type of DNA polymerase being used, the temperature and extension time will vary. Taq polymerase works best at temperatures between 75 and 80°C. Nonetheless, the enzyme is often utilized at 72°C. The length of the template affects the extension period as well.

The complete steps of denaturation, annealing, and extension/elongation are completed in a single PCR cycle. When all the right factors are present (ideal temperature ramping, substrates/reagents, and inhibitors), the amount of target DNA doubles at the end of each cycle, which leads to exponential amplification of that particular DNA segment.

- **Final Elongation:** At 70–74°C, this one-step process is optional. After the final PCR cycle, the final elongation can take 5 to 15 minutes, during which time any single-stranded DNA that is still present can be fully extended.
- **Final Hold:** The reaction can be temporarily stored by cooling the reaction chamber to 4–15°C for an extended period of time.

6.2.4. Validating PCR

The degree of success can be determined in many ways:

- Staining of amplified DNA products can be done with ethidium bromide (EtBr). Being a DNA intercalator, EtBr inserts itself between the base pairs in the double helix and has UV absorbance maxima at 300 and 360 nm and an emission maximum at 590 nm. DNA bound to ethidium bromide has a detection limit of 0.5–5.0 ng/band.
- For a more economical method of end-labeling PCR products, three primer combinations can be used: (i) fluorescently labeled universal primer, (ii) modified locus-specific primers, and (iii) 5′ universal primer sequence tails.
- **Agarose Gel Electrophoresis:** Uses gel as a molecular sieve and electric current to separate charged molecules like DNA, gel electrophoresis is the most widely used validating technique. Polyacrylamide (<500 bp) or agarose (for DNA >500 bp) can be used as gelling agents. Size-based separation of distinct DNA sequences occurs. To help visualize the DNA bands using a UV transilluminator, DNA staining dyes (such as EtBr) are applied to the gel. It is indicated that the target sequence was present and that the PCR amplified a correct product when a correct size DNA band is visible, as verified by a DNA ladder. When a DNA band of the wrong size is present, it suggests that a spurious product was produced; when none at all, it suggests that the target DNA was absent.
- Direct sequencing is not commonly used because DNA sequencers are expensive, difficult to access or take a long time to complete. However, restriction enzyme digestion can also be used to indirectly evaluate an amplicon's sequence.

6.2.5. Types of PCR

PCR methods come in various forms, each intended for a particular use. These are a few typical kinds:

6.2.5.1. Conventional PCR

Denaturation, annealing, and extension are the three primary steps in this fundamental PCR process. It amplifies fragments of DNA for use in a variety of downstream processes, including cloning, sequencing, and analysis.

6.2.5.2. Real-time PCR (qPCR)

Real-time PCR, sometimes referred to as quantitative PCR, makes it possible to track the amplification of a specific DNA molecule as it happens. It makes it possible to measure the starting concentration of target DNA. SNP genotyping, viral load measurement, and gene expression analysis are some examples of applications.

6.2.5.3. Reverse Transcription PCR (RT-PCR)

RNA is amplified and simultaneously quantified using RT-PCR. It entails RNA reverse transcription into complementary DNA (cDNA), which is then amplified using PCR. RNA virus quantification, viral detection, and gene expression research are three common applications for RT-PCR.

6.2.5.4. Nested PCR

Two sets of primers are used in nested PCR, which amplifies the target DNA in two consecutive PCR reactions. The second PCR is then conducted using the first PCR product as a template. When high specificity is needed or the target DNA is present in low amounts, nested PCR is employed.

6.2.5.5. Multiplex PCR

Using multiple primer pairs, multiplex PCR enables the simultaneous amplification of several target DNA sequences in a single PCR reaction. It is frequently used for genotyping, forensic analysis, and the detection of multiple pathogens in a single sample.

6.2.5.6. Long-range PCR

Long-range PCR is designed to amplify DNA fragments that are longer than those that are usually amplified by traditional PCR. It amplifies DNA fragments up to several kilobases in length using specialized DNA polymerases and buffer systems. Large genes, genomic regions, or entire genomes can be amplified using long-range PCR.

6.2.5.7. Digital PCR (dPCR)

A PCR reaction mixture is divided into thousands of separate reactions using digital PCR, each of which either contains one target DNA molecule or none at all. Standard curves are not necessary to determine the absolute quantity of the target DNA; instead, the number of positive partitions can be counted. Applications needing accurate nucleic acid quantification, like rare allele identification and absolute quantification of gene expression, benefit greatly from digital PCR (Bermingham and Luettich, 2003; Garibyan and Avashia, 2013).

6.3. DNA-SEQUENCING TECHNIQUES

In terms of molecules, knowing a DNA molecule's nucleotide sequence provides the final word on the molecule. It is frequently possible to infer a gene's function from its nucleotide sequence. For instance, one can compare the presumed amino acid sequence—which is derived from the nucleotide sequence—with the protein sequences of known genes. A protein with a comparable function is typically indicated by a significant sequence similarity. It is also possible to identify unique coding regions, such as transmembrane domains, DNA-binding sites, and receptor recognition sites. A gene's regulation may be inferred from the nucleotide sequences in noncoding regions, or areas that do not encode a protein or RNA molecule. Nucleotide sequence data is also necessary for characterizing gene activity and for molecular cloning investigations.

6.3.1. Sanger Sequencing (Chain Termination Method)

The sequencing of DNA fragments was made possible by Sanger sequencing, also referred to as the chain termination method, which completely changed the field of molecular biology. This technique, created in 1977 by Fred Sanger and associates, synthesizes DNA fragments *in vitro* and uses size-based separation to show the nucleotide sequence's order. Numerous later developments, such as the Human Genome Project and personalized medicine, were made possible by this process.

The synthesis of new DNA strands using DNA polymerase is the fundamental idea behind Sanger sequencing. Chain-terminating dideoxynucleotides (ddNTPs), which lack the 3'-OH group required for additional DNA strand elongation, are incorporated during this process. When regular deoxynucleotides (dNTPs) and ddNTPs are mixed together in the reaction, DNA synthesis can continue as usual until a ddNTP is incorporated at random, which effectively stops strand elongation. As a result, a number of different length DNA fragments are produced, each ending at a distinct nucleotide position.

A target DNA fragment is amplified using the polymerase chain reaction (PCR) or other techniques before Sanger sequencing can start. The DNA template is then split into four distinct reactions, each of which has a different fluorescently labeled ddNTP (ddATP, ddCTP, ddGTP, and ddTTP) but the same primer. Terminated fragments of various lengths are created during DNA synthesis, and each one is labeled with a distinct fluorescent dye that matches the terminating nucleotide.

After DNA synthesis, gel electrophoresis is used to separate the resulting fragments based on size. After loading the DNA samples into a polyacrylamide gel matrix, the fragments are moved through the gel using an electric field. The gel separates the fragments according to their lengths because smaller fragments migrate more quickly than larger ones.

A fluorescence imaging device, like a laser scanner, is used to identify the fluorescently labeled DNA fragments after they have been separated. An electropherogram is created by capturing and processing the released fluorescence, which shows each fluorescent signal's intensity as a function of fragment size. Based on the sequence of termination events, the original DNA template sequence can be inferred from the electropherogram.

Finding the peaks that correspond to each terminating nucleotide and figuring out their relative positions along the sequence are key components of electropherogram interpretation. From the shortest to the longest fragments, the sequence is read, with each peak denoting a distinct nucleotide in the DNA template. It is possible to reconstruct the entire sequence of the original DNA fragment by aligning the peaks from each of the four reactions.

Sanger sequencing has a number of benefits, such as great accuracy, compatibility with a wide range of DNA sources and templates, and comparatively long read lengths. It does, however, have certain drawbacks, including the requirement for separate reactions for every nucleotide and a lower throughput than more recent sequencing technologies, such as next-generation sequencing (NGS).

Sanger sequencing is still a useful technique in many fields of study and medicine, even with these drawbacks. It is especially useful for characterizing particular regions of interest, confirming NGS results, and locating genetic variants linked to certain illnesses. The Sanger sequencing workflow has also been simplified by developments in automation and miniaturization, which have reduced costs and increased accessibility for a variety of applications.

6.3.2. Next-Generation Sequencing (NGS)

Rapid sequencing of DNA and RNA molecules is possible thanks to the potent technology known as next-generation sequencing (NGS). NGS enables the massively parallel sequencing of millions of DNA fragments at once, in contrast to traditional Sanger sequencing, which entails sequencing one DNA fragment at a time. The cost and time needed for genome sequencing have been drastically lowered by this high-throughput method, opening up the technique for use in a variety of scientific and medical applications.

6.3.2.1. Historical Perspective

When the first commercial NGS platforms were released in the early 2000s, NGS began to take shape. While these platforms—Illumina, 454 Life Sciences, and SOLiD, among others—used various sequencing chemistries and methodologies, they were all based on the same parallel sequencing concept. NGS technologies have advanced quickly over time, offering increases in sequencing precision, efficiency, and affordability.

6.3.2.2. NGS Workflow

The NGS workflow consists of several key steps:

- **Library Preparation:** DNA or RNA samples are fragmented and adapters are ligated to the ends of the fragments. These adapters contain sequences necessary for attachment to the sequencing platform.

- **Clonal Amplification:** In some NGS platforms, such as Illumina, DNA fragments are amplified to generate clusters of identical DNA molecules on a solid surface (flow cell). Each cluster represents a single DNA fragment that will be sequenced in parallel.
- **Sequencing:** NGS platforms use various sequencing chemistries to determine the nucleotide sequence of DNA fragments. Common methods include sequencing by synthesis (e.g., Illumina), sequencing by ligation (e.g., SOLiD), and nanopore sequencing (e.g., Oxford Nanopore).
- **Data Analysis:** Raw sequencing data is processed to remove low-quality reads, adapter sequences, and other artifacts. The remaining reads are aligned to a reference genome or assembled de novo to reconstruct the original sequence.
- **Variant Calling:** Genetic variants, such as single nucleotide polymorphisms (SNPs) and insertions/deletions (indels), are identified by comparing the sequenced reads to the reference genome. Variant calling algorithms are used to distinguish true variants from sequencing errors.
- **Annotation and Interpretation:** Variants are annotated with information about their genomic location, functional impact, and relevance to diseases or traits. Bioinformatics tools are used to prioritize variants for further analysis and interpretation.

6.3.2.3. Applications of NGS

NGS has revolutionized various fields of biology and medicine, including:

- **Genomic Research:** NGS enables comprehensive analysis of genomes, transcriptomes, and epigenomes, facilitating studies on gene expression, regulation, and chromatin structure.
- **Clinical Diagnostics:** NGS is used for diagnosing genetic disorders, identifying disease-causing mutations, and predicting drug responses in personalized medicine.
- **Cancer Genomics:** NGS allows for the identification of somatic mutations, copy number alterations, and fusion genes associated with cancer development and progression.
- **Microbial Genomics:** NGS is used to study the diversity, evolution, and pathogenesis of microorganisms, including bacteria, viruses, and fungi.
- **Evolutionary Biology:** NGS provides insights into the evolutionary history, population genetics, and adaptation of species by sequencing whole genomes or targeted loci.
- **Forensic Genomics:** NGS is employed in forensic science for identifying individuals, determining paternity/maternity, and analyzing DNA evidence in criminal investigations.

6.3.3. Third-Generation Sequencing (Single-Molecule Sequencing)

Single-molecule sequencing, or third-generation sequencing (TGS) is a revolutionary development in the field of genomics. With TGS, single DNA molecules are directly sequenced as opposed to earlier generations of sequencing technologies, which relied on the amplification of DNA fragments and the detection of incorporated nucleotides. This strategy reduces bias and makes long-read sequencing of DNA fragments possible, which was previously difficult with other methods. It also does away with the requirement for PCR amplification and clonal amplification steps.

The core of TGS is the ability to watch in real-time as individual nucleotides are incorporated into a growing DNA strand. To do this, a number of methods have been developed, such as nanopore sequencing, nanodisc-based sequencing, and single-molecule real-time (SMRT) sequencing. All of these techniques have the same goal of directly sequencing DNA molecules without the need for amplification, but they each have different benefits and difficulties.

Single DNA strands are passed through a protein nanopore embedded in a synthetic membrane to perform nanopore sequencing, which was invented by Oxford Nanopore Technologies. Individual nucleotides disrupt the ionic current in ways unique to them as the DNA molecule translocates through the nanopore. Real-time sequencing of the DNA molecule can be achieved by monitoring these disruptions. Long read lengths (up to hundreds of kilobases), portability, and the capacity to sequence DNA in its native form without amplification or labeling are just a few benefits of nanopore sequencing.

Pacific Biosciences (PacBio) developed SMRT sequencing, which takes a different approach to single-molecule sequencing. This technique immobilizes DNA polymerase at the bottom of a tiny well known as a zero-mode waveguide (ZMW). After adding fluorescently labeled nucleotides to the reaction, a camera detects the fluorescence that is released as DNA polymerase incorporates the nucleotides into the developing DNA strand. The sequence of the DNA molecule can be ascertained by timing and intensities of these fluorescence signals. Long read lengths (up to tens of kilobases) and excellent accuracy are two well-known characteristics of SMRT sequencing, particularly when it comes to identifying DNA modifications like methylation.

A more recent method that incorporates elements of SMRT and nanopore sequencing is called nanodisc-based sequencing. This technique involves ensnaring DNA molecules in lipid nanodiscs, which are subsequently rendered immobile on a solid substrate. Real-time sequencing of the DNA molecules is made possible by monitoring changes in the electrical characteristics of the nanodiscs as nucleotides are added to the reaction. The use of nanodisc-based sequencing presents the possibility of high-throughput, highly accurate long-read sequencing.

Wide-ranging and multifaceted, TGS is revolutionizing fields like transcriptomics, epigenetics, and genomics. The ability of TGS to produce long sequencing reads, which are essential for deciphering complex genomic regions like repetitive sequences, structural variations, and haplotype phasing, is one of its greatest advantages. De novo genome assembly has advanced as a result of this capability, allowing high-quality genome assemblies to be produced using TGS data without the need for expensive and time-consuming scaffolding steps.

Apart from facilitating genome assembly, TGS plays a crucial role in providing unprecedented resolution in transcriptome characterization. Particularly for genes with intricate alternative splicing patterns, full-length transcripts are frequently difficult for conventional short-read RNA sequencing techniques to precisely reconstruct. By offering long reads that cover entire transcripts, TGS gets around this restriction and allows for more precise measurement of isoform diversity and gene expression.

Furthermore, by enabling the direct detection of DNA modifications like methylation and hydroxymethylation, TGS has completely changed the field of epigenetics research. Conventional techniques for identifying DNA modifications involve chemically modifying DNA and then sequencing it, which can result in biases and artifacts. TGS provides a real-time, label-free single-molecule method for identifying DNA modifications while shedding light on their dynamic regulation and functional implications.

TGS has applications in clinical diagnostics and personalized medicine in addition to basic research. More thorough profiling of disease-associated genomic regions has been made possible by long-read sequencing, which has improved the detection and management of infectious diseases, cancer, and genetic disorders. Additionally, TGS has potential applications in non-invasive prenatal testing, as short-read sequencing techniques may not always be able to detect fetal DNA fragments in mother blood.

Even with all of its benefits, TGS is not without problems and constraints. One of the main issues is that long-read sequencing technologies have relatively high error rates, which can make it more difficult to interpret and analyze sequencing data later on. Error correction techniques and base-calling algorithms are being worked on to increase sequencing accuracy.

One additional difficulty is the expense and intricacy of TGS tools and procedures, which may be unaffordable for numerous investigators and medical facilities. However, continuing technological developments and economies of scale are bringing TGS closer to the affordable range and making it available to a wider range of users.

6.3.4. Shotgun Sequencing

Shotgun sequencing, sometimes referred to as shotgun cloning or shotgun metagenomics, is a potent genomics technique that allows for the sequencing of entire genomes or DNA samples without the need for prior compositional knowledge. Shotgun sequencing separates DNA samples into tiny, randomly generated fragments, sequences them, and then assembles the fragments into a complete genome or genomic region, in contrast to traditional sequencing techniques that rely on sequencing particular target regions or amplifying DNA fragments. Comprehensive analyses of complex genomes, including those of organisms whose genetic sequences are unknown or highly variable, are made possible by this method.

First, DNA from the sample of interest—which could be a single organism, a mixed population of organisms (metagenomics), or even environmental DNA—must be isolated. This is the first step in the shotgun sequencing process. Following isolation, the DNA is broken up into smaller fragments, usually with a length of a few hundred to several thousand base pairs. One can accomplish fragmentation by utilizing physical techniques like sonication or enzymatic techniques like restriction enzymes.

High-throughput sequencing technologies, such as Illumina sequencing or next-generation sequencing (NGS) platforms, are used to sequence the DNA fragments after they are created. With just one sequencing run, these platforms can produce millions to billions of short DNA sequences, or reads, giving the original DNA sample thorough coverage. With overlapping areas between neighboring reads, each read represents a piece of the original DNA sequence.

Following sequencing, the next stage of shotgun sequencing involves using specialized bioinformatics software to put the short DNA reads together into longer contiguous sequences, or contigs. The overlapping areas between reads are used by assembly algorithms to put the fragments together and recreate the original DNA sequence. The aim of this process is to reconstruct the entire genome or genomic region from the fragmented sequencing data, which is similar to solving a jigsaw puzzle.

Differentiating between overlapping reads that come from different molecules or genomic regions and those that belong to the same DNA molecule is a major challenge in shotgun sequencing. This problem is more severe in metagenomic samples or complex genomes where there may be several identical sequences. Bioinformatics tools use a variety of algorithms and heuristics to resolve repetitive or ambiguous regions and accurately assemble the sequencing data to address this problem.

We now have a better understanding of genome structure, function, and evolution in a wide range of organisms and environments thanks to shotgun sequencing. Microbiology, evolutionary biology, environmental science, and personalized medicine are just a few of the disciplines that have made extensive use of it.

With shotgun metagenomics, microbial communities can be thoroughly analyzed in microbiology without requiring the use of culture-based techniques. The diversity of microorganisms found in complex ecosystems like soil, water, and the human gut microbiome can be identified and described by researchers through direct sequencing of DNA from environmental samples. Using this method has revealed new microbial species and provided insights into their functions in disease, nutrient cycling, and ecosystem functioning.

Shotgun sequencing has been applied to the study of the genomes of many organisms in evolutionary biology, including bacteria, viruses, plants, and animals. Researchers can determine genetic adaptations, deduce evolutionary relationships, and look into the molecular underpinnings of phenotypic traits by comparing the genomes of various species. Additionally, shotgun sequencing has provided information on the genetic diversity and population structure of endangered species, which has influenced management plans and conservation initiatives.

Shotgun metagenomics has transformed our knowledge of microbial ecology and biogeochemical processes in environmental science. Researchers can clarify the functional capabilities of microbial communities and their contributions to ecosystem processes like carbon cycling, nitrogen fixation, and pollutant degradation by sequencing DNA from environmental samples. Predicting and reducing the effects of environmental change, such as pollution and climate change, depend heavily on this data.

Shotgun sequencing has the potential to improve the diagnosis and treatment of infectious diseases, cancer, and genetic diseases in personalized medicine. Clinicians can determine disease-causing mutations, estimate the risk of a disease, and customize treatment plans based on each patient's distinct genetic profile by sequencing an individual's entire genome or exome. Additionally, somatic mutations and uncommon genetic variations linked to treatment response, drug resistance, and cancer progression can be found via shotgun sequencing.

Even with all of its benefits, shotgun sequencing is not without its drawbacks. The computational difficulty of compiling and interpreting vast amounts of sequencing data is one drawback, especially for complex genomes or metagenomic samples. This may be prohibitive for certain researchers or clinical laboratories due to the need for access to high-performance computing resources and bioinformatics expertise.

The possibility of sequencing errors, such as base-calling errors, insertion-deletion errors, and chimeric sequences, presents another difficulty because they can compromise the precision and dependability of the sequencing data. Although error correction algorithms and quality control procedures are used to lessen these problems, errors cannot be completely eliminated. In addition, shotgun sequencing might miss some genomic features that are critical to comprehending the function and regulation of the genome, such as structural variations, long-range chromosomal interactions, and epigenetic modifications. Shotgun sequencing can be combined with complementary sequencing methods, such as long-read sequencing, chromatin conformation capture (Hi-C), and epigenomic profiling, to provide a more thorough picture of the genome (Bermingham and Luettich, 2003; Garibyan and Avashia, 2013).

6.3.5. Metagenomic Sequencing

A groundbreaking method in genomics called metagenomic sequencing makes it possible to thoroughly examine microbial communities and their genetic potential. With metagenomic sequencing, scientists can directly sequence DNA from environmental samples, capturing the collective genomes of all the organisms present, in contrast to traditional sequencing techniques that concentrate on individual organisms or particular genes. This method offers insights into the diversity of microorganisms, the structure of communities, their potential for function, and ecological interactions in a range of habitats, from soil and water to the human body and beyond.

The first step in the metagenomic sequencing process is to extract and collect DNA from the relevant environmental sample. A variety of materials, including soil, sediment, water, air, animal tissues, and human body parts like the skin, gut, or oral cavity, could be used to gather this sample. The genetic material from all the organisms in the sample, including bacteria, viruses, fungi, archaea, and other microorganisms, is mixed together in the extracted DNA.

The process of preparing a library for metagenomic sequencing involves breaking up the DNA into smaller pieces and getting it ready for sequencing, which comes after DNA extraction. Both enzymatic and physical techniques, such as sonication and restriction enzymes, can be used to achieve fragmentation. After being broken up, DNA fragments are frequently filtered for particular size ranges that are appropriate for sequencing on the selected platform.

After the library is prepared, high-throughput sequencing technologies, such as those from Pacific Biosciences (PacBio), Oxford Nanopore, and Illumina, are used to sequence the DNA fragments. These platforms use the metagenomic DNA sample to produce millions to billions of short DNA sequences or reads. Every read is a piece of the original DNA sequence, giving an overview of the genetic makeup of the environmental sample. Following the sequencing process, the obtained data are subjected to bioinformatic analysis to reconstruct the genomes of individual microorganisms and assess their functional potential. There are various steps involved in this analysis, such as functional annotation, taxonomic classification, read assembly, and quality control. The short DNA reads are put together using assembly algorithms into longer contiguous sequences, or contigs that represent the genomes of distinct microorganisms.

Using taxonomic classification, the microbial taxa contained in the sample are identified based on how similar their gene sequences are to known reference sequences in databases like the Integrated Microbial Genomes (IMG) database and NCBI's GenBank. Predicting the biological processes that the genes in the metagenomic data, such as metabolic pathways, genes for antibiotic resistance, virulence factors, and other functional elements, encode is known as functional annotation.

Numerous fields, including environmental microbiology, microbial ecology, biotechnology, human health, and biomedicine, have found extensive uses for metagenomic sequencing. Metagenomic sequencing in environmental microbiology allows scientists to examine the microbial diversity and community structure in a range of ecosystems, such as freshwater, oceans, soil, harsh environments, and built environments. Understanding ecosystem dynamics, nutrient cycling, and the effects of environmental change requires knowledge of this information.

Within the field of microbial ecology, metagenomic sequencing offers valuable perspectives on the capacities of microbial communities and their functions in ecological processes like pollutant degradation, carbon cycling, and nitrogen fixation. Researchers can deduce the interactions and ecological roles of various microbial taxa within a community by analyzing the genes and metabolic pathways encoded by microbial genomes.

Metagenomic sequencing has made it easier to find new enzymes, biosynthetic routes, and bioactive substances with medical and industrial uses in biotechnology. Researchers can find enzymes with particular catalytic activities, like cellulases, lipases, and proteases, for use in biofuel synthesis, bioremediation, and biocatalysis by searching metagenomic libraries for genes of interest.

Metagenomic sequencing is transforming our understanding of the human microbiome and its role in health and disease in biomedicine and human health. Researchers can examine the role of the microbiome in several illnesses, such as obesity, inflammatory bowel disease, diabetes, cancer, and infectious diseases, by sequencing the microbial communities residing in various body sites, such as the gut, skin, oral cavity, and urogenital tract.

The microbiomes of various animals, plants, and ecosystems have also been studied using metagenomic sequencing, which has shed light on co-evolutionary dynamics, symbiosis, and host-microbe interactions. Through the analysis of microbial communities' genetic diversity and functional potential in various environments and host organisms, scientists can clarify the fundamental concepts guiding microbial ecology and evolution.

Metagenomic sequencing has limitations and challenges despite its many benefits. The possibility of exogenous DNA contamination during sample collection, extraction, and sequencing is one restriction. Laboratory reagents, apparatus, and the surrounding environment can all be contaminated, which can result in false-positive results and incorrect data interpretation. Bioinformatic analyses and strict quality control procedures are necessary to detect and reduce contamination in metagenomic datasets.

The computational complexity of analyzing vast amounts of metagenomic data—which can include billions of short DNA reads from various microbial taxa—presents another difficulty. This calls for the availability of bioinformatics knowledge, high-performance computing resources, and specialized software tools for functional annotation, taxonomic classification, and read assembly. Furthermore, accurate taxonomic classification and functional annotation of metagenomic data are hampered by the lack of reference genomes for many microbial taxa.

6.3.6. Cyclic Array Sequencing

Even though many of these steps have been automated, cloning large numbers of DNA fragments into plasmids, converting the plasmids into bacterial cells, and then selecting and extracting the plasmids from thousands to millions of colonies for sequencing using the Sanger dideoxynucleotide method is a very time-consuming and costly process. Cyclic array sequencing strategies, which immobilize the sequencing templates in a dense array on a surface so that very small volumes of reagents are required for the sequencing reactions, prepare libraries of DNA fragments for sequencing *in vitro*, and use a sequencing-by-synthesis approach to acquire hundreds of millions of sequences simultaneously (in parallel) have been developed to reduce the time and cost associated with large-scale sequencing. Cyclic array sequencing can produce the sequence of a human genome in two months as opposed to the eight months needed with the shotgun cloning–Sanger sequencing method.

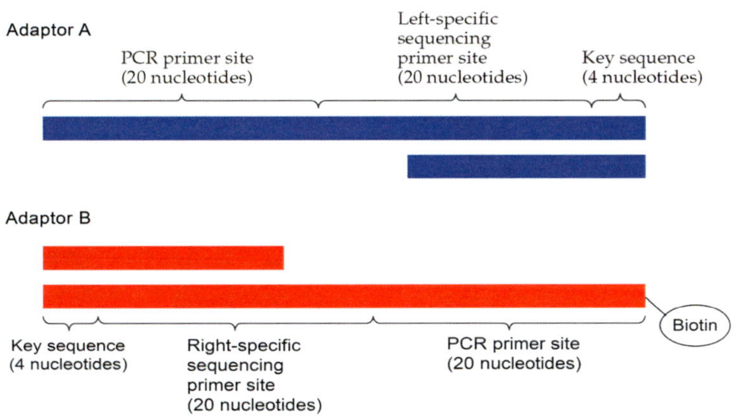

Figure 6.17. *Features of adaptors A and B that are used for template preparation, PCR amplification, and sequencing using a cyclic array-sequencing strategy with the 454 sequencing platform.*

Source: Bernard R. Glick, Jack J. Pasternak, and Cheryl L. Patten.; Molecular biotechnology: principles and applications of recombinant DNA; ISBN 978-1-55581-498-4.

Libraries of sequencing templates can now be created without the need to prepare clone libraries in bacterial cells thanks to the development of cell-free methods. PCR, on the other hand, is used to create clonal clusters that are isolated from other sequences and contain millions of copies of each template DNA molecule. Polymerase colonies are referred to as "polonies" because they are composed of multiple copies of the same nucleotide sequence that are contained within an entity, like a bead or a solid or gel surface. One method of cyclic array sequencing is to capture the PCR-amplified sequencing templates on a small bead's surface, arrange the beads in plate wells, and then use pyrosequencing to ascertain the sequences of the captured templates. Because of the company (454 Life Sciences) that developed the technology, this strategy is frequently referred to as 454 sequencing.

The original DNA is broken up into pieces that are typically 350 bp in size. The 5′ ends are phosphorylated, and the frayed ends are blunt ended. The polished, phosphorylated genomic DNA fragments are ligated to two nonphosphorylated adaptor sequences (A and B) (Fig 6.17). The adaptors contain sequences for priming the sequencing reaction, calibrating the signal output of the sequencing reaction, and amplifying the genomic sequence by PCR. A 4-nucleotide calibration (key) sequence, a 20-nucleotide sequencing primer site (left-template-specific primer; left-specific sequencing primer), and a 20-nucleotide PCR primer site (primer A) make up the components of adaptor A's long arm, which is read from the 3′ end. A key sequence, a PCR primer site (primer B), a sequencing primer site (right-template specific primer; right-specific sequencing primer), and a biotin tag on the 5′ end are all present in adaptor B (Figure 6.17). The 5′ phosphate end of the genomic DNA is joined to the 3′ hydroxyl group of an adaptor during ligation, leaving a nick (gap) at the 3′ hydroxyl end of the genomic DNA because the adaptors are not phosphorylated. A DNA polymerase without exonuclease activity fills in the genomic DNA's 3′ ends to form complete double-stranded DNA molecules. During replication, the long arms of the adaptors serve as templates, and the short adaptor strands are displaced (Figure 6.18A).

A genomic DNA fragment may have (1) an A and B adaptor, (2) only adaptor A, or (3) only adaptor B flanking it on either end following ligation and filling in (Fig 6.18B). With adaptors A and B on either end of the genomic DNA fragment, the following step is intended to recover single-stranded DNA, i.e., A-genomic DNA-B. After adding streptavidin beads to the DNA sample, adaptor B's biotin binds to the streptavidin beads. As a result, the molecules of A-genomic DNA-B and B-genomic DNA-B are immobilized. Conversely, the A-genomic DNA-A molecules wash away because they are unable to bind. Following denaturation, only the A-genomic DNA-B molecules' nonbiotinylated strands are liberated and gathered. Under these circumstances, the biotinylated strands that are still attached to the streptavidin beads are disposed of (Figure 6.18B).

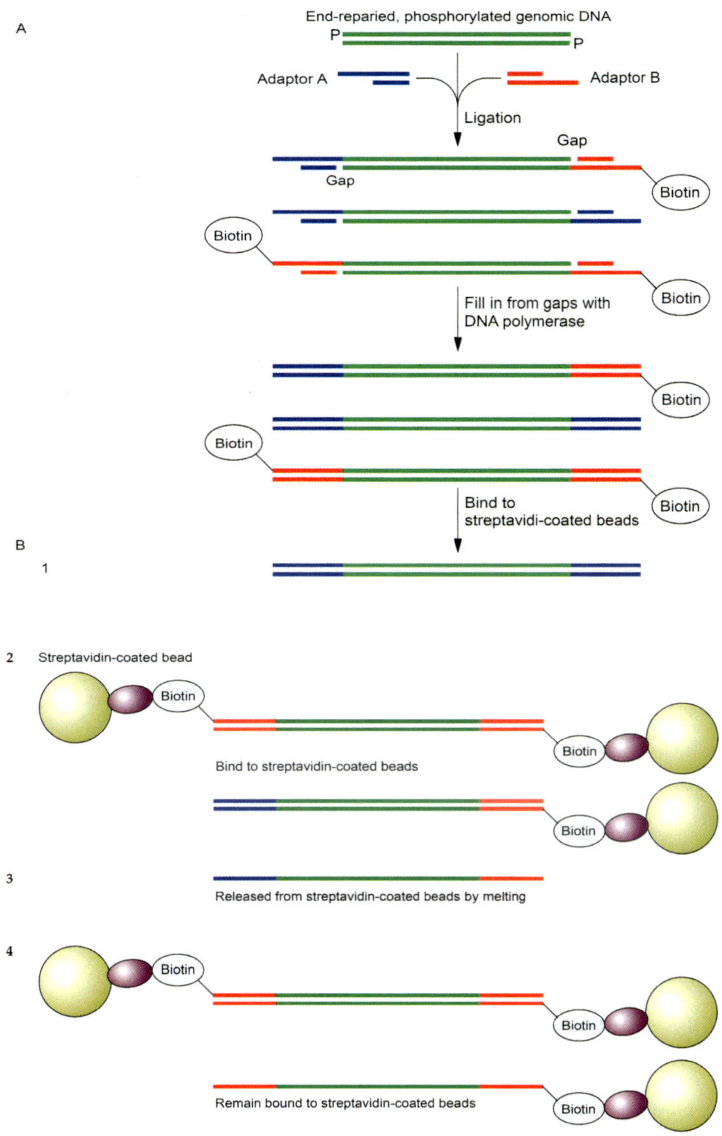

Figure 6.18. *Preparation of single-stranded genomic DNA fragments for the 454 sequencing platform.*

Note: (A) The frayed ends of genomic fragments are polished (end repaired), phosphorylated, and ligated with adaptors A and B. (B) The ligation products are mixed with streptavidin-coated beads. (1) The adaptor A–genomic DNA–adaptor A molecules do not bind to the streptavidin-coated beads and are washed out. (2) The DNA molecules with biotin tags bind to the streptavidin-coated beads. (3) The adaptor A–genomic DNA–adaptor B strands without a biotin tag are released by melting, concentrated, and retained for sequencing. (4) DNA molecules with biotin tags remain bound to the streptavidin-coated beads and are discarded.

Source: Bernard R. Glick, Jack J. Pasternak, and Cheryl L. Patten.; Molecular biotechnology: principles and applications of recombinant DNA; ISBN 978-1-55581-498-4.

The next step of the process involves amplifying each genomic sequence that has been gathered separately. To do this, 20-nucleotide oligomers whose sequences are complementary to the PCR primer site of adaptor B are bound to beads (DNA capture beads) at their 5′ ends. More than 107 of these oligomers are carried by each DNA capture bead. After that, only one DNA strand can base pair with an oligomer on a bead when the isolated A-genomic DNA-B strands are combined with the DNA capture beads in a certain ratio (Figure 6.19). To create a water-in-oil emulsion, the beads and PCR reagents—PCR primers that anneal to sequences that are part of adaptors A and B—are then vigorously mixed with oil. The setup is such that a single bead and PCR components are present in an oil globule's water droplet (Figure 6.20A). Stated differently, every oil globule functions as a distinct reaction chamber. These PCR "microreactors" could number up to 1,000 per microliter. Emulsion PCR is the name given to this type of PCR.

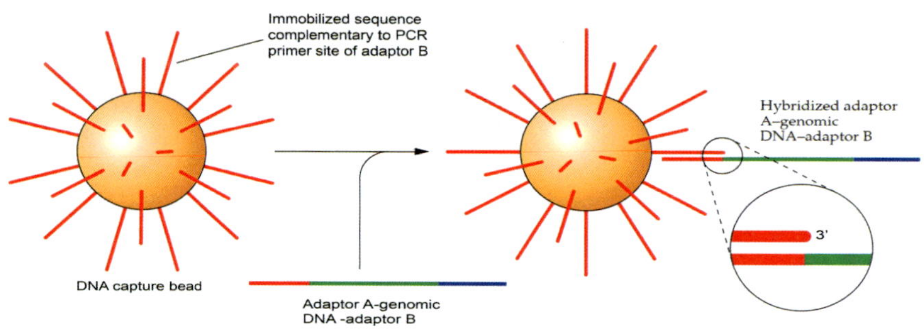

Figure 6.19. *DNA capture bead. Oligomers that are complementary to the PCR amplification sequence of adaptor B are attached at their 5′ ends to a bead. Each DNA capture bead hybridizes with only one adaptor A–genomic DNA–adaptor B strand. The inset highlights the available 3′ end of the immobilized oligomer.*

Source: Bernard R. Glick, Jack J. Pasternak, and Cheryl L. Patten.; Molecular biotechnology: principles and applications of recombinant DNA; ISBN 978-1-55581-498-4.

Strands that have the same sequence as the isolated A-genomic DNA-B molecule are created during PCR cycling in each microreactor, and they base pair with the immobilized oligomers. The synthesis of an entire complementary strand from the 3′ ends of the oligomers comes next (Figure 6.20B to D). Following 20 to 30 cycles, each bead contains roughly 10^7 complementary DNA molecules bound to the original A-genomic DNA-B strand hybridization. After PCR, all of the free DNA molecules are removed by washing, breaking up the emulsion and gathering the beads. Melting them allows you to extract DNA strands that are base paired with the bound DNA strands. It is essential to concentrate the beads with attached full-length DNA extensions because not all DNA capture beads enclose themselves in a PCR microreactor. By hybridizing to oligomers that are attached to magnetic beads and complementary to the adaptor A sequence at the free 3′ end of every full length immobilized DNA molecule, these beads are enriched. Every DNA capture bead that doesn't hybridize is thrown away. Melting them releases the DNA capture beads that are bound by the oligomer attached to the magnetic beads.

Using a magnet, the magnetic beads are finally taken out. Now, to prepare the DNA capture beads for sequencing, a primer that matches the left-template specific sequence of adaptor A, a protein that binds to single-stranded DNA, and DNA polymerase are added. Even though sample preparation seems like a lot of work, it really just takes two or three days. On the other hand, traditional whole-genome shotgun sequencing takes 20 to 30 days due to cloning, bacterial growth, and vector DNA preparation.

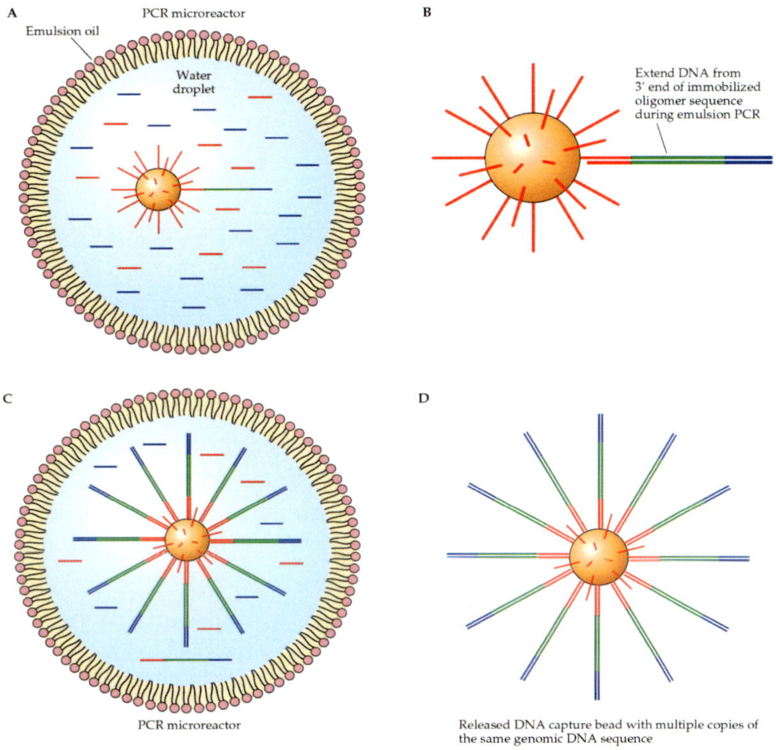

Figure 6.20. *Emulsion PCR.*

Note: (A) Representation of a DNA capture bead with a hybridized A–genomic DNA–B strand and PCR amplification primers in a water-in-oil droplet (PCR microreactor) before the initiation of the PCR cycles. (B) DNA capture bead with the original hybridized adaptor A–genomic DNA–adaptor B strand and the complementary sequence extended from the 3′ end of the immobilized oligomer. Blue, adaptor A; green, genomic DNA; red, adaptor B. (C) Representation of a DNA capture bead during PCR cycling with many immobilized full-length sequences that are complementary to the originally hybridized DNA strand. (D) A DNA capture bead with many copies of the same genomic sequence.

Source: Bernard R. Glick, Jack J. Pasternak, and Cheryl L. Patten.; Molecular biotechnology: principles and applications of recombinant DNA; ISBN 978-1-55581-498-4.

By adding single nucleotides in cycles and then detecting light emission to identify which incorporated nucleotide releases pyrophosphate, a technique known as pyrosequencing can be used to determine the nucleotide sequence. Wells with picoliter capacities are used to conduct the sequencing reactions. There are one millionth of a liter in a picoliter. To improve the efficiency of the sequencing reaction, two additional types of beads are used in addition to the DNA capture beads (~28-µm diameter). To begin pyrosequencing, luciferase and ATP sulfurylase are first attached to tiny beads with a diameter of 2.8 µm. Second, uniformity within the reaction wells and support for the DNA capture beads are achieved by using microparticles (0.8-µm diameter) free of immobilized molecules. After mixing, the beads are added to a plate with either 2.6 × 10^6 or 0.8 × 10^6 wells, depending on the plate's size. Only one DNA capture bead can fit inside each hexagonal well (Figure 6.21). The light signal corresponding to the incorporation of a specific nucleotide is transmitted from each well to a sensor via optic fibers positioned adjacent to the plate. Each well's sequence of incorporated nucleotides is represented by a series of signal outputs (flow signals), which are recorded and saved in a computer.

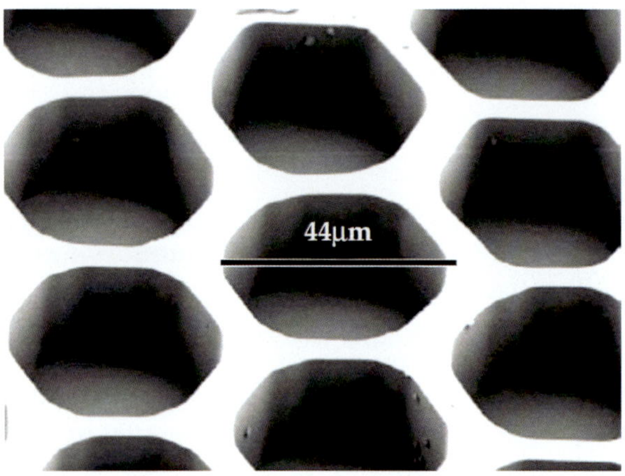

Figure 6.21. *Wells of a picotiter plate used for 454 sequencing. Reproduced with the kind permission of Roche Diagnostics North America, Indianapolis, IN.*

Source: Bernard R. Glick, Jack J. Pasternak, and Cheryl L. Patten.; Molecular biotechnology: principles and applications of recombinant DNA; ISBN 978-1-55581-498-4.

Pyrosequencing reagents, apyrase, and one of the four deoxyribonucleotides are sequentially poured into the wells during a single sequencing round. This procedure is carried out once for every nucleotide. A round lasts roughly 60 seconds. The signal output is calibrated and the wells containing DNA capture beads are found using the "key" sequence on the adaptors. Before the accuracy of the read starts to decline, high-quality sequence information is typically generated from 200 to 300 bases. The most likely sets of contiguous nucleotides are identified by an assembler once the flow signal data from all of the wells have been combined at the conclusion of a run. In 4 hours and 5 minutes, this platform produces approximately 25 × 10^6 raw base reads, which is approximately 10 times the coverage of a genome with 2 × 10^6 bp. The effectiveness of this approach has been shown by a number of large-scale sequencing projects, such as the sequencing of the human genome (using Nobel laureate James Watson's genome as a test case) and the sequencing of DNA in complex environmental samples that contain multiple microbial genomes in an effort to identify potential pathogens causing the unexplained collapse of honeybee colonies. The Neanderthal genome sequencing project demonstrated the value of PCR amplification of sequencing templates, particularly in situations where low-quality templates are scarce, as is typically the case with ancient DNA samples.

Some cyclic array-sequencing techniques combine base extension with ligation or reversible chain terminators with emulsion PCR, or another PCR-based technique, for clonal amplification of sequencing templates. There is also research being done on techniques for sequencing individual DNA molecules without amplification. In one of these methods, a DNA molecule is moved through a nanopore, a very small channel, and the nucleotide sequence is deduced from the successive changes in electrical conductance that result from particular base pairs. One benefit of nanopore sequencing is that it does not necessitate extensive template preparation. Furthermore, sequence information might theoretically be produced in a matter of microseconds. Last but not least, even with the progress made in large-scale DNA sequencing, it will be amazing if any team ever prevails in the difficult $10 million competition known as the X Prize. The goal is to create a platform that can sequence 100 human genomes with 99.999% accuracy in 10 days for a maximum of $10,000 per genome. (Bermingham and Luettich, 2003; Garibyan and Avashia, 2013)

CASE STUDY

SEQUENCING OF DNA TO COMBAT ANTIMICROBIAL RESISTANCE

Background: Global Health Institute (GHI) is a research organization focused on addressing the growing threat of antimicrobial resistance (AMR). AMR occurs when bacteria evolve to resist the effects of antibiotics, making infections harder to treat. One of the projects involves studying a particularly resistant strain of *Escherichia coli* (*E. coli*) that has been causing outbreaks in hospitals.

Problem Statement: The resistant *E. coli* strain has been spreading rapidly and is resistant to multiple antibiotics. Traditional methods of identifying resistance mechanisms are slow and often do not provide a complete picture. GHI aims to use DNA sequencing to rapidly identify resistance genes and understand the genetic basis of resistance in this strain.

Solution: GHI decides to employ whole-genome sequencing (WGS) to analyze the DNA of the resistant *E. coli* strain. This approach allows for a comprehensive analysis of the entire bacterial genome, identifying all potential resistance genes and providing insights into their mechanisms.

Implementation

Sample Collection and DNA Extraction:

- GHI collects samples from infected patients and isolates the resistant *E. coli* strain.
- DNA is extracted from the bacterial cultures using standard protocols ensuring high-quality genomic DNA suitable for sequencing.

Whole-Genome Sequencing:

- **Sequencing Technology:** GHI uses a next-generation sequencing (NGS) platform, such as Illumina or Oxford Nanopore, to sequence the DNA of the resistant *E. coli* strain.
- **Library Preparation:** DNA samples are prepared for sequencing by fragmenting the DNA and adding adaptors. Libraries are then amplified and loaded onto the sequencing platform.

Bioinformatics Analysis:

- **Genome Assembly:** Raw sequencing data is processed to assemble the complete genome sequence of the resistant *E. coli* strain. This involves aligning sequencing reads to a reference genome or de novo assembly.
- **Gene Annotation:** The assembled genome is annotated to identify genes, with

a particular focus on known antibiotic resistance genes and mobile genetic elements, such as plasmids and transposons, that can facilitate the spread of resistance.

- **Comparative Genomics:** The resistant strain's genome is compared to genomes of susceptible *E. coli* strains to pinpoint genetic differences and identify novel resistance mechanisms.

Functional Validation:

- **Experimental Validation:** Key resistance genes identified through sequencing are experimentally validated. This may include cloning the genes into susceptible bacterial strains to confirm their role in conferring resistance.
- **Phenotypic Testing:** The resistant *E. coli* strain is subjected to various antibiotics to correlate genomic findings with phenotypic resistance profiles.

Implementation in Clinical Settings:

- **Diagnostic Development:** Based on sequencing results, GHI develops rapid diagnostic tests to detect the presence of the identified resistance genes in clinical samples.
- **Infection Control:** Hospitals use this information to enhance infection control measures, such as isolating patients carrying the resistant strain and tailoring antibiotic treatment based on the resistance profile.

- **Results:** The whole-genome sequencing of the resistant *E. coli* strain reveals multiple resistance genes, including some previously unknown to confer resistance. The data provide insights into the genetic mechanisms of resistance and the evolutionary pathways of the strain. The rapid diagnostic tests developed from this information allow for timely identification and management of infections in clinical settings.
- **Conclusion**: This case study demonstrates the power of DNA sequencing in addressing the critical issue of antimicrobial resistance. By employing whole-genome sequencing and comprehensive bioinformatics analysis, GHI was able to identify resistance mechanisms, develop diagnostic tools, and inform treatment and infection control strategies. This approach underscores the importance of genomics in modern microbiology and public health.

SUMMARY

- Chemically synthesized single-stranded DNA oligonucleotides are used for assembling whole genes, amplifying specific DNA sequences, introducing mutations into cloned genes, screening gene libraries, sequencing DNA, and facilitating gene cloning.
- Solid-phase synthesis, i.e., attachment of the growing DNA strand to a solid support, is used so that all the reactions can be conducted in one reaction vessel, the reagents from one reaction step can be readily washed away before the reagents for the next step are added, and the reagents can be used in excess in an attempt to drive the reactions to completion.
- Chemically synthesized oligonucleotides (<100-mers) have a myriad of functions. Single-stranded hybridization oligonucleotide probes (20- to 40-mers) can be formulated by deducing the codons from the amino acid sequence of a protein and then used to screen a genomic library for the gene.
- Polymerase chain reaction (PCR) is one of the most commonly used method in modern molecular biology. The technique was developed by the Nobel laureate, Kary Mullis, in 1984. It is an *in vitro* process to multiply a target molecule of DNA with extreme precision, making it easy to be handled and examined by routine molecular biological methods.
- The double-stranded DNA molecule amplified by PCR is called the target or template DNA. The template defines the sequence in which new nucleotides are added during the PCR process. This process is carried out *in vitro* by cyclically varying temperature, enabling separation of DNA strands, hybridization of primers, and polymerization.
- PCR requires four different deoxynucleoside triphosphates or dNTPs to synthesize new DNA strands: adenine (A), guanine(G), cytosine(C), thymine(T). The dNTPs are usually provided at a concentration of 200 μM in the reaction mixture. The concentration of these four dNTPs must be equal in the reaction mixture, as unequal concentration of even a single dNTPs leads to misincorporation of nucleotides by the DNA polymerase.
- Metagenomic sequencing is a revolutionary approach in genomics that enables the comprehensive study of microbial communities and their genetic potential. Unlike traditional sequencing methods that focus on individual organisms or specific genes, metagenomic sequencing allows researchers to sequence DNA directly from environmental samples, capturing the collective genomes of all organisms present.
- Cloning of large numbers of DNA fragments into plasmids, transforming the plasmids into bacterial cells, and then picking and extracting the plasmids from thousands to millions of colonies for sequencing using the Sanger dideoxynucleotide method is very time-consuming and expensive, even though many of these steps have been automated.
- A single sequencing round consists of flooding the wells, in succession, with one of the four deoxyribonucleotides, pyrosequencing reagents, and finally

apyrase. This process is repeated for each nucleotide. The duration of a round is about 60 seconds.

- The double-stranded DNA molecule amplified by PCR is called the target or template DNA. The template defines the sequence in which new nucleotides are added during the PCR process. This process is carried out *in vitro* by cyclically varying temperature, enabling separation of DNA strands, hybridization of primers, and polymerization.

REVIEW QUESTIONS

1. Describe the phosphoramidite method for chemical synthesis of DNA. What are the main steps involved in this process?
2. What are the essential components of a PCR used to amplify a specific sequence of DNA?
3. Explain the types of PCR.
4. What are the advantages of cycle array sequencing compared to the shotgun cloning method for large-scale sequencing?
5. Explain the principle of chain-termination in Sanger sequencing. How does this principle enable the determination of DNA sequences?
6. Discuss the role of bioinformatics in the analysis of DNA sequencing data. What are some common tools and techniques used for this purpose?

REFERENCES

1. Bermingham, N., & Luettich, K. (2003). Polymerase chain reaction and its applications. *Current Diagnostic Pathology, 9*(3), 159–164. https://doi.org/10.1016/S0968-0896(03)00031-4.
2. Bowne, S. J., Sullivan, L. S., Koboldt, D. C., et al. (2011). Identification of disease-causing mutations in autosomal dominant retinitis pigmentosa (adRP) using next-generation DNA sequencing. *Investigative Ophthalmology & Visual Science, 52*(1), 494–503. https://doi.org/10.1167/iovs.10-6031.
3. Dewey, F. E., Pan, S., Wheeler, M. T., et al. (2012). DNA sequencing: Clinical applications of new DNA sequencing technologies. *Circulation, 125*(7), 931–944. https://doi.org/10.1161/CIRCULATIONAHA.111.044077.
4. Garibyan, L., & Avashia, N. (2013). Research techniques made simple: Polymerase chain reaction (PCR). *The Journal of Investigative Dermatology, 133*(3), e6. https://doi.org/10.1038/jid.2013.1.
5. Innis, M., & Gelfand, D. (1999). Optimization of PCR: Conversations between Michael and David. In *PCR Applications* (pp. 3–22). Academic Press.
6. Logsdon, G. A., Vollger, M. R., & Eichler, E. E. (2020). Long-read human genome sequencing and its applications. *Nature Reviews Genetics, 21*(10), 597–614. https://doi.org/10.1038/s41576-020-0118-8.

7. Mutz, K. O., Heilkenbrinker, A., Lönne, M., et al. (2013). Transcriptome analysis using next-generation sequencing. *Current Opinion in Biotechnology, 24*(1), 22–30. https://doi.org/10.1016/j.copbio.2012.09.004.

8. Robledo, D., Palaiokostas, C., Bargelloni, L., et al. (2018). Applications of genotyping by sequencing in aquaculture breeding and genetics. *Reviews in Aquaculture, 10*(3), 670–682. https://doi.org/10.1111/raq.12185.

9. Shendure, J., & Aiden, E. L. (2012). The expanding scope of DNA sequencing. *Nature Biotechnology, 30*(11), 1084–1094. https://doi.org/10.1038/nbt.2448.

10. Shendure, J., Balasubramanian, S., Church, G. M., et al. (2017). DNA sequencing at 40: Past, present and future. *Nature, 550*(7676), 345–353. https://doi.org/10.1038/nature24286.

CHAPTER

7

Molecular Diagnostics

LEARNING OBJECTIVES

After studying this chapter, you will be able to:

- Describe immunological diagnostic procedures;
- Understand the monoclonal antibodies;
- Define biofluorescent and bioluminescent systems;
- Focus on molecular diagnosis of genetic disease;
- Discuss the diagnostic techniques in molecular biology;
- Explain the applications of molecular diagnostic techniques.

KEY TERMS FROM THIS CHAPTER

Allele-specific oligonucleotide	Biomarker
CRISPR	Digital PCR (dPCR)
DNA microarray	Genotyping
In situ hybridization	Microarray
Molecular probe	Pathogen detection
Point-of-care testing	qPCR (quantitative PCR)
RT-PCR (reverse transcription PCR)	

INTRODUCTION

The study of biological markers at the molecular level, such as DNA, RNA, and proteins, to identify and track diseases, evaluate genetic predispositions, and inform treatment choices is the rapidly developing field of molecular diagnostics in medicine. This field uses cutting-edge methods and tools, such as microarrays, next-generation sequencing (NGS), and polymerase chain reaction (PCR), to pinpoint precise genetic mutations, variations, and expressions that may be a sign of a range of medical disorders. Molecular diagnostics offers an accurate and individualized approach to healthcare by analyzing these molecular markers. This allows for early disease detection, precise diagnosis, and customized treatment plans that enhance patient outcomes.

The domains of infectious diseases, genetic disorders, and oncology have demonstrated the unique importance of molecular diagnostics. Molecular diagnostics in oncology can detect genetic mutations that fuel the growth of cancer, opening the door to targeted therapies that attack cancer cells only, sparing healthy tissues as little as possible. Molecular testing has the ability to identify pathogens in infectious diseases with high sensitivity and specificity, even in the early stages of infection, allowing for prompt and effective treatment. Furthermore, these diagnostics can identify inherited conditions in the field of genetic disorders, facilitating early intervention and family planning choices.

7.1. IMMUNOLOGICAL DIAGNOSTIC PROCEDURES

Numerous immunological detection techniques are straightforward, precise, and sensitive. They have many uses, such as pathogen identification, drug testing, evaluation and tracking of different cancers, detection of particular metabolites, and infectious agent monitoring. There are restrictions, though. When using antibodies, for instance, it is necessary that the genes that contribute to the target site's presence be expressed and that the target site not be obscured or otherwise blocked so as to hinder the antibody's ability to bind to it. This is especially true if the target is a protein.

Generally speaking, conventional methods of diagnosing infectious agents depend on one of two things: either a distinct feature that is easily identified or a discrete set of characteristics that are typical of the pathogenic agent. The goal of the clinical microbiologist's search is to identify the fewest biological traits that can positively identify and positively identify a pathogenic agent. Certain infectious agents, for instance, generate unique biochemical molecules. The challenge lies in figuring out when the distinctive element appears in a biological specimen. A specialized biochemical assay that is highly specific for the marker molecule can often be used to directly identify such a molecule. This strategy has the drawback of possibly encouraging the development of numerous, highly customized detection systems for various harmful organisms. It is desirable to have a standardized technique for identifying any key marker molecule, regardless of its chemical makeup. Assays that are based only on identifying specific antibody–antigen complexes have eliminated the need to develop a unique identification process for each individual marker molecule because antibodies bind with high specificity to discrete target sites (antigens).

7.1.1. Enzyme-Linked Immunosorbent Assay (ELISA)

If an antibody has bonded to its target antigen, this can be ascertained in a variety of methods. ELISA, or enzyme-linked immunosorbent assay, is one technique that is commonly employed in diagnostic detection. There are two possible ELISA procedures: indirect (Figure 7.1A) and direct (Figure 7.1B). The following steps make up a generalized indirect ELISA protocol (Figure 7.1A):

Bind the sample being tested for the presence of a specific molecule or organism to a solid support, such as a plastic microtiter plate, which usually contains 96 sample wells. Wash the support to remove unbound molecules. Add a marker-specific antibody (primary antibody directed against the target antigen) to the bound material, and then wash the support to remove the unbound primary antibody. Add a second antibody (secondary antibody) that binds specifically to the primary antibody and not to the target molecule. Bound (conjugated) to the secondary antibody is an enzyme, such as alkaline phosphatase, peroxidase, or urease, that can catalyze a reaction that converts a colorless substrate into a colored product. Wash the mixture to remove any unbound secondary antibody–enzyme conjugate. Add the colorless substrate. Observe or measure the amount of colored product.

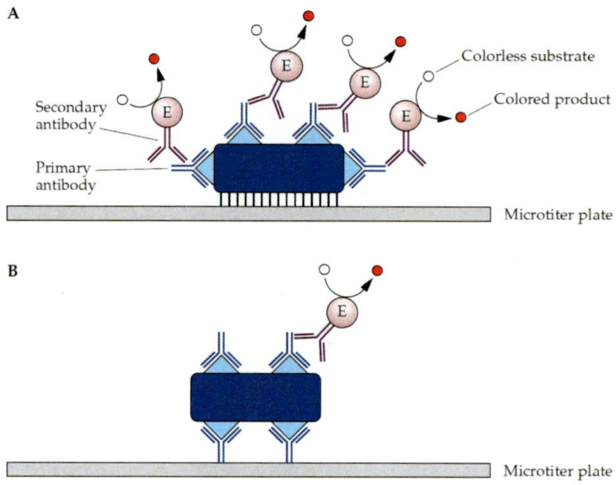

Figure 7.1. *Generalized ELISA protocol for detecting a target antigen. The primary antibody is often obtained from rabbits that have been immunized with the target antigen, while the secondary antibody is from goats immunized with rabbit antibodies. The enzyme (E) is conjugated to the secondary antibody. (A) Indirect ELISA; (B) direct ELISA.*

Source: Bernard R. Glick, Jack J. Pasternak, and Cheryl L. Patten.; Molecular biotechnology: principles and applications of recombinant DNA; ISBN 978-1-55581-498-4.

The second washing step eliminates the primary antibody from the sample if it is unable to bind to a target site. As a result, the final mixture is colorless and the secondary antibody–enzyme conjugate has nothing to bind to. It is thus eliminated during the third washing step. In contrast, if the target site is present in the sample, the attached enzyme catalyzes the reaction to form an easily detectable colored product as the primary antibody binds to it and the secondary antibody binds to the primary antibody. Every new diagnostic test just needs a single, distinct primary antibody because commercially available secondary antibodies that have been complexed with an enzyme are available. Additionally, the signal is amplified when multiple primary antibody molecules bind to multiple secondary antibody molecules, each of which is attached to multiple enzyme molecules.

A monoclonal antibody specific to the target antigen is first bound to the microtiter plate surface using a direct ELISA protocol (Figure 7.1B). The sample is put into the microtiter plate well and allowed to interact with the bound antibody to determine the concentration of a specific antigen in the sample. A wash is then performed to get rid of any unbound molecules. Before the presence of bound antigen being visible, the primary antibody and the secondary antibody conjugated to an enzyme are added. The primary antibody's specific binding to the target site is the main characteristic of an ELISA system. The purified protein preparation of the target molecule, for instance, is typically utilized in the production of the antibodies that will be employed to identify it. Several distinct antibodies that would each bind to a different antigenic determinant (epitope) on the target molecule make up the resultant antibody mixture, which is found in the serum (antiserum) of an inoculated animal, typically a rabbit. An amalgamation of these antibodies is referred to as a polyclonal preparation. The use of polyclonal antibodies for certain diagnostic assays has two limitations: (1) the concentrations of various antibodies in a polyclonal preparation may differ from batch to batch, and (2) polyclonal antibodies are unable to discriminate between two targets that are similar to one another, e.g., when there is only one determinant separating the pathogenic form (target) from the nonpathogenic one (nontarget). These issues can be resolved, though, because it is now feasible to produce a monoclonal antibody, which is an antibody preparation that targets just one antigenic determinant. Additionally, polyclonal antibody-based diagnostic assays are widely used for a variety of purposes despite these drawbacks.

7.2. MONOCLONAL ANTIBODIES

Mammals have developed an intricate web of cellular defenses against pathogen invasion and poisonous substances. Lymphatic system cells can be made to produce particular proteins, known as antibodies, as part of the defensive response. These antibodies bind to foreign substances, known as antigens, and, with the aid of other immune system proteins, such as the complement system, neutralize their biological effects. Each antibody-producing cell produces and secretes a single antibody in response to an immune challenge, which has the high affinity to recognize a specific region of the immunizing antigen known as the epitope, also known as the antigenic determinant. Since an antigen typically contains multiple distinct epitopes, multiple immune system cells will typically produce distinct antibodies directed against different antigen epitopes. A group of antibodies that react to the same antigen collectively is known as a polyclonal antibody (Figure 7.2).

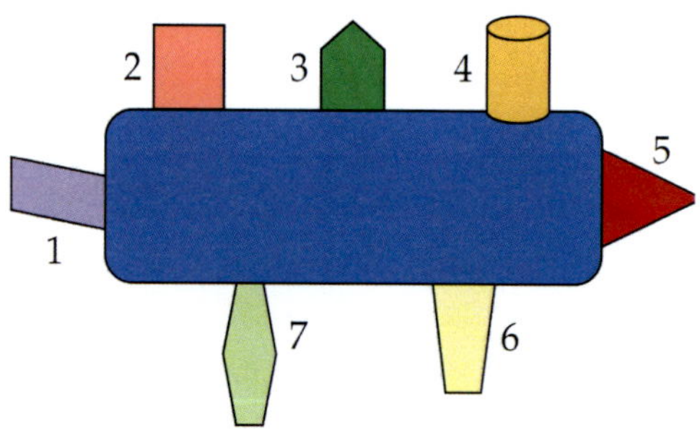

Figure 7.2. *Schematic representation of a target antigen. The surface of the antigen depicted has seven (numbered 1 to 7) different antigenic determinants (epitopes). When this antigen is used to immunize an animal, each antigenic determinant elicits the synthesis of a different antibody. Together, the different antibodies that interact with an antigen constitute a polyclonal antibody directed against that antigen.*

Source: Bernard R. Glick, Jack J. Pasternak, and Cheryl L. Patten.; Molecular biotechnology: principles and applications of recombinant DNA; ISBN 978-1-55581-498-4.

Despite the fact that the polyclonal nature of antibodies was not understood at the beginning of the 20th century, it was discovered that antibody specificity could be utilized to stop infections. Subsequently, antibodies were employed as diagnostic tools to identify harmful compounds in clinical samples. Regrettably, the efficacy of a polyclonal antibody preparation differs between batches due to the fact that, in certain vaccinations, specific antigenic determinants of a given antigen are potent stimulators of cells that produce antibodies, while in other vaccinations, the immune system reacts more vigorously to distinct epitopes of the same antigen. Additionally, different animals frequently react to a given antigen in different ways. Since different epitopes have varying potencies, this variation may have an impact on how well various preparations neutralize antigens. As a result, a single batch of polyclonal antibodies might not be as effective as a previous antibody preparation due to a low level of antibody molecules directed against a major epitope.

Consequently, developing a cell line that could be cultivated in culture and produce a single type of antibody molecule (monoclonal antibody) with a high affinity for a particular target antigen was a fundamental goal for the applied use of antibodies, whether as diagnostic tools or as components of therapeutic agents. A cell line of this kind would offer a steady supply of identical antibody molecules. The B lymphocytes, or B cells, responsible for producing antibodies regrettably cannot proliferate in a culture. Nonetheless, it was thought that this issue might be resolved by developing a hybrid cell type. This hybrid would have the cell division capabilities of a compatible cell type to allow the cells to grow in culture, as well as the B-cell genetic components for producing antibodies. It was known that healthy B cells could occasionally develop into cancerous cells called myelomas, which have the capacity to proliferate in culture while still carrying many characteristics of B cells. Myeloma cells were therefore candidates for fusion with B cells that produced antibodies, particularly those that did not produce antibody molecules. The mid-1970s saw the realization of these concepts.

7.2.1. Formation and Selection of Hybrid Cells

Mice are vaccinated with an antigen as the first step towards creating a hybrid cell line that generates a single antibody. The animals are tested, usually with an ELISA or a comparable device, after multiple inoculations spread over a few weeks to see if they have produced an immune response. If so, they are killed, and their spleens are taken out, cleaned, minced, and given a mild shake to liberate individual cells, some of which are B cells that produce antibodies. A suspension of myeloma cells lacking the enzyme hypoxanthine-guanine phosphoribosyl transferase ($HGPRT^-$) genetically is combined with the splenic cell suspension. After a brief period of mixing with 35% polyethylene glycol, the combined cell suspensions are moved to a growth medium that contains thymidine, aminopterin, and hypoxanthine (HAT medium).

The application of polyethylene glycol promotes cell fusion. However, the fusion events are sporadic and unpredictable. Within the mixture, there will be myeloma cells, spleen cells, myeloma–myeloma fusion cells, and spleen–spleen fusion cells. Nevertheless, none of the other cell types can proliferate in the HAT medium, so the only cells that can grow there are the myeloma-spleen fusion cells. No culture medium will allow spleen cells or spleen-spleen fusion cells to proliferate. Hypoxanthine is not able to be used by HGPRT-myeloma or myeloma-myeloma fusion cells as a precursor for the biosynthesis of purines, guanine and adenine, which are necessary for the synthesis of nucleic acids. They do, however, have a second purine biosynthesis pathway that occurs naturally and makes use of the enzyme dihydrofolate reductase. Since aminopterin inhibits the activity of dihydrofolate reductase, it is added to the medium. As a result, HGPRT-myeloma and myeloma-myeloma fusion cells perish in HAT medium due to their inability to synthesize purines (Figure 7.3).

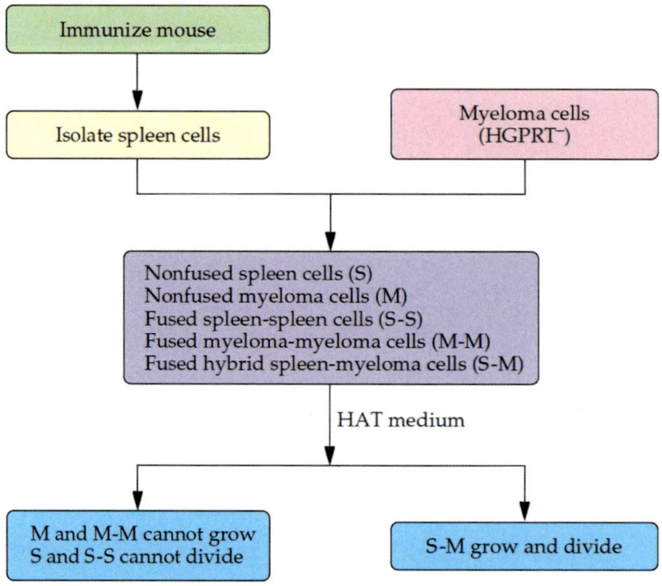

Figure 7.3. *The HAT procedure for selecting hybrid spleen–myeloma (hybridoma) cells.*

Source: Bernard R. Glick, Jack J. Pasternak, and Cheryl L. Patten.; Molecular biotechnology: principles and applications of recombinant DNA; ISBN 978-1-55581-498-4.

The myeloma cell's cell division functions are active, and the spleen cell contributes a functional HGPRT that can use the exogenous hypoxanthine in the medium even though aminopterin blocks purine production through dihydrofolate reductase, which is how the spleen–myeloma fusion cells survive in HAT medium. To overcome the block in pyrimidine production brought on by aminopterin's inhibition of dihydrofolate reductase, thymidine is supplied. After the fusion treatment, only spleen-myeloma fusion cells remained viable and proliferated in the HAT medium for approximately 10 to 14 days. After that, these cells are seeded into the plastic microtiter plate wells and grown on full culture medium without HAT.

7.2.2. Identification of Specific Antibody-Producing Hybrid Cell Lines

Finding the hybrid cells that make antibodies against the immunization antigen is the next step. A widely used screening method makes use of the culture medium, which has released antibodies in it. A well of a different microtiter plate that has been precoated with the target antigen is filled with the medium that was taken from the wells containing growing cells. A primary antibody present in the culture medium that identifies an epitope on the antigen will attach to the antigen and remain there after further washing. The test plate wells are filled with a second antibody, also known as a secondary antibody that is particular to mouse antibodies. Any primary antibody that is attached to the antigen will bind to it.

The secondary antibody is conjugated with an enzyme that can change a colorless substrate into a colored compound before use in the immunoassay. An antibody specific to the antigen was present in the original culture medium, as evidenced by the color in one of the test wells (Figure 7.4). The first wash will get rid of the primary antibody if there isn't an antibody in the culture medium that binds to the antigen. As a result, the second washing step removes the secondary antibody, which has nothing to bind to when it is added. The substrate solution in a well where this series of events takes place stays colorless.

A mixture of cell fusions may be present in those wells of the original microtiter plate whose media produce a positive (color) response in the immunoassay. To create cell lines from single cells, these cells are diluted with culture medium and seeded into new wells (clones). The media used to cultivate the clones is retested to identify which cell lines—also known as hybridomas, or spleen-myeloma cells—produce monoclonal antibody molecules that are capable of recognizing the target antigen. If multiple distinct hybridomas are identified, additional investigations are carried out to ascertain whether the distinct clones generate antibodies targeting the same antigenic determinant. Every clone that produces monoclonal antibodies can be kept in culture for a very long time. Furthermore, specimens can be preserved in liquid nitrogen to furnish a reservoir of cells for prospective application. Using a monoclonal antibody instead of a polyclonal one can significantly increase the specificity of an ELISA protocol because the former binds to a single discrete site.

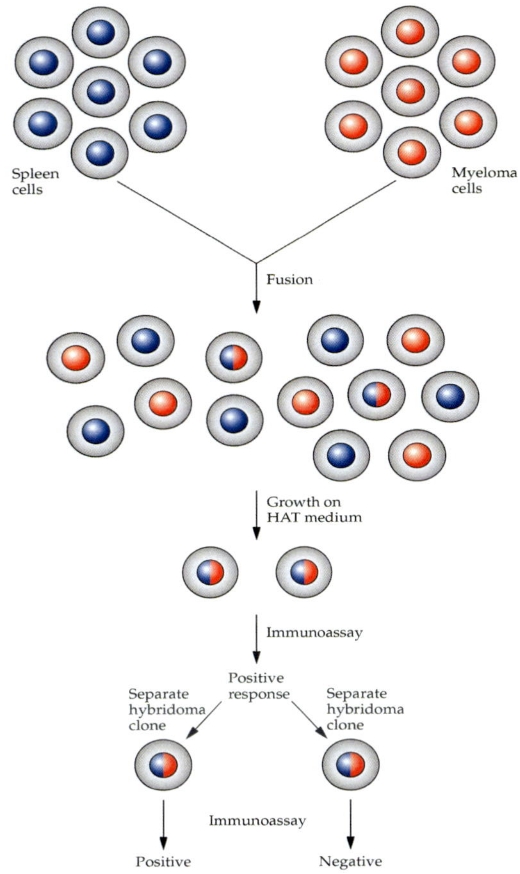

Figure 7.4. *Screening for the production of a monoclonal antibody. Spleen cells from a mouse that was immunized with a specific antigen are isolated and fused in culture with myeloma cells that do not produce antibodies. Fused cells are selected for the ability to grow on HAT medium, which contains hypoxanthine, aminopterin, and thymidine. Cells that produce a specific antibody to the immunizing antigen (hybridomas) are identified by an immunoassay and individually subcultured. A hybridoma, which grows in culture and secretes a single type of antibody molecules, is the source of a monoclonal antibody.*

Source: Bernard R. Glick, Jack J. Pasternak, and Cheryl L. Patten.; Molecular biotechnology: principles and applications of recombinant DNA; ISBN 978-1-55581-498-4.

For use as immunodiagnostic agents against a range of substances and pathogenic organisms, numerous monoclonal antibodies have been created (Table 7.1). It is possible to choose and produce monoclonal antibodies and portions of antibodies (Fab or Fv fragments) directed against a target antigen in *Escherichia coli* as an alternative to isolating and synthesizing monoclonal antibodies in hybridoma cells in culture.

Table 7.1. *Targets for Diagnostic Monoclonal Antibodies*

Polypeptide Hormones	Tumor markers	Cytokines	Drug Monitoring	Miscellaneous targets	Infectious disease
Chorionic gonadotropin	Carcinoe-mbryonic antigen	Interleukins 1–8	Theophylline	Thyroxine	Chlamydia
Growth hormone	Prostate-specific antigen	Colony-stimulating factor	Gentamicin	Vitamin B_{12}	Herpes
Luteinizing hormone	Interleukin-2 receptor		Cyclosporin	Ferritin	Rubella
Follicle-stimulating hormone	Epidermal growth factor receptor			Fibrin degradation products	Hepatitis B
Thyroid-stimulating hormone				Tau protein	Legionella
Prolactin					Human immunodeficiency virus

Source: Bernard R. Glick, Jack J. Pasternak, and Cheryl L. Patten.; Molecular biotechnology : principles and applications of recombinant DNA; ISBN 978-1-55581-498-4.

7.3. BIOFLUORESCENT AND BIOLUMINESCENT SYSTEMS

There are several applications for proteins that fluoresce or glow naturally or that are readily induced to do so as biological reporters. To engineer cells to produce a measurable signal in response to a specific chemical or physical agent in their environment, for instance, genes encoding these bioreporter proteins may be employed. In one implementation of this system, a fluorescent or luminescent protein gene is driven by a promoter that is responsive to specific environmental cues, resulting in the production of a fluorescent or luminescent signal upon promoter activation (Figure 7.5).

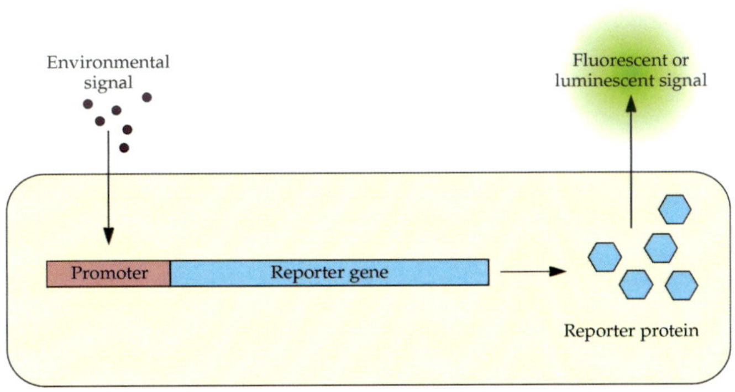

Figure 7.5. *Schematic representation of a bacterial cell, engineered to respond to a particular environmental compound, producing a detectable fluorescent or bioluminescent signal.*

Source: Bernard R. Glick, Jack J. Pasternak, and Cheryl L. Patten.; Molecular biotechnology: principles and applications of recombinant DNA; ISBN 978-1-55581-498-4.

7.3.1. Colored Fluorescent Proteins

7.3.1.1. Green Fluorescent Protein

When exposed to ultraviolet light, the 238 amino acid long photoprotein green fluorescent protein—which was isolated from the jellyfish Aequorea Victoria—fluoresces green. Green fluorescent protein, which is incorporated into cells, enables the real-time monitoring of intact living cells, despite the fact that many fluorescent dyes are phototoxic. Fluorescence microscopy has undergone a revolution thanks to the use of this reporter molecule. Green fluorescent protein has been used by researchers for a wide range of purposes, including tracking tumor cells in gene therapy protocols, evaluating the reactions of particular cell types to different therapeutic drugs and treatments, tracking protein-protein interactions, and tracking the outcomes of individual proteins in various therapies.

7.3.1.2. Red Fluorescent Protein

Scientists started looking for other colored fluorescent proteins in nature and through targeted mutagenesis after the successful discovery and use of green fluorescent protein. It would be possible to simultaneously monitor multiple biological processes if fluorescent proteins came in different colors. It is imperative that these proteins be as bright and stable as possible for practical applications. The propensity of many naturally occurring colored fluorescent proteins to form homodimers or homotetramers is one of their drawbacks. These multimeric structures may have a negative effect on the proteins' subcellular localization, which may result in intracellular aggregation and other anomalies. A team of scientists extracted a red fluorescent protein gene from Discosoma coral and, using a series of haphazard mutations, created a mutant that could only exist as a monomer as opposed to a tetramer. Proteins that produced bright, stable red fluorescence were chosen after each iterative cycle of random mutagenesis. 33 mutations were needed to produce monomeric red fluorescent protein. Despite its success, monomeric red fluorescent protein had a number of disadvantages over its native tetrameric form, including lower photostability and brightness.

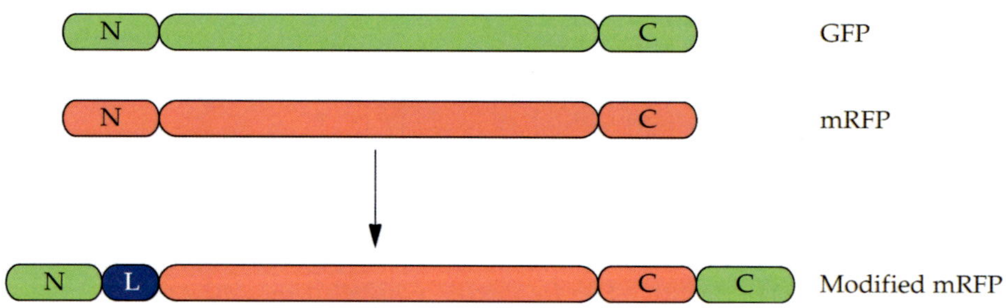

Figure 7.6. *Construction of a modified monomeric red fluorescent protein (modified mRFP). The regions of the gene encoding the N and C termini of the green fluorescent protein (GFP) were spliced onto the gene for mRFP following the removal of the portion of the mRFP gene encoding the N terminus and the addition of an oligonucleotide encoding a short linker peptide (L).*

Source: Bernard R. Glick, Jack J. Pasternak, and Cheryl L. Patten.; Molecular biotechnology: principles and applications of recombinant DNA; ISBN 978-1-55581-498-4.

Further rounds of mutagenesis were carried out to expand the repertoire of available fluorescent proteins and to address some of the issues with a monomeric red fluorescent protein that have been mentioned above. First, the red fluorescent protein gene was mutated to include DNA encoding seven amino acids from the green fluorescent protein's

N terminus (Figure 7.6). Subsequently, the red fluorescent protein gene was inserted with DNA corresponding to the six amino acids found in the green fluorescent protein's C terminus. Following that, this construct served as the basis for multiple rounds of directed evolution and random mutagenesis. Seven distinct monomeric colored fluorescent proteins were ultimately generated (Figure 7.7). There is a debate regarding the best colored fluorescent protein. Certain types differ in terms of brightness, photostability, and sensitivity to pH changes. As a result, different fluorescent proteins can be applied in different ways. Furthermore, the utility of this approach can be increased by labeling multiple different cellular components or cell types simultaneously using two or three of these proteins.

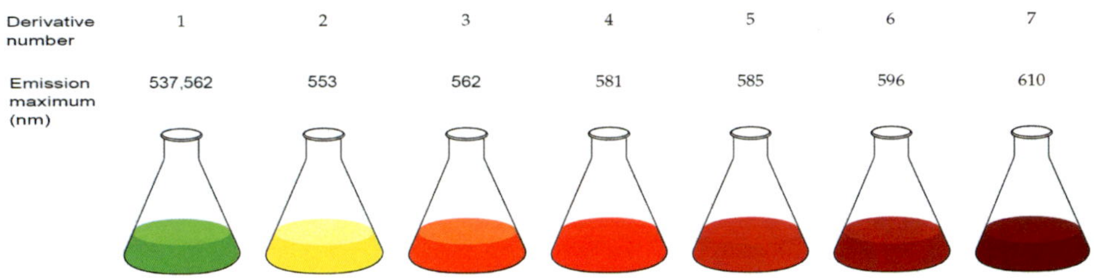

Figure 7.7. *Various colored monomeric fluorescent proteins derived from monomeric red fluorescent protein showing their emission wavelengths and color maxima following excitation. The colors have been called, from derivative 1 through 7, honeydew, banana, orange, tomato, tangerine, strawberry, and cherry, respectively.*

Source: Shaner et al., Nat. Biotechnol. 22:1567–1572, 2004.

7.3.2. Luciferase

Numerous different organisms, including bacteria, algae, fungi, jellyfish, insects, shrimp, and squid, can produce the luciferase enzyme, which catalyzes a reaction that emits light. The term "lux gene" refers to luciferase genes found in bacteria, whereas "Luc gene" refers to those found in other organisms, the most studied and used of which is the firefly. The lux system generates a light peak at 490 nm and consists of five genes, luxCDABE. All five lux genes are used in some applications to track the concentrations and presence of different environmental chemicals, such as metals (cobalt, copper, iron, lead, mercury, nickel, and zinc) or organic compounds (phenol, salicylate, benzene, trichloroethylene, ammonia, xylene, toluene, and ethylbenezene). The light-generating system doesn't need any additional chemicals when all five genes are utilized. Therefore, within a few minutes to a few hours after adding a contaminant-containing sample to bacterial cells carrying luxCDABE, a quantifiable amount of bioluminescence is produced. Sometimes, only the luxAB genes are present in the reporter system. The substrate decanal must be added to this system during the assay procedure, even though luxA and luxB work together to generate the light signal.

Light is produced at 550–575 nm via a reaction catalyzed by firefly luciferase. In addition, the system needs to include luciferin, an organic compound with a low molecular weight, as a substrate for the light reaction. Luc genes are typically employed in conjunction with eukaryotic cells.

7.3.3. Microbial Biosensors

Techniques that can quickly and simply identify the numerous potentially harmful substances that pollute the environment are required. Numerous extremely advanced analytical techniques are available to identify and quantify specific pollutants once the contaminated sites have been located and their range has been defined. Constitutively bioluminescent bacteria (i.e., bioluminescence does not need to be induced, in contrast to the previously described scenario) make them suitable candidates for pollutant detectors. Pollutants are clearly visible when they are present because the bioluminescence is reduced in their presence. Naturally bioluminescent bacteria are not helpful for testing terrestrial groundwater because they need saline conditions and a specific pH range, such as the marine bacterium *Vibrio fischeri*.

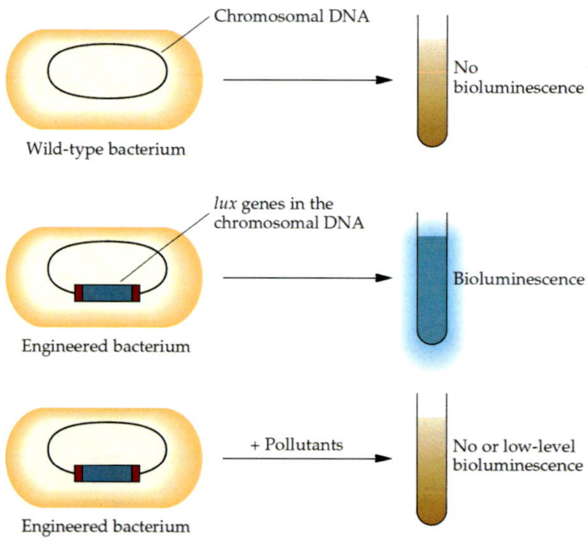

Figure 7.8. *Assaying for the presence of pollutants with genetically engineered bioluminescent P. fluorescens.*

Source: Bernard R. Glick, Jack J. Pasternak, and Cheryl L. Patten.; Molecular biotechnology: principles and applications of recombinant DNA; ISBN 978-1-55581-498-4.

However, soil bacteria like *Pseudomonas fluorescens* may have random insertions of structural genes encoding bioluminescence-causing enzymes (luxCDABE) into their chromosomal DNA. Since these genes lack a transcriptional promoter, the only luminescent colonies that can be seen in a darkroom after being inserted into the *P fluorescens* chromosomal DNA are those in which the lux genes are inserted downstream from a constitutive *P fluorescens* promoter (without interfering with any crucial bacterial genes). For testing with different environmental pollutants, the cells that luminesce the most and grow at a rate comparable to the wild-type strains are chosen. A suspension of bioluminescent *P fluorescens* is combined with the solution under test, and after a 15-minute incubation period, the luminescence of the suspension is measured in a luminometer to screen water samples for the presence of different pollutants (both metals and organic compounds) (Figure 7.8). The inhibition of cell luminescence in test samples containing low to moderate levels of specific pollutants is likely due to the pollutant's direct interference with bacterial metabolism. This process is a good first screen for determining whether pollution is present at a specific site because it is quick, easy, and affordable. Other techniques can be used to identify the real pollutants once a bacterial biosensor yields a positive result.

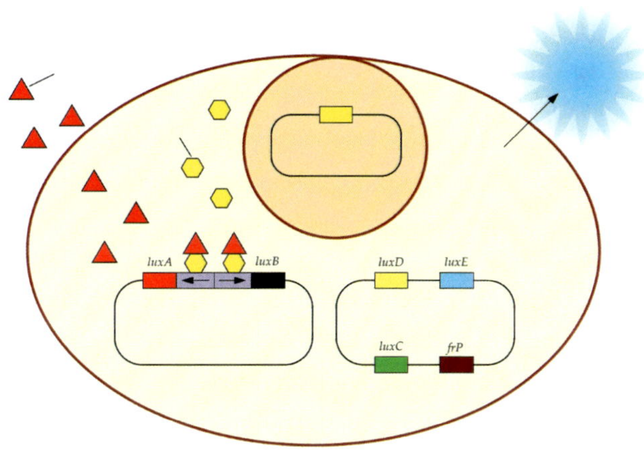

Figure 7.9. *Schematic representation of a strain of the yeast Saccharomyces cerevisiae that has been engineered to detect low levels of estrogenic substances in the environment. Each yeast cell contains the human estrogen receptor (hER) gene integrated into its chromosomal DNA, one plasmid (pUTK404) that contains constitutively expressed bacterial luxCDE genes and the flavin oxidoreductase (frp) gene from the bacterium Vibrio harveyi, and one plasmid (pUTK407) that contains luxA and luxB genes under the regulatory control of estrogen response elements (ERE) and constitutive bacterial promoters (not shown). Following the binding of the hER protein to an estrogenic compound, the complex binds to an ERE and activates luxAB transcription. Expression of the luxAB, luxCDE, and frp genes leads to the production of a measurable bioluminescent signal.*

Source: Sanseverino et al., Appl. Environ. Microbiol. 71:4455–4460, 2005; Bernard R. Glick, Jack J. Pasternak, and Cheryl L. Patten.; Molecular biotechnology: principles and applications of recombinant DNA; ISBN 978-1-55581-498-4.

It is estimated that about 87,000 distinct chemical compounds in the US require testing for estrogenic activity, i.e., steroid-like activity that has the potential to upset vertebrates' endocrine systems. These compounds can be tested using a variety of techniques, but most of them are too slow for this kind of extensive screening. As a result, researchers have created a straightforward and accurate method for quickly screening these 87,000 chemicals. Yeast (*Saccharomyces cerevisiae*) cells have been genetically modified to generate detectable amounts of light when exposed to incredibly low concentrations of estrogenic compounds (Figure 7.9). It could take as little as an hour to detect light production when testing for estrogenic compounds using this method, which induces luminescence. Furthermore, after 6 hours of incubation, the assay reached its peak bioluminescence when 5×10^{-11} M of the common estrogen 17β-estradiol was added to the engineered yeast cells. Naturally, larger concentrations were needed to identify many estrogenic compounds. This method's primary flaw is that the yeast cell wall and transport system allow certain substances to enter the cell more easily while preventing the absorption of other substances. This can distort the findings and, in certain cases, imply that a substance is not estrogenic if it cannot penetrate yeast cells with sufficient efficiency. However, many estrogenic compounds that were thought to be non-strogenic before being tested using this protocol may be found using this technique.

While some scientific teams are concentrating on developing and refining cells as biosensors, others have directed their attention toward automating these systems. An automated system may consist of genetically modified cells that produce blue-green light (~490 nm) in reaction to particular compounds, a light-tight enclosure, an environment that supports the cells, and an integrated circuit for signal processing and light detection. Even though some basic prototypes have been built, using living cells still presents significant challenges.

7.4. MOLECULAR DIAGNOSIS OF GENETIC DISEASE

It is now feasible to determine whether a person or their progeny is at risk for developing a particular inherited disease by using genetic diagnostics. DNA analysis can be used for early diagnosis before the onset of symptoms, prenatal diagnosis of serious genetic conditions, and identification of carriers of hereditary disorders.

Certain genetic mutations can only be definitively identified through DNA testing. Previously, biochemical assays that determined whether a gene product was present or absent were nearly exclusively used in genetic testing. However, expression of the mutant gene is not necessary for a DNA-based test to detect it, which means that screening assays for all single-gene diseases could potentially be developed.

7.4.1. Screening for Cystic Fibrosis

Genetic disease screening is frequently a complicated process. This reflects the fact that many diseases are caused by any one of a large number of genetic alterations to the normal DNA for that gene, rather than a single alteration to the wild-type DNA, as is the case with sickle-cell anemia. For example, mutations to the cystic fibrosis transmembrane conductance regulator (CFTR) gene cause defects in chloride ion transport, which leads to cystic fibrosis, one of the most common lethal autosomal recessive disorders in Europeans and their descendants, with an incidence of approximately 1 in every 2,500 live births and a carrier frequency of approximately 1 in 29. As of right now, there are about 1,400 known mutations in the CFTR gene that can cause cystic fibrosis. It is a difficult task to screen people who might be at risk for cystic fibrosis for 1,400 different mutations. Thankfully, compared to other mutations that cause cystic fibrosis, some of them are far more common. Indeed, more than 90% of patients with cystic fibrosis have at least one ΔF508 allele, and approximately 50% of cases of the disease are homozygous for ΔF508. It is estimated that screening for ΔF508 and the next 20 most common mutations should identify approximately 98% of carriers and affected individuals with cystic fibrosis, even though separate tests are needed for each mutation.

Several methods are used in the diagnostic testing for cystic fibrosis nowadays. Allele-specific oligonucleotide dot blots, also known as allele-specific hybridization, are among the most popular techniques. This method involves amplifying an individual's genomic DNA or cDNA via PCR, transferring the resultant material to a membrane, and then hybridizing the resultant material (individually) to labeled oligonucleotide probes for the mutant (typically ΔF508) and wildtype genes (Figure 7.10). This makes it feasible to differentiate between people who are healthy, people who carry cystic fibrosis, and people who have the disease (Figure 7.11). Using this method, the probe or the probe-target complex can be labeled in a number of ways, such as by using fluorescent dyes, radioactivity, or enzymes that change color when they interact with specific substrates. This method, which can identify 12 common and 17 uncommon cystic fibrosis mutations, can be automated and is presently offered for sale as a kit.

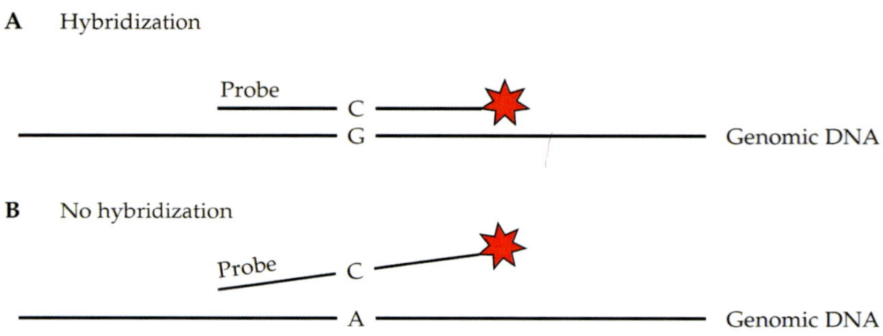

Figure 7.10. *Labeled oligonucleotide probe hybridizes to a completely complementary genomic DNA but, under stringent conditions (A), not to a DNA sequence in which one of the bases in the middle has been altered (B). In this case, the probe was designed to hybridize to wild-type DNA. Other probes, designed to hybridize to the genomic DNA of known mutations, would not bind to the wild-type DNA.*

Source: Bernard R. Glick, Jack J. Pasternak, and Cheryl L. Patten.; Molecular biotechnology: principles and applications of recombinant DNA; ISBN 978-1-55581-498-4.

A different approach based on the PCR amplification of particular alleles has been sold as a kit. With this protocol, multiple PCRs are carried out concurrently for every DNA sample; the primers anneal to distinct regions corresponding to distinct mutations. The existence of a specific size DNA band after amplification suggests the presence of a particular mutation.

Different-sized DNA fragments are usually separated in this situation using either capillary electrophoresis or gel electrophoresis. This is a very quick test that can identify many different types of mutations. It cannot, however, distinguish between homozygotes and heterozygotes, so further testing is necessary to ascertain whether a positive result indicates a carrier of the cystic fibrosis gene or an affected individual.One common method for finding mutations causing cystic fibrosis is the PCR/OLA technique.

This method has the highest detection rate of all the diagnostic tests for this disease and is regarded as being extremely accurate when compared to a great deal of other protocols. It also lends itself easily to automation. Even though all of the aforementioned procedures have been successful, researchers are still working to improve and develop these as well as other methods for diagnosing cystic fibrosis.

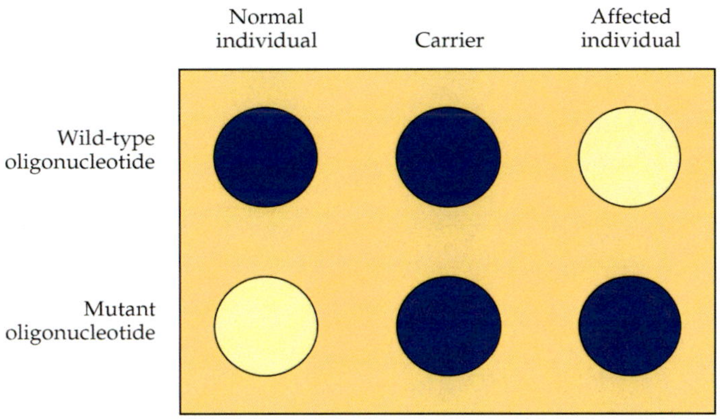

Figure 7.11. *Allele-specific oligonucleotide dot blot to diagnose individuals who are either carriers of a mutant CFTR gene (heterozygotes) or affected by the disease (homozygotes). The dark dot blot indicates that the labeled oligonucleotide has bound to the individual's DNA.*

Source: Bernard R. Glick, Jack J. Pasternak, and Cheryl L. Patten.; Molecular biotechnology: principles and applications of recombinant DNA; ISBN 978-1-55581-498-4.

7.4.2. Sickle-Cell Anemia

The genetic disease known as sickle-cell anemia arises from a single nucleotide alteration in the codon encoding the sixth amino acid of the β chain of the hemoglobin molecule. Because of a single amino acid shift from glutamic acid to valine, the hemoglobin molecule's conformation is distorted in individuals homozygous for the defect (S/S), resulting in irregular (sickle-shaped) red blood cells. This genetic mutation has severe anemia as well as progressive damage to the heart, lungs, brain, joints, and major organ systems as biological effects. Because of the mutation, the hemoglobin is unable to carry enough oxygen, which results in anemia. S/S homozygotes have a very short life expectancy. Individuals who are genetic carriers, or heterozygous (A/S), have red blood cells with a normal shape and do not exhibit any symptoms unless they are exposed to extreme weather or high altitudes that reduce their oxygen supply. The probability of a child having sickle-cell anemia is 25% if both parents are heterozygous, i.e., shall be homozygote for S/S. Black Africans and their descendants, as well as Hispanic populations, carry a high frequency of the sickle-cell anemia gene. In the US, carrier screening for the sickle-cell anemia gene is a regular practice aimed at identifying those who carry the gene and may pass it on to their progeny.

The CvnI restriction endonuclease site is eliminated due to a single-nucleotide alteration in the β-globin gene that accidentally causes sickle-cell anemia. This restriction enzyme cleaves the DNA between the C and the T after recognizing the sequence CCTNAGG. (The letter N designates that this position can be occupied by any one of the four nucleotides). The DNA sequence for the normal gene is CCTGAGG, but the sickle-cell anemia gene has CCTGTGG. A DNA diagnostic assay is based on this difference (Figure 7.12A).

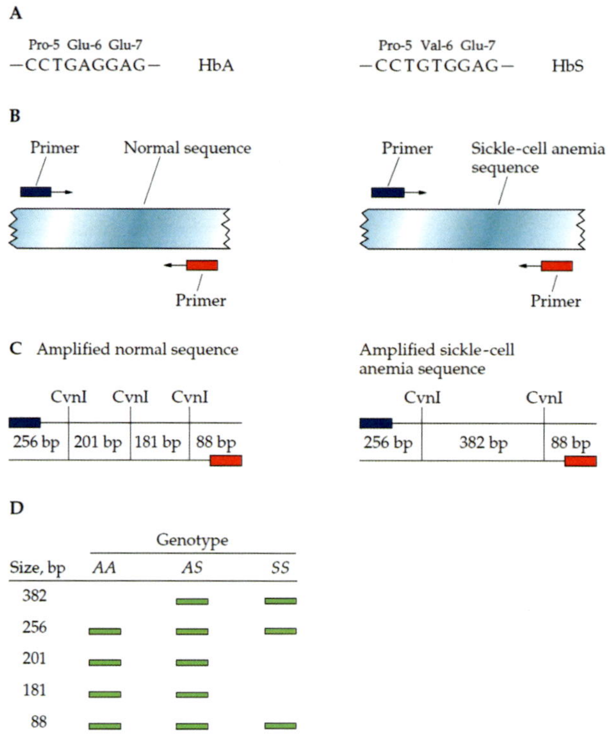

Figure 7.12. *Detection of the sickle-cell anemia gene at the DNA level.*

Note: (A) A portion of the sequence of the wild-type (HbA) and sickle-cell (HbS) human β-globin gene. The amino acids (numbered from the N-terminal end of the peptide chain) encoded by this portion of the DNA are shown above the DNA sequence. (B) PCR amplification of the portion of the β-globin gene containing the CvnI recognition site that is altered in the mutant gene. (C) CvnI digestion of the PCR products. The normal (wild-type) gene has three CvnI sites between the PCR primers, and the mutant gene has two. (D) Size distribution of fragments following gel electrophoresis of CvnI-digested PCR-amplified β-globin DNA. AA, homozygous condition for the normal β-globin gene; AS, heterozygous condition; SS, homozygous condition for the sickle-cell anemia β-globin gene.

Source: Bernard R. Glick, Jack J. Pasternak, and Cheryl L. Patten.; Molecular biotechnology: principles and applications of recombinant DNA; ISBN 978-1-55581-498-4.

7.4.3. Padlock Probes

An oligonucleotide that is complementary to a target sequence (DNA or RNA) at both of its 5′ and 3′ ends but not in the middle is known as a padlock probe (Figure 7.13). The middle portion of a padlock probe loops out when it hybridizes to its target sequence, while the 5′ and 3′ ends of the probe approach each other closely. DNA ligase can be used to join the 5′ and 3′ ends of the probe to one another after hybridization if the probe is precisely complementary to the target sequence. A high "specificity of detection" is ensured by the requirement that two sequences—actually, two ends of the same oligonucleotide—bind to the target. It is easy to detect allelic sequence variants because both sequences must perfectly hybridize to the target for DNA ligation to occur. No ligation takes place if there is a mismatch at either end. The middle (linker) portion of the padlock probe contains reporter molecules, like biotin or digoxigenin, which allow the probe-target hybrid to be detected after the ligation reaction. Padlock probes usually consist of a middle region of about 50 nucleotides and complementary sequences at the 5′ and 3′ ends that are roughly 15 to 20 nucleotides long. Because it is less complicated and requires fewer steps than the OLA procedure, this process has gained popularity among researchers. Furthermore, the process can be automated and only needs one oligonucleotide, as opposed to two for OLA.

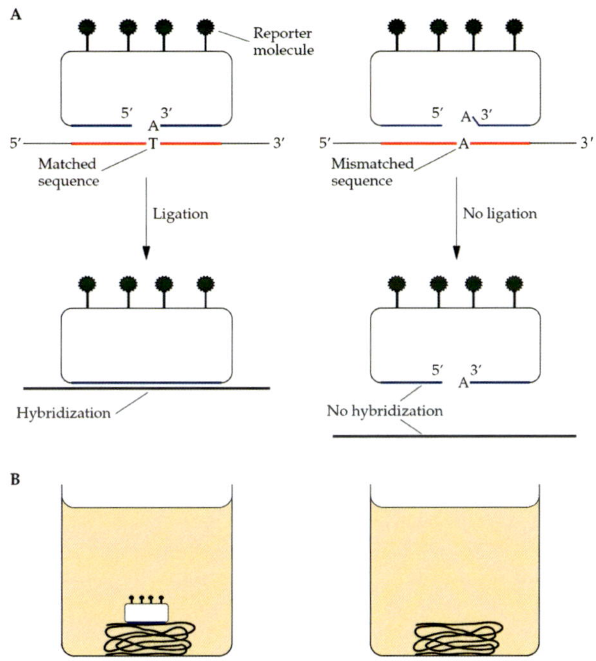

Figure 7.13. *Schematic representation of the functioning of a padlock probe.*

Note: (A) When the bases at the 5′ and 3′ ends of the probes are completely paired to the target DNA, ligation can take place. When there is a single-base mismatch at the 3′ end of the probe, ligation cannot occur and the probe assumes a conformation that does not allow hybridization. (B) Under stringent conditions, the ligated probe remains bound to the target DNA, which is bound to the surface of a 96-well microtiter plate. The nonligated probe is removed during washing. The bound probe is detected by interaction with the reporter molecules. If the reporter is biotin, then avidin and a biotinylated enzyme, such as alkaline phosphatase, are added sequentially. A colored well indicates that the probe is present and bound to the target DNA or RNA.

Source: Bernard R. Glick, Jack J. Pasternak, and Cheryl L. Patten.; Molecular biotechnology: principles and applications of recombinant DNA; ISBN 978-1-55581-498-4.

7.4.4. Genotyping with Fluorescence-Labeled PCR Primers

Color-based detection systems that are nonradioactive can be developed using PCR primers labeled with various fluorescent dyes. DNAs of the mutant and wild types are distinguished using PCR, which is carried out using two distinct primers. One has rhodamine (red) labeled at the 5′ end and is exactly complementary to the wild-type DNA. The other, which is complementary to the mutant DNA, has fluorescein (green) labeled at its 5′ end (Figure 7.14). A third, unlabeled primer that is complementary to the opposing strand controls amplification in both situations. The presence of these three primers in the same reaction mixture will result in the amplification of either the wild-type DNA, the mutant DNA, or both, depending on which target DNAs are initially present to act as PCR templates. This is because PCR amplification can only take place when the primer is exactly complementary to the target DNA. Following PCR and unincorporated primer removal, the reaction mixture will fluoresce red if an individual is homozygous for wild-type DNA, green if they are homozygous for mutant DNA, and green if they have both mutant and wild-type DNA (i.e., is heterozygous), yellow fluorescence will be seen in the reaction mixture. Any single-nucleotide target site of any sequenced gene can be used with this automated assay. This method's limitation lies in the fact that it can only identify SNPs. It is not possible to analyze multiple loci because a large number of cross-reactions between primer pairs could result from having a lot of different PCR primers in one reaction tube. This would lead to a lot of non-specific PCR products being formed.

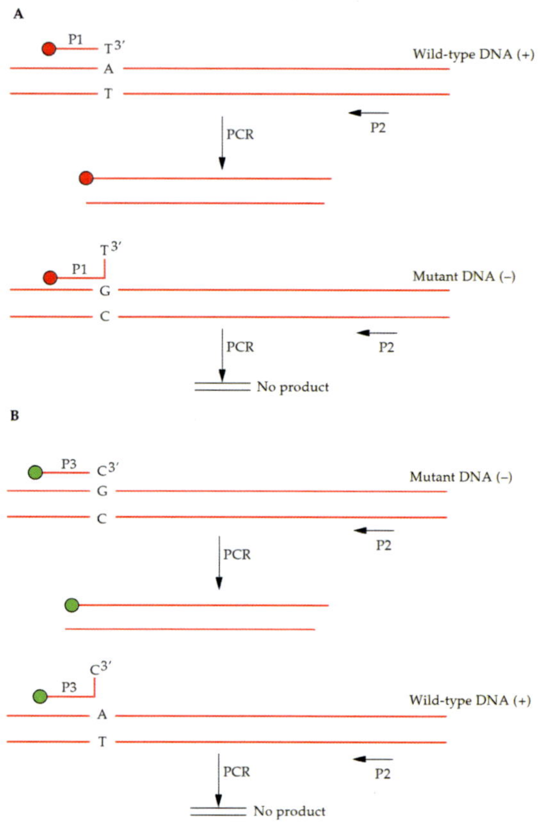

Figure 7.14. *Detection of a single-base mutation with fluorescence-labeled PCR primers.*

Note: (A) Primers P1 and P2 amplify DNA from the wild-type sequence. The same primers cannot amplify DNA from the mutant sequence because primer P1 is mismatched with this DNA. Primer P1 is labeled at its 5′ end with rhodamine (red). Primer P2 is unlabeled; (B) Primers P3 and P2 amplify DNA from the mutant but not the wild-type sequence. Primer P3 is labeled at its 5′ end with fluorescein (green). Primer P2 is unlabeled. The plus and minus signs denote wild-type and mutant sites, respectively. The genotypes 1/1, 1/2, and 2/2 produce PCR products that contain rhodamine only, rhodamine and fluorescein only and that fluoresce red, yellow, and green, respectively.

Source: Bernard R. Glick, Jack J. Pasternak, and Cheryl L. Patten.; Molecular biotechnology: principles and applications of recombinant DNA; ISBN 978-1-55581-498-4.

7.4.5. TaqMan Assay

Individuals are tested for the presence of SNPs indicative of any of a number of genetic diseases using the TaqMan protocol (Figure 7.15). It is based on the 5′ nuclease activity of Taq polymerase, which is frequently used to amplify DNA in PCR applications and was made popular by a specific company. Two TaqMan probes are used to monitor wild-type and mutant alleles at the same time. Every probe has a distinct fluorescent dye attached to its 5′ end and is precisely complementary to both the mutant and wild-type DNA sequences. Because the quencher molecule is present at the 3′ end of the probe, intact probes—whether bound or unbound to cDNA—do not fluoresce. The TaqMan probe is displaced by the growing DNA strand as PCR proceeds from primers flanking the probe hybridization site. The Taq polymerase's 5′ nuclease activity then degrades the TaqMan probe's 5′ end, releasing the fluorescent dye and removing it from the quencher molecule's vicinity. TaqMan probes will only degrade and fluoresce if they were previously bound to the target DNA. Any mismatched probes won't fluoresce because they will be displaced but not cleaved as a result of mutations in the TaqMan probe binding region. The detection of fluorescence at two distinct wavelengths—one for each TaqMan probe—allows for the differentiation of three types of organisms: heterozygotes, which possess one mutant gene and one wild-type gene, and homozygous individuals for the target mutation. Actually, two or three mutations can be assayed for simultaneously using this method. The technique can only be successfully applied if two conditions are met: (1) the target DNAs' exact DNA sequences must be known, and (2) the fluorescent dyes' fluorescence maxima must be well-separated and nonoverlapping.

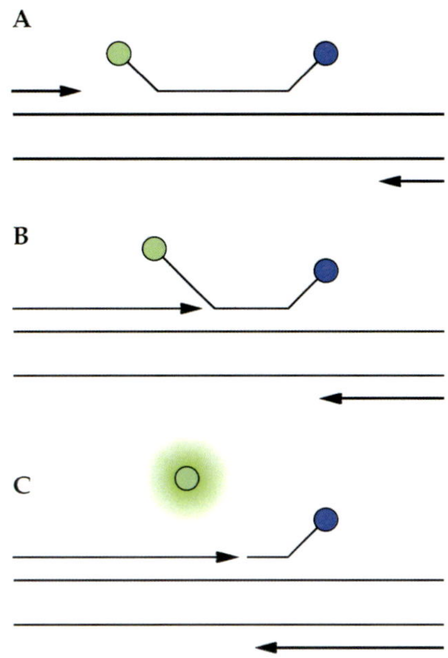

Figure 7.15. *TaqMan assay.*

Note: (A) A TaqMan probe that is complementary to the wild-type DNA is added before PCR amplification of the DNA sequence. The probe contains a fluorescent dye attached to its 5′ end (green) and a quencher attached to its 3′ end (blue). When the probe is intact, the quencher interacts with the fluorophore, quenching its fluorescence. The PCR primers are indicated by arrows; (B) In the extension phase of PCR, the TaqMan probe is displaced by the growing DNA strand; and (C) Subsequently, the 5′ fluorescent dye is cleaved from the probe by the 5′ nuclease activity of the Taq polymerase, leading to a dramatic increase in the fluorescence of the reporter dye (shown as a starburst of color from the dye).

Source: Bernard R. Glick, Jack J. Pasternak, and Cheryl L. Patten.; Molecular biotechnology: principles and applications of recombinant DNA; ISBN 978-1-55581-498-4.

7.5. DIAGNOSTIC TECHNIQUES IN MOLECULAR BIOLOGY

A variety of methods are employed in clinical microbiology labs to assist in clinical decision-making regarding the diagnosis, management, and prognosis of infections. There are two main methods for diagnosing. The gold standard is the pathogen's direct detection in clinical samples taken from the cases. Many techniques are employed, including polymerase chain reaction (PCR) pathogen genome detection, electron microscopy, and culture. Table 7.2 provides a summary of the main performance concerns. The identification of the patient's immune response to infection is frequently used to indirectly diagnose infections. One frequently used technique for verifying acute infection is the detection of pathogen-specific IgM. Clinical microbiology diagnostics nowadays depend more and more on molecular methods. The main elements influencing the clinical application of molecular tests will be the subject of this succinct review. The most popular methods are explained. The benefits and drawbacks are listed in Table 7.3.

Table 7.2. *Comparison of Laboratory Methods for Direct Detection of Infections*

Technique	Sensitivity	Turnaround	Availability
Culture	Moderate	> 3 days	Expensive Virus culture expertise disappearing, now little use in clinical laboratories for primary diagnosis Reference activity for antigenic monitoring, e.g. for influenza
Electron microscopy	Low	<1 day	Expensive Expertise disappearing, little used clinically Reference tests for 'new' infections

Antigen detection	Moderate	< 1 day	Laboratory-based tests for HIV, rotavirus available Wider use of point-of-care tests within clinics likely Use in clinical setting requires quality control
Direct immunofluorescence	Moderate	<1 day	Moderate cost Still in use for respiratory infections
Polymerase chain reaction	High	<1 day	Highly sensitive and specific, moderate cost Most widely used approach in clinical setting Used for quantification to guide treatment and to monitor progress Combined with sequencing to define antiviral sensitivity and for genotyping

Source: https://www.ncbi.nlm.nih.gov/pmc/articles/PMC7108329/.

Table 7.3. *Advantages and Disadvantages of Different Molecular Diagnostic Techniques*

Technique	Advantages	Disadvantages
PCR and its modifications	Suitable for the detection of wide variety of pathogens with various modifications Suitable for a wide variety of specimen materials In comparison to antibody or antigen detection far more sensitive and specific	Specimen contamination will lead to false-positive results More expensive than, e.g., serological methods Requires meticulous working techniques and quality control Requires special facilities and instrumentation

Isothermal amplification	In addition to PCR: Very efficient method NASBA is not affected by possible DNA contamination of the sample	In addition to PCR: Expensive
Fluorescence *in situ* hybridization	Allows the detection of nucleic acid directly from the sample without prior nucleic acid amplification Low cost Modest technical requirements The technical part of the analysis is quick Laboratory contamination of the sample is not an issue Usually very specific	Poor sensitivity Screening of slides slow, unless an expensive automated microscope is available
Genotyping	Suitable for identifying antimicrobial resistance and high-risk strains Direct sequencing provides accurate and extensive data Can be used for phylogenetic analyses	Direct sequencing requires expensive equipment and technical expertise Laborious method
Mass spectrometry	Suitable for typing Allows detection of both known and new mutations Very fast compared to DNA sequencing Suitable for analysis of large number of samples	Not suitable for the analysis of long oligonucleotides due to decreased sensitivity Cannot be used for large-scale genomic sequence analysis MALDI is susceptible to metal contaminants
Microarrays	Can detect both nucleic acids and antibodies Very good sensitivity Ability to detect several pathogens simultaneously High sample throughput	Too expensive to be adapted for routine diagnostics Usually requires nucleic acid amplification before analysis

Source: https://www.ncbi.nlm.nih.gov/pmc/articles/PMC7108329/.

7.5.1. Nucleic Acid Amplification Techniques

7.5.1.1. Cycling Amplification

7.5.1.1.1. Polymerase Chain Reaction (PCR)

The most popular molecular diagnostic methods for identifying infectious pathogens are the different types of PCR. Because of its exceptional sensitivity and specificity, PCR has revolutionized laboratory investigations in certain clinical situations, such as the diagnosis of viral encephalitis or the monitoring of cytomegalovirus (CMV) and Epstein-Barr virus (EBV) in organ transplant patients. In these cases, PCR is unquestionably the method of choice. Millions of identical DNA copies can be amplified from a small initial amount of pathogen genome in a clinical sample using PCR. The target DNA is essentially extracted first, then heated to a high temperature to denature it. After annealing specific oligonucleotide primers to the DNA at a lower temperature, the DNA polymerase enzyme copies the template strand during an extension phase. Millions of identical DNA copies are produced by repeating this cycle, typically 30 to 40 times. Refer to Figure 7.16 for PCR's basic principles.

The use of PCR in the diagnosis of bacteria, fungi, viruses, and parasites is growing. For a wide range of infections, including enteroviruses, parvoviruses, respiratory viruses, SARS-CoV, and poxviruses, as well as general or specific bacterial screening for bacteria like *Neisseria meningitidis*, *Streptococcus pneumoniae*, *Bordetella pertussis*, and *Borrelia* spp., qualitative PCR is widely used. Quantitative PCR can be used to measure nucleic acid copy numbers of various viruses, including CMV, EBV, HBV, HCV, and HIV, by comparing the sample's copy number to a control with a known copy level. Information on the course and outlook of the illness, as well as the efficacy of antiviral therapy, can be obtained through quantification.

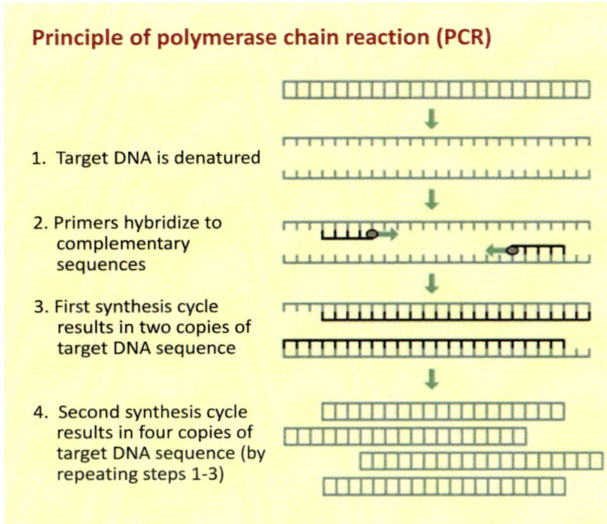

Figure 7.16. *Principle of polymerase chain reaction (PCR).*

Source: https://www.ncbi.nlm.nih.gov/pmc/articles/PMC7108329/.

Real-time PCR is the preferred technique these days. A fluorescent reporter, which provides signal in direct proportion to the amount of PCR amplicon, is used in real-time PCR. Following each heat cycle, the technique enables the real-time detection and quantification of PCR amplicons. Melting curve analysis is a technique used to distinguish between PCR amplicons based on their length and varying guanine-cytosine contents. This can be used in place of sequencing to identify the various pathogen strains or gene mutations linked to antibiotic resistance, e.g. Vancomycin-resistant enterococci and meticillin-resistant *Staphylococcus aureus*). When compared to traditional methods, real-time PCR is typically more sensitive and specific due to its high sample throughput and speed.

Multiplex format is an option for both real-time and conventional PCR, enabling the simultaneous detection of multiple pathogens. Nevertheless, optimizing sensitivity and specificity in multiplex PCR is typically more challenging, and the technique might be more favorable to some amplification products. Screening positive blood cultures for sepsis-causing bacterial pathogens and screening respiratory viruses from nasopharyngeal swabs are two clinical applications of multiplex PCR (Kim, Lee, Ki, 2009).

7.5.1.1.2. Ligase Chain Reaction (LCR)

LCR is a PCR modification. This technique involves the hybridization of two adjacent probes to a single target DNA strand. A highly specific thermostable DNA ligase recognizes the small gap between the two adjacent primers and ligates the DNA to form a single probe. The amplification procedure then uses the ligated products as templates. LCR is highly specific and only detects mutations involving a single base pair. The most popular use of LCR is for detecting Chlamydia trachomatis in urine and cervical specimens.

7.5.1.2. Isothermal Amplification

Isothermal amplification methods, in contrast to PCR, rely on a unique enzyme and don't need thermocycling, which involves repeated cycles of heating and cooling that can take several hours. Isothermal amplification can be completed quickly; it typically takes less than an hour. Nevertheless, the method is currently rather costly. Here, we go over some of the most well-known techniques for isothermal amplification.

Nucleic Acid Sequence-Based Amplification (NASBA): T7 RNA polymerase, RNaseH, and a primer containing a T7 promoter transform the target RNA into double-stranded DNA in the isothermal NASBA analysis (Figure 7.17). Using the DNA as a template, multiple copies of RNA with the target's opposite polarity can be produced. These RNA copies can then be used to create more DNA templates. If some minor adjustments are made in the initial stages of the process, NASBA can also be applied to DNA. A real-time multiplex NASBA assay for the detection of Mycoplasma pneumoniae, Chlamydophila pneumoniae, and *Legionella* spp. in respiratory specimens is a recent example of NASBA technology in clinical diagnostics.

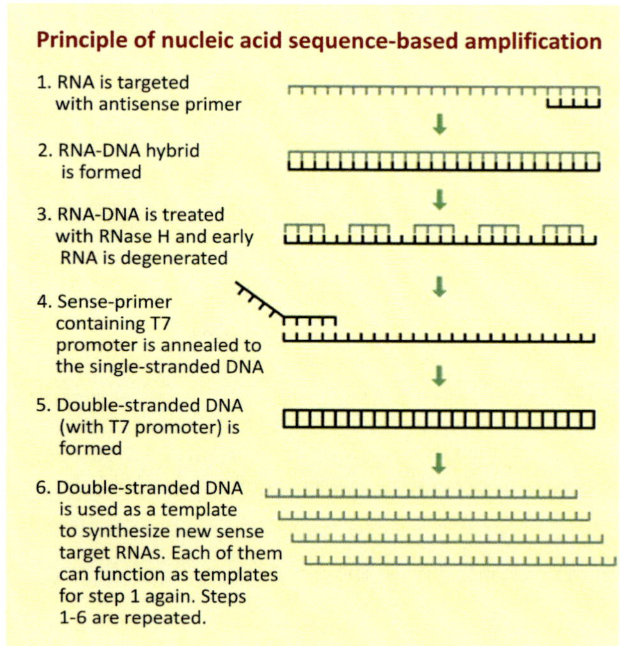

Figure 7.17. *Principle of nucleic acid sequence-based amplification.*

Source: https://www.ncbi.nlm.nih.gov/pmc/articles/PMC7108329/.

- **Transcription-Mediated Amplification (TMA):** The Theory of TMA shares a resemblance with NASBA, yet it employs the RNaseH function of reverse transcriptase instead of utilizing a distinct enzyme for this purpose. Since the emergence of West Nile virus (WNV) in the U.S. in 1999, screening for primary WNV infection in blood donors has been introduced. Currently, TMA is widely recognized as the most sensitive nucleic acid test commercially available for WNV infection and is extensively utilized for blood donor screening in the U.S. Another significant clinical application of TMA technology on a global scale is its use in detecting Chlamydia trachomatis and Neisseria gonorrhoeae in urine and urogenital swab samples.

- **Strand Displacement Amplification (SDA):** SDA is an additional technique for isothermal amplification. An annealed DNA template is used to anneal the first set of restriction site primers. The amplification process is then initiated by annealing second primers next to the first ones, which nicks the synthesized DNA. Next, the restriction enzyme HincII is added. As the newly synthesized strands are displaced, amplification is restarted by an exonuclease-deficient variant of the Klenow fragment of *Escherichia coli* DNA polymerase I. The amplification process of SDA is highly efficient, much like other isothermal amplification techniques. Using female endocervical swabs and male first-void urine samples, for instance, SDA is used to diagnose N. goorrhoeae infection.

7.5.2. Nucleic Acid Hybridization Techniques

7.5.2.1. Fluorescence in Situ Hybridization (FISH)

Without first undergoing nucleic acid amplification, the FISH technique enables the direct detection of microbial nucleic acid from the sample (or cultured sample). In summary, the method involves fixing the specimen on a microscope slide, hybridizing the prepared sample with a particular fluorescent-labeled probe, and using a fluorescent microscope to visually detect the hybridization. Medical microbiology employs this technique, e.g., in the identification of Helicobacter pylori from stomach biopsy specimens and the assessment of antibiotic resistance. Recognizing bacteria in CSF or blood cultures containing Staphylococcus aureus (Can, Yilmaz, and Demirbilek, 2005).

7.5.2.2. Line Probe Assay

Another nucleic acid hybridization test is called the line probe assay (LiPA), in which biotin-labeled PCR products are hybridized with particular oligonucleotide probes that are affixed at predetermined spots on a nitrocellulose strip as parallel lines. The quick identification of rifampicin resistance in Mycobacterium tuberculosis is one of the popular uses for LiPA. The detection of HBV mutations resistant to antiviral drugs is another use for the LiPA technique.

7.5.3. Microarrays

Microarrays are made up of a two-dimensional matrix of biomolecules printed or synthesized on a membrane made of nylon, silicon, glass, or plastic. Using an oligonucleotide or protein as an example, an immobilized biomolecule can be recognized by microarrays using both nucleic acids and antibodies that bind to it. Target labeling with fluorescent probes or antibodies allows highly sophisticated scanners to identify positive reactions. One of the microarray modifications is schematically illustrated in Figure 7.18. There are many different ways to modify microarrays: from immobilizing DNA probes in three-dimensional gel drops to detecting hybridization electrochemically without first amplification of the PCR product. Both serological microarrays and DNA are expected to be significant components of the technologies used in clinical microbiology labs in the future. Applications such as the identification of chloroquine-resistant Plasmodium falciparum, the detection of herpesvirus from CSF, and the identification of rifampicin- and isoniazid-resistant M tuberculosis have already been suggested and introduced.

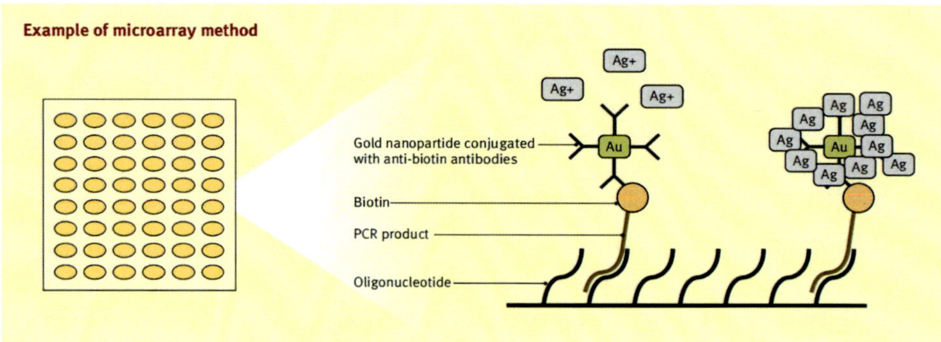

Figure 7.18. *This example is a modification of a microarray for the detection of human papilloma virus. In this example, biotin-labeled PCR products are hybridized with immobilized oligonucleotides. The chip is incubated in an anti-biotin antibody-conjugated gold nanoparticle solution. The silver enhancement by the gold nanoparticles bound to the biotin precipitates silver metal particles at the chip surfaces. This blocks light irradiated from above, which can be measured, allowing detection and quantitative analysis of the target DNA.*

Source: https://www.ncbi.nlm.nih.gov/pmc/articles/PMC7108329/.

7.5.4. Genotyping

The process of identifying variations in distinct pathogen strains or subtypes at the nucleic acid level is known as genotyping. Direct sequencing, reverse hybridization to genotype-specific probes, restriction fragment length polymorphism, and mass spectrometry detection of single nucleotide polymorphisms are the methods available for genotyping PCR amplicons. Since direct sequencing yields comprehensive and precise data, it is typically regarded as the gold standard. Since different genotypes of HCV infection respond to treatment differently—patients with genotypes 2 or 3 respond better to treatment than those with genotype 1 infection—genotyping is crucial to the management of HCV infection. For HCV genotyping, both direct sequencing and probe hybridization techniques are frequently employed. One instance of genotyping that is currently becoming commonplace is the typing of the human papilloma virus (HPV). Around the world, eight oncogenic HPV types (16, 18, 31, 33, 35, 45, 52, and 58) account for 80% of cervical cancers and intraepithelial neoplasia; however, regional variations are observed. Women with slightly abnormal Pap tests can benefit from early intervention, tailored clinical management, and risk assessment thanks to the oncogenic HPV type genotyping. In addition to its widespread application in molecular epidemiology, genotyping is also used to identify high-risk strains and antimicrobial resistance.

7.5.5. Mass Spectrometry

Mass spectrometry makes it possible to measure a sample's molecular mass, which is useful in a variety of clinical diagnostic contexts. With this method, the sample is first ionized, then the ions are separated based on their mass-to-charge ratios, and lastly the ion current is used to detect the separated ions. There are several methods for ionizing the sample, but one popular method for genotyping single nucleotide polymorphisms is matrix-assisted laser desorption/ionization (MALDI). The MALDI mass spectrometry is a valuable instrument for the typing of bacteria and viruses, e.g., in the identification of strains resistant to antibiotics or hypervirulent. The method has been made public, e.g., for HCV genotyping and quasispecies surveillance (a cloud of variant RNA viruses that develops over time within a viral isolate) in chronic HCV infection. Only five of the 13 serogroups are commonly linked to invasive disease; it is also utilized in the detection of hypervirulent N meningitidis strains. Recently, the global spread of novel influenza virus genotypes has been monitored using an approach based on electrospray ionization mass spectrometry followed by broad-range RT-PCR. (Urwin et al., 2004).

7.5.6. Key Factors Affecting Test Performance and Reliability

Microbiological test results are never conclusive, just like the results of any other laboratory test. The clinical significance of the test result should always be interpreted in light of the patient's clinical presentation. This also applies to molecular diagnostic techniques. Undoubtedly, the high sensitivity of PCR raises new challenges for the clinical interpretation of test findings. Both clinical and asymptomatic infections are caused by a variety of infectious agents; for instance, the norovirus, which causes gastroenteritis, frequently results in asymptomatic infections. It is challenging to assign causality to a single agent because mixed infections are common (5–10 percent), according to a systematic application of PCR to respiratory and enteric illnesses.

7.5.6.1. Sensitivity, Specificity, and Predictive Values

Predictive values—perhaps the most significant factor—along with specificity and sensitivity characterize a laboratory method's performance. The ability of the assay to correctly identify affected individuals is referred to as sensitivity, while the ability to correctly identify non-affected individuals is referred to as specificity. In molecular diagnostic techniques, false-positive and/or false-negative results can arise from inadequate assay performance, contaminated samples, incorrect specimen material, or improper sample collection timing. Furthermore, a positive result might not have any clinical significance in certain cases because of these techniques' low detection threshold.

The test's performance at the population level is indicated by the assay's positive and negative predictive values, i.e., the likelihood that someone who tests positively is actually impacted, and someone who tests negatively is likely not impacted at all. The test's predictive values are contingent not only on the assay's sensitivity and specificity but also on the disease's prevalence within the population. When used in high-risk groups, for instance, the performance of any given HIV antibody test differs from that of the general population; the positive predictive value of the test will decrease as the disease's prevalence decreases.

7.5.6.2. Contamination

One potential source of error is cross-contamination of the specimen, which can happen when sampling, when laboratory reagents get contaminated, or as a result of specimen-to-specimen contamination during sample processing. The primary concern with PCR is preventing contamination from the amplified products of PCR, which have a high concentration. This was a significant problem when PCR was initially introduced, and as a result, the various PCR steps' respective laboratory work areas have been physically separated to prevent contamination. Under ideal circumstances, different employees will never handle clean reagents and amplified products. It takes careful and aseptic handling procedures to prevent specimen-to-specimen contamination. Because real-time PCR requires less handling, the risk of contamination has decreased since its introduction. Frequent quality control of the assays is essential, and this includes continual trend analyses, internal positive and negative controls, and analysis of the pathogen's epidemiology.

7.5.6.3. Specimen Selection

Numerous specimens, such as serum, whole blood, cerebrospinal fluid (CSF), urine, genital swabs, respiratory swabs, stool, and various tissue biopsies, can be subjected to molecular analyses. The type of specimen and the pathology site must be taken into consideration when interpreting the results clinically. Finding a possible infection (e.g., rotavirus) in a meningitis patient's feces sample might not be noteworthy. The quantity of the pathogen is also crucial. For instance, the presence of CMV nucleic acid in blood indicates an ongoing infection, but it may not have any clinical significance because reactivations of the infection can also occur in healthy individuals. For this reason, quantitative CMV PCR is the preferred method for use in patients with compromised immune systems to direct treatment.

7.5.6.4. Specimen Collection

To prevent false-negative results, the specimen must be collected, handled, and delivered to the laboratory appropriately. The timeliness and quality of the sample collection process are crucial. Testing for C trachomatis in urine using PCR is a good illustration of this. Since shedding can be detected for much longer by PCR than by culture, too early post-treatment testing may lead to clinically false-positive results due to detectable traces of nucleic acid by PCR without presence of infectious pathogen. Inaccurate timing can also result in false-negative results. One example of this is when stool and respiratory samples are tested by PCR after the virus has already cleared, late in the illness process. Serological testing in this instance might aid in the diagnosis.

7.6. APPLICATIONS OF MOLECULAR DIAGNOSTIC TECHNIQUES

7.6.1. Prenatal Tests

Traditional prenatal diagnostics examine the chromosome count and appearance (karyotype). The availability of cell-free DNA in plasma has led to the adoption of noninvasive prenatal testing utilizing fetal DNA in mother plasma. Prenatal diagnosis through direct or indirect methods of detection has been made possible by the occurrence of disease mutations and inheritance patterns. While accuracy in indirect mutation analysis depends on the distance between the disease locus and the DNA marker, detection accuracy in direct mutation analysis is very accurate. For chromosomal abnormalities like Down syndrome, this has been used.

7.6.2. Infectious Diseases

Molecular diagnostics is used to identify particular strains of viruses, like the H1N1 virus, as well as infectious diseases like chlamydia, influenza, and tuberculosis. FISH and flow cytometry have been used to quickly and independently detect Salmonella species, and they have been used in conjunction with PCR to accurately detect Staphylococcus and Listeria species. A loop-mediated isothermal amplification test can be used to identify infectious agents genetically, as demonstrated by the diagnosis of the malaria parasite. Nevertheless, proteome, bacteriophage, or chromatographic profile analysis are still more frequently used to identify infections, even with these advancements in genome analysis. Drug resistance genes can be found using molecular diagnostics to determine the particular pathogen strain.

7.6.3. Disease Risk Management

A hereditary or random mutation in a patient's genome may alter their likelihood of contracting an illness in the future. For instance, Lynch syndrome is a hereditary condition that increases a patient's risk of developing colorectal and other cancers. Early identification of the condition can result in close monitoring, which raises the patient's prospects of success. Biological markers can indicate cardiovascular risk, and screening can assess the likelihood that a child will be born with a genetic condition like cystic fibrosis. It is complex from an ethical standpoint to test for genetics because patients might not want to know their risk. Insurance rates may increase for a known risk in nations without universal healthcare.

Many chromosomal aberrations, both stable and unstable, are caused by ionizing radiation (IR). Using standard chromosome staining methods, one can easily identify unstable aberrations, which are characterized by a significant compromise in chromosome morphology. FISH probes can target a specific sub-region or the entire chromosome(s). The technique can be used to identify the chromosome or specific DNA sequence(s) involved in a given aberration formation in addition to visualizing stable aberrations. Multiple fluorescence in situ hybridization (mFISH) and Spectral Karyotyping (SKY) are two highly sensitive cytogenetics techniques to identify inter-chromosomal stable aberrations that form in the bone marrow cells after exposure to total body irradiation. While fluorescently labeled DNA probes are used in both techniques, there are differences in the way the fluorescent signals are detected and acquired as images. Radiation biology, cancer cytogenetics, retrospective radiation biodosimetry, clinical cytogenetics, evolutionary cytogenetics, and comparative cytogenetics are among the research fields that have adopted the use of these two approaches. Variable FISH techniques have been found to bridge the gap between traditional chromosome banding analysis and molecular genetic DNA studies of risk factors by improving the thorough interpretation of numerical and complex chromosome aberrations (Hardenbol et al., 2003).

CASE STUDY

IMPLEMENTATION OF MOLECULAR DIAGNOSTICS FOR INFECTIOUS DISEASE MANAGEMENT

Background: XYZ Hospital, a major healthcare facility, is dedicated to improving the accuracy and speed of diagnosing infectious diseases. Traditional diagnostic methods, such as culture techniques and serological assays, often take several days to yield results and may lack sensitivity and specificity. To address these limitations, XYZ Hospital decides to implement molecular diagnostic techniques, which offer rapid, sensitive, and specific detection of pathogens.

Challenges

- **Delayed Diagnosis:** Traditional diagnostic methods can be time-consuming, leading to delays in initiating appropriate treatment for infectious diseases.
- **Sensitivity and Specificity:** Conventional methods may not detect low levels of pathogens or distinguish between closely related organisms, resulting in false negatives or false positives.
- **Antimicrobial Resistance:** The rise of antimicrobial resistance necessitates accurate and timely identification of resistant strains to guide effective treatment.
- **Operational Costs:** Implementing molecular diagnostics requires significant investment in equipment, training, and operational costs.

Solution: XYZ Hospital decides to adopt molecular diagnostic techniques, particularly polymerase chain reaction (PCR) and next-generation sequencing (NGS), to enhance the detection and management of infectious diseases.

Implementation Steps

- Infrastructure Development:
  - **Laboratory Setup:** The hospital invests in modern laboratory infrastructure, including PCR machines, sequencers, and biosafety equipment, to facilitate molecular testing.
  - **Staff Training:** Comprehensive training programs are conducted for laboratory personnel to ensure proficiency in molecular diagnostic techniques and data interpretation.

Diagnostic Workflow Integration:

- **Sample Collection and Handling:** Standardized protocols for sample collection, handling, and processing are established to maintain the integrity of specimens.

- **Molecular Testing Implementation:** PCR and NGS tests are integrated into the diagnostic workflow for rapid identification of pathogens. Tests include multiplex PCR for simultaneous detection of multiple pathogens and NGS for comprehensive genomic analysis.

Quality Control and Assurance:

- **Validation and Calibration:** All molecular diagnostic tests undergo rigorous validation to ensure accuracy, sensitivity, and specificity. Regular calibration and maintenance of equipment are performed.
- **Internal and External Controls:** Internal controls are included in each test run, and participation in external quality assurance programs ensures ongoing accuracy and reliability.

Clinical Application and Impact:

- **Rapid Diagnosis:** Molecular diagnostics significantly reduce the time to diagnosis, allowing for quicker initiation of appropriate treatments and improving patient outcomes.
- **Antimicrobial Stewardship:** Accurate identification of pathogens and their resistance profiles supports targeted antimicrobial therapy, reducing the use of broad-spectrum antibiotics and combating antimicrobial resistance.
- **Outbreak Management:** The hospital uses molecular diagnostics for rapid detection and containment of infectious disease outbreaks, enhancing public health response capabilities.

Results

- **Improved Diagnostic Accuracy and Speed:** The adoption of molecular diagnostics reduces the time to diagnosis from days to hours, significantly improving the accuracy and speed of pathogen detection.
- **Enhanced Patient Outcomes:** Rapid and precise diagnosis enables timely and targeted treatment, leading to better patient outcomes, shorter hospital stays, and reduced healthcare costs.
- **Antimicrobial Resistance Management:** By accurately identifying resistant strains, the hospital implements more effective antimicrobial stewardship programs, curbing the spread of resistance.
- **Efficient Outbreak Control:** Molecular diagnostics facilitate rapid identification and containment of infectious disease outbreaks, protecting public health and preventing widespread transmission.

Conclusion: The implementation of molecular diagnostics at XYZ Hospital transforms the approach to diagnosing and managing infectious diseases. By leveraging advanced technologies such as PCR and NGS, the hospital achieves faster, more accurate, and comprehensive detection of pathogens. This enhances patient care, supports antimicrobial stewardship, and strengthens the hospital's capacity to respond to infectious disease outbreaks. The investment in molecular diagnostics proves to be a valuable and impactful advancement in the hospital's diagnostic capabilities and overall public health strategy.

CLASS ACTIVITY

"Explore the role of Molecular Diagnostics in identifying genetic markers for diseases. Discuss how techniques like PCR and DNA sequencing revolutionize medical diagnostics by detecting genetic variations quickly and accurately."

SUMMARY

- Molecular Diagnostics is the application of molecular biology techniques to study human diseases, including infectious disease, inherited conditions and cancer. Molecular diagnostics research is focused on genomic and proteomic analysis with the goal of identifying disease biomarkers, creating better diagnostic assays and ultimately finding new treatments and potential cures. Techniques used in molecular diagnostic applications include core molecular biology methods such as nucleic acid isolation and quantification, PCR amplification, sequencing, and STR analysis.
- Many immunological detection methods are sensitive, specific, and simple. They can be used for a wide range of applications, including drug testing, assessment and monitoring of various cancers, detection of specific metabolites, pathogen identification, and monitoring infectious agents.
- ELISAs have been used for detecting various proteins, identifying viruses and bacteria, and determining the presence of low-molecular-weight compounds in a wide range of biological samples. To increase the specificity of the primary antibody and to ensure the reliability of the antibody preparation, monoclonal antibodies are often used for diagnostic ELISAs.
- The polyethylene glycol treatment facilitates fusion between cells. Nevertheless, the fusion events are rare and random. There will be myeloma cells, spleen cells, myeloma–spleen fusion cells, myeloma– myeloma fusion cells, and spleen–spleen fusion cells in the mixture. The HAT medium, however, allows only the myeloma–spleen fusion cells to grow, because none of the other cell types can proliferate in this medium. Unfused spleen cells and spleen–spleen fusion cells cannot grow in any culture medium.
- The luciferase enzyme, which catalyzes a light-emitting reaction, may be produced by a variety of different organisms, including bacteria, algae, fungi, jellyfish, insects, shrimp, and squid. Luciferase genes from bacteria are termed lux genes, while those from other organisms—the most widely studied and utilized being the firefly—are termed luc genes.
- The ability to diagnose the occurrence of specific inherited diseases in humans at the genetic level makes it possible for individuals to discover whether they or their offspring are at risk. DNA analysis can be used for the identification of carriers of hereditary disorders, for prenatal diagnosis of serious genetic conditions, and for early diagnosis before the onset of symptoms.
- The PCR/OLA procedure is also commonly used to detect cystic fibrosis mutations. This technique is considered to be highly accurate compared too many other protocols and has the highest detection rate of any of the diagnostic tests for this disease. Moreover, it is readily amenable to automation. Notwithstanding the success with all of the procedures mentioned above, researchers continue to refine and develop these and other approaches to the diagnosis of cystic fibrosis.

- Sickle-cell anemia is a genetic disease that is the result of a single-nucleotide change in the codon for the sixth amino acid of the β chain of the hemoglobin molecule. In individuals homozygous for the defect (S/S), the shape of the red blood cells is irregular (sickle shaped) because the conformation of the hemoglobin molecule is distorted by a single amino acid change from glutamic acid to valine.
- The TaqMan protocol is used to check individuals for the presence of SNPs that are indicative of any of a variety of genetic diseases. Made popular by one particular company, it is based on the 5′ nuclease activity of Taq polymerase, which is commonly used to amplify DNA in PCR applications.

REVIEW QUESTIONS

1. Briefly describe how the change in the human β-globin gene that gives rise to sickle-cell anemia can be detected by using PCR.
2. What is an ELISA? How does it work?
3. For diagnostic assays, what is meant by sensitivity, specificity, and simplicity?
4. What are the main types of ELISA, and how does each type differ in terms of procedure and applications?
5. Explain the difference between biofluorescence and bioluminescence. Provide examples of organisms or systems where each phenomenon is utilized.
6. What is polymerase chain reaction (PCR), and how is it used in molecular diagnostics?
7. Describe the process of real-time PCR and its advantages over conventional PCR.
8. What is the principle behind nucleic acid amplification techniques like PCR, LAMP, and NASBA?

REFERENCES

1. Barker, R. H., Suebsaeng, L., Rooney, W., Alecrim, G. C., Dourado, H. V., & Wirth, D. F. (1986). Specific DNA probe for the diagnosis of *Plasmodium falciparum* malaria. *Science, 231*(4743), 1434–1436. https://doi.org/10.1126/science.3520020.
2. Campbell, R. E., Tour, O., Palmer, A. E., Steinbach, P. A., Baird, G. S., Zacharias, D. A., & Tsien, R. Y. (2002). A monomeric red fluorescent protein. *Proceedings of the National Academy of Sciences, 99*(12), 7877–7882. https://doi.org/10.1073/pnas.112205299.
3. Can, F., Yilmaz, Z., & Demirbilek, M. (2005). Diagnosis of *Helicobacter pylori* infection and determination of clarithromycin resistance by fluorescence *in situ* hybridization from formalin-fixed, paraffin-embedded gastric biopsy specimens. *Canadian Journal of Microbiology, 51*(7), 569–573. https://doi.org/10.1139/w05-048.
4. Carlson, D. P., Superko, C., Mackey, J., Gaskill, M. E., & Hansen, P. (1990). Chemiluminescent detection of nucleic acid hybridization. *Focus, 12*(1), 9–12.

5. Chehab, F. F., & Kan, Y. W. (1989). Detection of specific DNA sequences by fluorescence amplification: A color complementation assay. *Proceedings of the National Academy of Sciences, 86*(22), 9178–9182. https://doi.org/10.1073/pnas.86.22.9178.

6. Erlich, H. A., Gelfand, D., & Sninsky, J. J. (1991). Recent advances in the polymerase chain reaction. *Science, 252*(5013), 1643–1651. https://doi.org/10.1126/science.1900565.

7. Eshaque, B., & Dixon, B. (2006). Technology platforms for molecular diagnosis of cystic fibrosis. *Biotechnology Advances, 24*(1), 86–93. https://doi.org/10.1016/j.biotechadv.2005.12.005.

8. Hardenbol, P., Banér, J., Jain, M., Nilsson, M., Namsaraev, E. A., Karlin-Neumann, G. A., Fakhari-Rad, H., Ronaghi, M., Willis, T. D., Landegren, U., & Davis, R. W. (2003). Multiplexed genotyping with sequence-tagged molecular inversion probes. *Nature Biotechnology, 21*(6), 673–678. https://doi.org/10.1038/nbt817.

9. Kim, S. R., Ki, C. S., & Lee, N. Y. (2009). Rapid detection and identification of 12 respiratory viruses using a dual priming oligonucleotide system-based multiplex PCR assay. *Journal of Virological Methods, 156*(1–2), 111–116. https://doi.org/10.1016/j.jviromet.2008.10.014.

10. Sayler, G. S., & Layton, A. C. (1990). Environmental application of nucleic acid hybridization. *Annual Review of Microbiology, 44*, 625–648. https://doi.org/10.1146/annurev.mi.44.100190.003205.

11. Urwin, R., Russell, J. E., Thompson, E. A., Holmes, E. C., Feavers, I. M., & Maiden, M. C. (2004). Distribution of surface protein variants among hyperinvasive meningococci: Implications for vaccine design. *Infection and Immunity, 72*(10), 5955–5962. https://doi.org/10.1128/IAI.72.10.5955-5962.2004.

12. Weitz, H. J., Ritchie, J. M., Bailey, D. A., Horsburgh, A. M., Killham, K., & Glover, L. A. (2001). Construction of a modified mini-Tn5 luxCDABE transposon for the development of bacterial biosensors for ecotoxicity testing. *FEMS Microbiology Letters, 197*(2), 159–165. https://doi.org/10.1111/j.1574-6968.2001.tb10688.x.

CHAPTER

8 Protein Synthesis

LEARNING OBJECTIVES

After studying this chapter, you will be able to:

- Discuss translation of mRNA;
- Explain the protein structure;
- Define protein folding and processing;
- Explain the cell free protein synthesis.

KEY TERMS FROM THIS CHAPTER

5' cap	Amino acid	Anticodon
Elongation	Initiation complex	Introns
mRNA	Peptide bond	Poly-a tail
Ribosome	RNA polymerase	RNA processing
Splicing	Termination	Transcription
Translation	tRNA	

INTRODUCTION

Protein synthesis is a fundamental biological process that allows cells to build specific proteins. It involves a series of intricate steps that begin with the transcription of DNA and conclude with the translation of mRNA into a functional protein.

The process initiates in the nucleus of a eukaryotic cell, where DNA, the genetic blueprint of an organism, resides. Here, a section of the DNA containing the gene that codes for a particular protein is transcribed into messenger RNA (mRNA). This step is facilitated by enzymes that unwind the DNA double helix and construct a complementary strand of mRNA according to the genetic code.

Once the mRNA is synthesized, it undergoes processing to ensure its integrity and functionality. This includes the addition of a 5' cap and a poly-A tail, as well as the removal of non-coding sequences called introns, leaving only the protein-coding exons. The processed mRNA then exits the nucleus and travels to the cytoplasm, where protein synthesis continues.

In the cytoplasm, the mRNA binds to a ribosome, the cellular machinery responsible for translation. Transfer RNA (tRNA) molecules bring specific amino acids to the ribosome, guided by the mRNA sequence. The sequence of nucleotides on the mRNA, known as codons, dictates the order in which amino acids are added to the growing polypeptide chain.

As each tRNA molecule delivers its amino acid, the ribosome catalyzes the formation of peptide bonds between adjacent amino acids. This process continues until a stop codon is reached on the mRNA, signaling the ribosome to release the completed polypeptide chain. This chain then folds into its functional three-dimensional structure, guided by various cellular factors.

Ultimately, protein synthesis is vital for the growth, maintenance, and function of cells and organisms. It ensures that specific proteins, each with unique structures and functions, are produced in the right place and at the right time, contributing to the complexity and diversity of biological systems.

8.1. TRANSLATION OF MRNA

The process by which proteins are made from mRNA templates has been remarkably conserved over the course of evolution. Polypeptide chains are synthesized from the amino to the carboxy terminus, and all mRNAs are read from 5′ to 3′. A nearly universal genetic code specifies each amino acid using three bases (a codon) in the mRNA. All cells use the same basic mechanisms for synthesizing proteins: ribosomes are used for translation, and transferrin (tRNA) acts as an intermediary between the mRNA template and the amino acids that are incorporated into the protein. Thus, interactions between three different RNA molecules—mRNA templates, tRNAs, and rRNAs—as well as different proteins that are necessary for translation are involved in protein synthesis.

8.1.1. Transfer RNAs

All 20 amino acids must line up with their matching codons on the mRNA template in order for translation to occur. Numerous tRNAs act as adaptors for this process in every cell. It makes sense that various tRNAs would have comparable general structures considering their shared role in protein synthesis. They do, however, also have distinct identifying sequences that enable the right amino acid to be joined and aligned with the right codon in mRNA.

Due to complementary base pairing between distinct molecule regions, transfer RNAs exhibit distinctive cloverleaf structures and range in length from 70 to 80 nucleotides (Figure 8.1). All tRNAs fold into comparable compact L shapes, as further demonstrated by X-ray crystallography studies. These folds are probably necessary for the tRNAs to fit onto ribosomes during the translation process. Two distinct areas of the molecule are involved in the tRNAs' adaptor function. Amino acids are covalently linked to the ribose of the terminal adenosine of all tRNAs, which have the sequence CCA at their 3′ terminus. The anticodon loop, which is found at the opposite end of the folded tRNA, subsequently identifies the mRNA template by complementary base pairing with the appropriate codon.

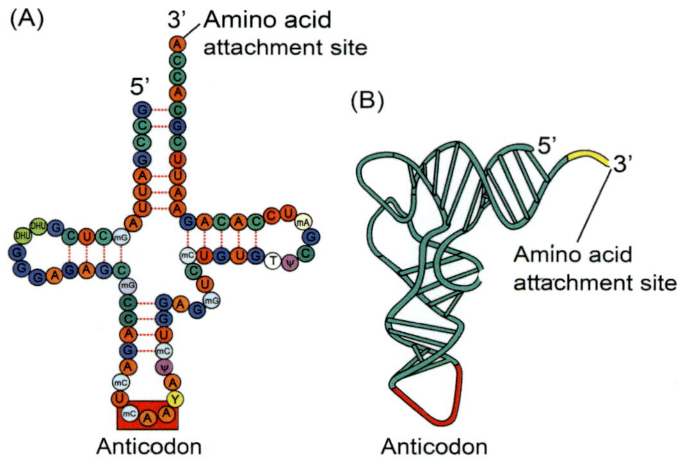

Figure 8.1. *The structure of tRNA is illustrated in open "cloverleaf" form (A) to show complementary base pairing. Modified bases are indicated as mG, methylguanosine; mC, methylcytosine; DHU, dihydrouridine; T, ribothymidine; Y, a modified purine (usually adenosine); and ⊠, pseudouridine. The folded form of the molecule is shown in (B) and a space-filling model.*

Source: https://courses.lumenlearning.com/suny-microbiology/chapter/structure-and-function-of-rna/.

The specificity of codon-anticodon base pairing and the attachment of each amino acid to the proper tRNA are necessary for the integration of the appropriately encoded amino acids into proteins.

Aminoacyl tRNA synthetases are a class of enzymes that Paul Zamecnik and Mahlon Hoagland identified in 1957. They mediate the attachment of amino acids to particular tRNAs.

These enzymes are each capable of identifying a single amino acid and the appropriate tRNA (or tRNAs) that it should be attached to. Two steps are involved in the reaction (Figure 8.2). Initially, an aminoacyl AMP synthetase intermediate is formed by the amino acid's reaction with ATP. The 3′ terminus of the tRNA is then connected to the activated amino acid.

The highly selective enzymes known as aminoacyl tRNA synthetases are required to identify the correct acceptor tRNAs by recognizing specific base sequences in addition to individual amino acids.

A proofreading mechanism that causes incorrect aminoacyl AMPs to hydrolyze instead of joining tRNA during the reaction's second step contributes to the high fidelity of amino acid recognition. The aminoacyl tRNA synthetase is likewise very selective in identifying the correct tRNA; it recognizes particular nucleotide sequences that, in most cases, include the anticodon and uniquely identify each species of tRNA.

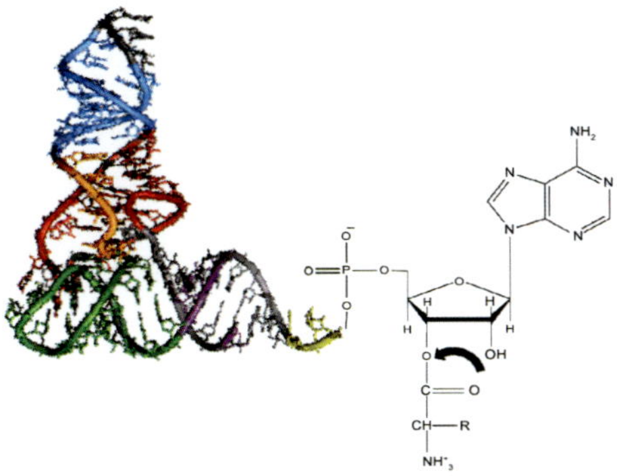

Figure 8.2. *Attachment of amino acids to tRNAs.*

Source: By Scientific29 – Own work, CC BY-SA 3.0, https://commons.wikimedia.org/w/index.php?curid=23570164.

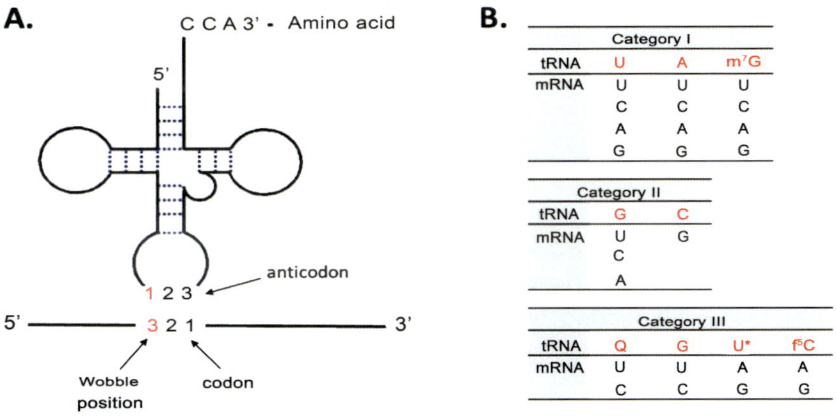

Figure 8.3. *Base pairing at the third codon position is relaxed, allowing G to pair with U, and inosine (I) in the anticodon to pair with U, C, or A. Two examples of abnormal base pairing, allowing phenylalanyl (Phe) tRNA to recognize either UUC or UUU codons and alanyl (Ala) tRNA to recognize GCU, GCC, or GCA, are illustrated.*

Source: Bezerra AR, Guimarães AR, Santos MAS. Non-Standard Genetic Codes Define New Concepts for Protein Engineering. Life. 2015; 5(4):1610–1628. https://doi.org/10.3390/life5041610.

After being attached to tRNA, an amino acid is aligned on the mRNA template by complementary base pairing between the mRNA codon and the anticodon of the tRNA. The significance of this unusual base pairing in codon-anticodon recognition relates to the redundancy of the genetic code. Of the 64 possible codons, three are stop codons that signal the termination of translation; the other 61 encode amino acids. Thus, most of the amino acids are specified by more than one codon. In part, this redundancy results from the attachment of many amino acids to more than one species of tRNA. *E. coli*, for example, contain about 40 different tRNAs that serve as acceptors for the 20 different amino acids. In addition, some tRNAs are able to recognize more than one codon in mRNA, as a result of nonstandard base pairing (called wobble) between the tRNA anticodon and the third position of some complementary codons (Figure 8.3). Relaxed base pairing at this position results partly from the formation of G-U base pairs and partly from the modification of guanosine to inosine in the anticodons of several tRNAs during processing. Inosine can base-pair with either C, U, or A in the third position, so its inclusion in the anticodon allows a single tRNA to recognize three different codons in mRNA templates.

8.1.2. The Ribosome

In both prokaryotic and eukaryotic cells, the sites of protein synthesis are called ribosomes. Ribosomes were originally identified as particles that were found through the ultracentrifugation of cell lysates. They are typically categorized based on the rate at which they sediment, with bacterial ribosomes being 70S and eukaryotic cells' somewhat larger ribosomes being 80S. Prokaryotic and eukaryotic ribosomes are made up of two different subunits, each of which has rRNAs and distinctive proteins. The fact that cells normally have a large number of ribosomes illustrates how crucial protein synthesis is to cell metabolism. In *E. coli,* about 20,000 ribosomes, or roughly 25% of the dry weight of the cell, while approximately 10 million ribosomes are found in rapidly proliferating mammalian cells.

The small subunit (designated 30S) of *E. coli* ribosomes consists of the 16S rRNA and 21 proteins; the large subunit (50S) is composed of the 23S and 5S rRNAs and 34 proteins. Each ribosome contains one copy of the rRNAs and one copy of each of the ribosomal proteins, with one exception: One protein of the 50S subunit is present in four copies. The subunits of eukaryotic ribosomes are larger and contain more proteins than their prokaryotic counterparts have. The small subunit (40S) of eukaryotic ribosomes is composed of the 18S rRNA and approximately 30 proteins; the large subunit (60S) contains the 28S, 5.8S, and 5S rRNAs and about 45 proteins.

The ability of ribosomes to form *in vitro* through the self-assembly of their protein and RNA components is a notable characteristic. Purified ribosomal proteins and rRNAs can be combined, as Masayasu Nomura originally reported in 1968, and under the right circumstances, will reform a functional ribosome. While ribosome assembly *in vivo* is significantly more complex, especially in eukaryotic cells, the ability of ribosomes to self-assemble *in vitro* has been a valuable tool for experiments, enabling the examination of the functions of individual proteins and rRNAs.

Similar to tRNAs, rRNAs use complementary base pairing to create distinctive secondary structures. The rRNAs fold into unique three-dimensional structures when they interact with ribosomal proteins. At first, it was believed that rRNAs had a structural function, acting as a platform for the assembly of ribosomal proteins. But as more RNA molecules were found to possess catalytic activity, the potential catalytic function of rRNA started to receive attention. It was discovered that rRNAs are absolutely necessary for the *in vitro* assembly of functional ribosomes, supporting this theory. Conversely, ribosome activity was reduced but remained intact when numerous ribosomal proteins were absent.

The first direct proof of rRNA's catalytic activity came from studies conducted in 1992 by Harry Noller and associates. These researchers showed that even after standard protein extraction procedures have removed about 95% of the ribosomal proteins, the large ribosomal subunit can still catalyze the formation of peptide bonds (the peptidyl transferase reaction). On the other hand, RNase treatment totally prevents the formation of peptide bonds, offering compelling evidence in favor of the theory that RNA catalyzes this process. These findings have been validated and expanded upon by additional research, which shows that synthetic 23S rRNA fragments can catalyze the peptidyl transferase reaction even in the complete lack of ribosomal protein. Thus, ribosomal RNA catalyzes the basic reaction of protein synthesis. It is now believed that ribosomal proteins, rather than being the main catalytic components of ribosomes, aid in correct rRNA folding and improve ribosome function by correctly arranging the tRNAs.

Significant evolutionary ramifications result from rRNA involvement in the peptidyl transferase reaction. It is believed that RNAs were the first macromolecules to replicate themselves. The ability of ribozymes to catalyze reactions involving RNA substrates, such as RNase P and self-splicing introns, to support this idea is compelling. The function of rRNA in peptide bond formation expands the range of RNA's catalytic activities beyond self-replication and includes direct participation in protein synthesis. Further research supports the idea that the original aminoacyl tRNA synthetases were RNAs rather than proteins since it shows that the Tetrahymena rRNA ribozyme can catalyze the attachment of amino acids to RNA. The capacity of RNA molecules to catalyze the processes necessary for both self-replication and protein synthesis may offer a crucial link in our understanding of the early evolution of cells (Dobson, 1999; Hartl and Hayer-Hartl, 2002)

8.1.3. The Organization of mRNAs and the Initiation of Translation

Prokaryotic and eukaryotic cells use similar mechanisms for protein synthesis, but they also differ, especially in terms of the signals that dictate the locations at which polypeptide chain synthesis starts on an mRNA template (Figure 8.4). Translation starts at particular initiation sites rather than at the 5′ end of the mRNA.

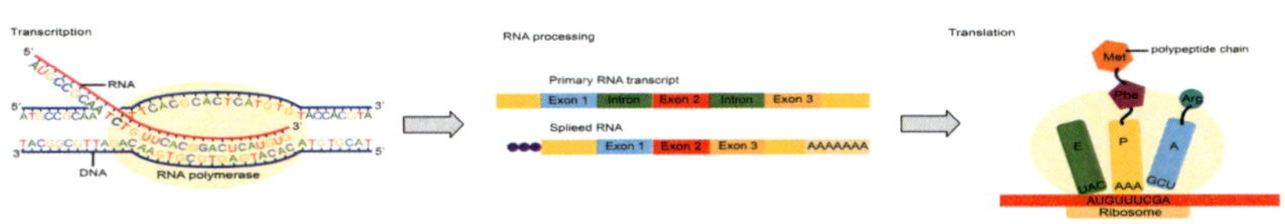

Figure 8.4. *Prokaryotic and eukaryotic mRNAs.*

Source: https://courses.lumenlearning.com/wm-biology1/chapter/the-central-dogma/.

Therefore, the 5′ terminal regions—also known as 5′ untranslated regions—of both prokaryotic and eukaryotic mRNAs are noncoding sequences. Prokaryotic mRNAs often encode multiple polypeptides that are synthesized independently from different initiation sites, whereas eukaryotic mRNAs typically only encode one polypeptide chain. For instance, the *E. coli* three genes make up the lac operon, and they are all translated from the same mRNA. While monocistronic messenger RNAs only encode one polypeptide chain, polycistronic messenger RNAs encode multiple polypeptides. Ultimately, noncoding 3′untranslated regions are the last stop on both prokaryotic and eukaryotic mRNAs. Translation in prokaryotic and eukaryotic cells always begins with the amino acid methionine, which is typically encoded by AUG. Bacteria occasionally use alternative initiation codons, like GUG. However, when these codons appear at the start of a polypeptide chain, they direct the incorporation of methionine instead of the amino acid they normally encode (GUG normally encodes valine). While unmodified methionines initiate protein synthesis in eukaryotes (apart from mitochondria and chloroplasts, whose ribosomes resemble those of bacteria), modified methionine residues (N-formylmethionine) initiate protein synthesis in the majority of bacteria.

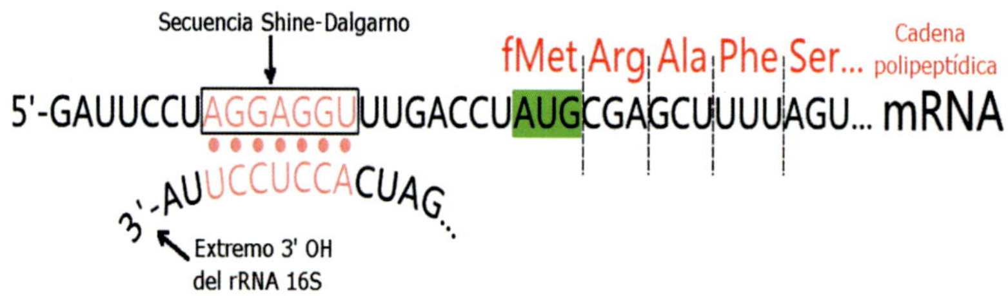

Figure 8.5. *Initiation sites in prokaryotic mRNAs are characterized by a Shine-Delgarno sequence that precedes the AUG initiation codon. Base pairing between the Shine-Delgarno sequence and a complementary sequence near the 3′ terminus of 16S rRNA aligns the mRNA on the ribosome. In contrast, eukaryotic mRNAs are bound to the 40S ribosomal subunit by their 5′ 7-methylguanosine caps. The ribosome then scans along the mRNA until it encounters an AUG initiation codon.*

Source: https://commons.wikimedia.org/wiki/File:Shine-Dalgarno_sequence-es.png.

The signals that distinguish initiation codons in prokaryotic and eukaryotic cells differ, which is in line with the different roles that monocistronic and polycistronic mRNAs play (Figure 8.5). The sequence known as a Shine-Delgarno sequence, named for its discoverers, comes before the initial codons in bacterial mRNAs. This sequence aligns the mRNA on the ribosome for translation by base-pairing with a complementary sequence close to the 3′ terminus of 16S rRNA. Because of this base-pairing interaction, bacterial ribosomes can start translation at the internal initiation sites of polycistronic messages as well as at the 5′ end of an mRNA. On the other hand, ribosomes attach to the 5′ terminus of most eukaryotic mRNAs to recognize them. After that, the ribosomes continue to travel downstream of the 5′ cap until they come across an AUG initiation codon. The efficiency of initiation is influenced by sequences surrounding AUGs, so translation frequently starts at an AUG downstream rather than the first one in the mRNA. But unlike prokaryotic mRNAs, eukaryotic mRNAs lack a sequence similar to the Shine-Delgarno sequence. Eukaryotic mRNAs function as monocistronic messages that encode only single polypeptides, and translation of these messages begins at a site that is determined by scanning from the 5′ terminus.

8.1.4. The Process of Translation

The three stages of translation are initiation, elongation, and termination (Figure 8.6). The first stage of the initiation stage occurs in both prokaryotes and eukaryotes when a particular initiator, methionyl tRNA, and the mRNA bind to the small ribosomal subunit. The polypeptide chain then continues to elongate on the functional ribosome that is formed when the large ribosomal subunit joins the complex. The different phases of the translation process also require many distinct nonribosomal proteins (Table 8.1).

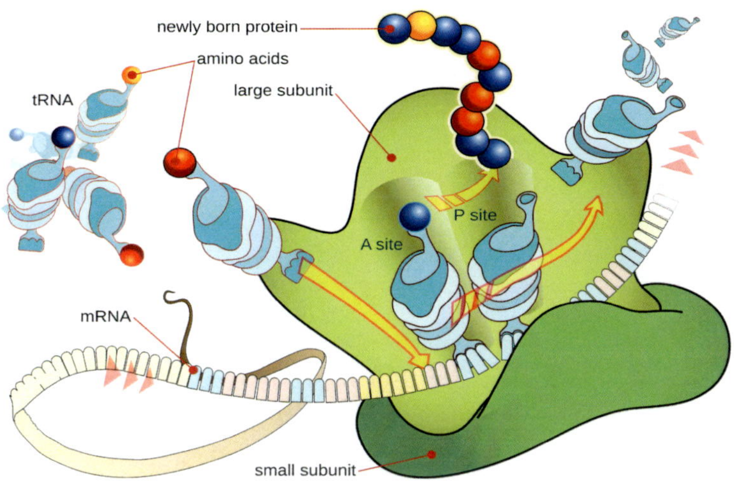

Figure 8.6. *Overview of translation.*

Source: https://courses.lumenlearning.com/suny-wmopen-biology1/chapter/translation/.

Table 8.1. *Translation Factors*

Species	**aa-tRNA incorporation**			**tRNA translocation**		**Other translation elongation factors**				
Bacteria	**EF-Tu**	**EF-Ts**	**SelB**	**EF-G**	**EF4**	**EF-P**	—	**Tet(O)/(M)**	**RelA**	**BipA**
Archaea	aEF1A	—	—	aEF2	—	aIF5A	—	—	—	—
Eukaryotes	eEF1A	eEF1B	EFsec	eEF2	mtEF4	eIF5A	eEF3	—	—	—
Mitochondria	mtEF-Tu	mtEF-Ts	—	mtEF-G1	mtEF4	—	—	—	—	—

Source: Xu B, Liu L and Song G (2022) Functions and Regulation of Translation Elongation Factors. Front. Mol. Biosci. 8:816398. doi: 10.3389/fmolb.2021.816398.

Three initiation factors (IF-1, IF-2, and IF-3) bind to the 30S ribosomal subunit to initiate translation in bacteria (Figure 8.7). The complex is then formed by the mRNA and initiator N-formylmethionyl tRNA, with the initiator tRNA being specifically recognized by IF-2, a GTP-bound protein. Following the release of IF-3, a 50S ribosomal subunit can attach itself to the complex. Because of this association, GTP bound to IF-2 hydrolyzes, releasing IF-1 and IF-2 (bound to GDP). As a result, during the elongation stage of translation, a 70S initiation complex is formed, ready to start forming peptide bonds. This complex is composed of initiator tRNA and mRNA that are bound to the ribosome.

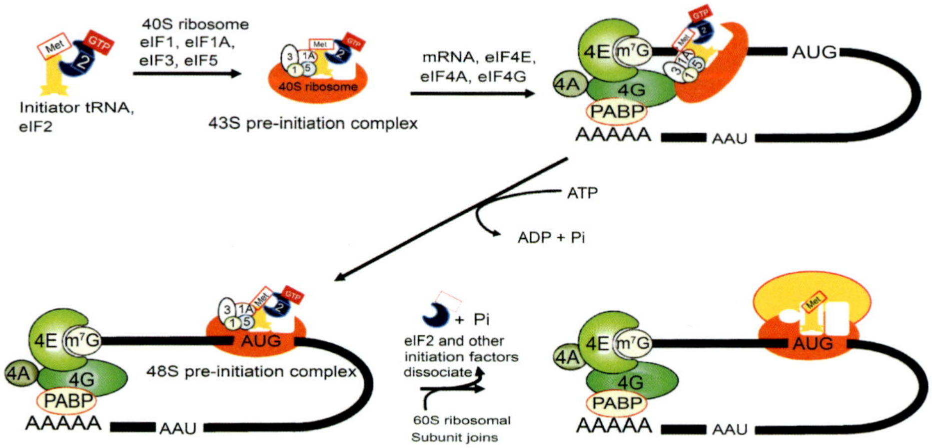

Figure 8.7. *Three initiation factors (IF-1, IF-2, and IF-3) first bind to the 30S ribosomal subunit. This step is followed by binding of the mRNA and the initiator N-formylmethionyl (fMet) tRNA, which is recognized by IF-2 bound to GTP. IF-3 is then released, and a 50S subunit binds to the complex, triggering the hydrolysis of bound GTP, followed by the release of IF-1 and IF-2 bound to GDP.*

Source: Kim H-J. Cell Fate Control by Translation: mRNA Translation Initiation as a Therapeutic Target for Cancer Development and Stem Cell Fate Control. Biomolecules. 2019; 9(11):665. https://doi.org/10.3390/biom9110665.

In eukaryotes, initiation is more complex and requires a minimum of 10 proteins called eIFs (eukaryotic initiation factors; Table 8.1), each of which is made up of several polypeptide chains. The 40S ribosomal subunit is bound by the factors eIF-1, eIF-1A, and eIF-3, while the initiator methionyl tRNA is linked to eIF-2 (in a complex with GTP) (Figure 8.8). The eIF-4 family of factors recognizes and transports the mRNA to the ribosome. eIF-4E is capable of identifying the 5′ cap of the mRNA. A different factor called eIF-4G binds to eIF-4E as well as a protein called PABP, which is connected to the poly-A tail at the 3' end of the mRNA. Because eukaryotic initiation factors are able to detect both the5' and3' ends of mRNAs, they are able to explain how polyadenylation stimulates translation. The mRNA is then brought to the 40S ribosomal subunit by the initiation factors eIF-4E and eIF-4G, in conjunction with eIF-4A and eIF-4B, with eIF-4G interacting with eIF-3. The 40S ribosomal subunit then searches the mRNA for the AUG initiation codon by working with the bound methionyl tRNA and eIFs. GTP attached to eIF-2 is hydrolyzed by eIF-5 when the AUG codon is reached. The 80S initiation complex of eukaryotic cells is formed by the binding of a 60S subunit to a 40S subunit following the release of initiation factors, which includes eIF-2 bound to GDP.

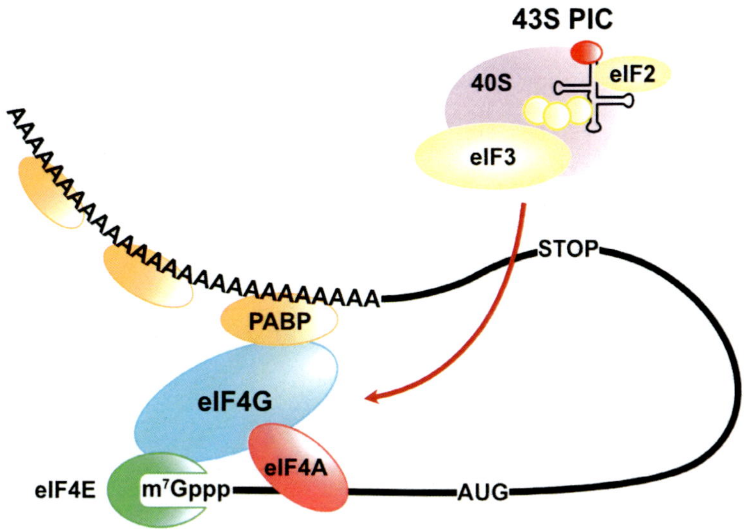

Figure 8.8. *Initiation of translation in eukaryotic cells.*

Source: Fukao A, Tomohiro T, Fujiwara T. Translation Initiation Regulated by RNA-Binding Protein in Mammals: The Modulation of Translation Initiation Complex by Trans-Acting Factors. Cells. 2021; 10(7):1711. https://doi.org/10.3390/cells10071711.

The polypeptide chain is extended during translation to carry out the process once the initiation complex has formed. Prokaryotic and eukaryotic cells use elongation mechanisms that are remarkably similar (Figure 8.9). Three locations on the ribosome are known as the P (peptidyl), A (aminoacyl), and E (exit) sites for tRNA binding. At the P site, methionyl tRNA, the initiator, is bound. The subsequent aminoacyl tRNA binds to the A site by pairing with the second codon of the mRNA, which is the initial step in elongation. An elongation factor (eEF-1α in eukaryotes, EF-Tu in prokaryotes) complexed with GTP escorts the aminoacyl tRNA to the ribosome.

The elongation factor bound to GDP is released, and the correct aminoacyl tRNA is inserted into the ribosome's A site, causing the GTP to hydrolyze to GDP. A mistaken aminoacyl tRNA, which would bind less firmly to the mRNA codon, can dissociate from the ribosome instead of being used for protein synthesis during the time interval provided by the requirement for GTP hydrolysis before EF-Tu or eEF-1α is released from the ribosome. This is the rate-limiting step in elongation. Because it gives time for the codon-anticodon pairing to be checked before the peptide bond forms, the use of a high-energy GTP at this stage is crucial to accurate protein synthesis.

A peptide bond can form between the initiator methionyl tRNA at the P site and the second aminoacyl tRNA at the A site once EF-Tu (or eEF-1α) has exited the ribosome. The large ribosomal subunit catalyzes this reaction, and rRNA is essential for it. Methionine is transferred to the aminoacyl tRNA at the ribosome's A site, where it forms a peptidyl tRNA, leaving the uncharged initiator tRNA at the P site. Translocation, the subsequent stage of elongation, is linked to GTP hydrolysis once more and calls for the addition of an additional elongation factor (eEF-2 in eukaryotes, EF-G in prokaryotes). The ribosome relocates three nucleotides along the mRNA during translocation, putting the subsequent codon in an unoccupied A site. In this step, the uncharged tRNA is translocated from the P site to the E site and the peptidyl tRNA is moved from the A site to the P site. After that, the ribosome has an empty A site and a peptidyl tRNA bound to the P site. The uncharged tRNA is released from the E site when a new aminoacyl tRNA binds to the A site, preparing the ribosome for the insertion of the subsequent amino acid in the expanding polypeptide chain.

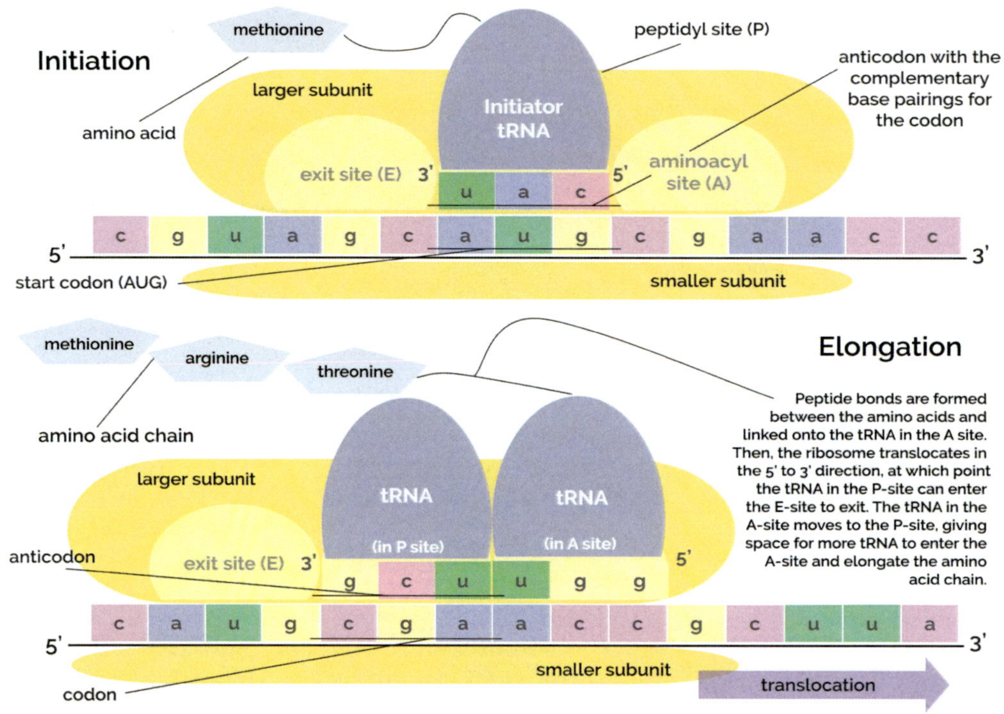

Figure 8.9. *Elongation stage of translation.*

Source: By Jordan Nguyen – Own work, CC BY-SA 4.0, https://commons.wikimedia.org/w/index.php?curid=79164021.

The EF-Tu (or eEF-1α) that is released from the ribosome bound to GDP needs to be reconverted to its GTP form as elongation proceeds (Figure 8.10). A third elongation factor, known as EF-Ts (eEF-1β⊠ in eukaryotes), is necessary for this conversion. It attaches itself to the EF-Tu/GDP complex and facilitates the exchange of bound GDP for GTP. After this exchange, EF-Tu/GTP regenerates and becomes prepared to escort a fresh aminoacyl tRNA to the ribosome's A site, initiating a fresh cycle of elongation. One common method of controlling protein activities is demonstrated by the regulation of EF-Tu through GTP binding and hydrolysis. Numerous proteins involved in the regulation of cell growth and differentiation, protein transport and secretion, and other processes are governed by similar mechanisms.

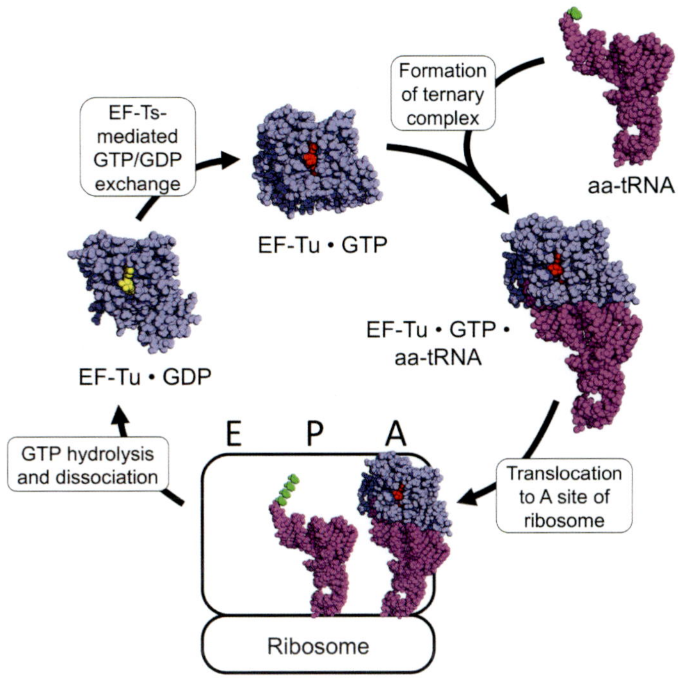

Figure 8.10. *EF-Tu complexed to GTP escorts the aminoacyl tRNA to the ribosome. The bound GTP is hydrolyzed as the correct tRNA is inserted, so EF-Tu complexed to GDP is released. The EF-Tu/GDP complex is inactive and unable to bind another tRNA. In order for translation to continue, the active EF-Tu/GTP complex must be regenerated by another factor, EF-Ts, which stimulates the exchange of the bound GDP for free GTP.*

Source: By Awchen – Own work, CC BY-SA 4.0, https://commons.wikimedia.org/w/index.php?curid=56937773.

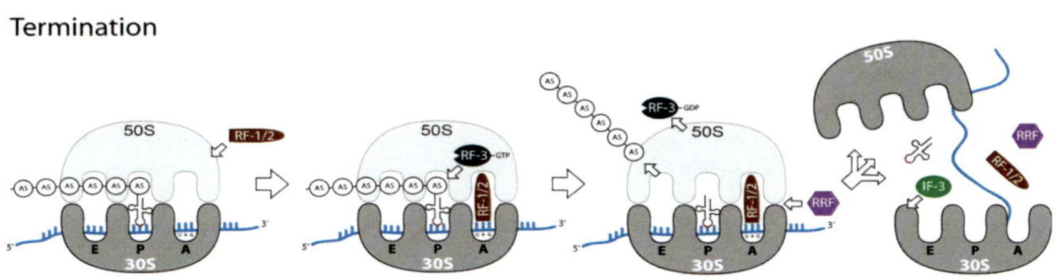

Figure 8.11. *A termination codon (e.g., UAA) at the A site is recognized by a release factor rather than by a tRNA. The result is the release of the completed polypeptide chain, followed by the dissociation of tRNA and mRNA from the ribosome.*

Source: https://upload.wikimedia.org/wikipedia/commons/9/9d/Termination_der_Translation.png.

The polypeptide chain keeps getting longer until a stop codon—UGA, UAG, or UGA—is moved into the ribosome's A site. Instead of having tRNAs with anticodons complementary to these termination signals, cells have release factors that identify the signals and stop the synthesis of new proteins (Figure 8.11). Two release factors, RF-1 and RF-2, are present in prokaryotic cells and are capable of identifying termination codons (UAA or UAG; Table 8.1). All three termination codons in eukaryotic cells are recognized by a single release factor, eRF-1. Additionally, prokaryotic and eukaryotic cells have release factors (RF-3 and eRF-3, respectively) that function in concert with RF-1 (or eRF-1) and RF-2 but do not recognize particular termination codons. A termination codon at the A site is bound by the release factors, which then cause the bond between the tRNA and the polypeptide chain at the P site to hydrolyze, releasing the completed polypeptide from the ribosome. The ribosomal subunits and the mRNA template separate as a result of the release of the tRNA.

Multiple ribosomes can translate messenger RNAs concurrently in both prokaryotic and eukaryotic cells. A new polypeptide chain can start to synthesize when a ribosome attaches to the mRNA after the first one has left the initiation site. A sequence of ribosomes that are separated by roughly 100 to 200 nucleotides typically translate mRNAs. A polyribosome, also known as a polysome, is a collection of ribosomes attached to an mRNA molecule. Every ribosome in the group carries out its own independent function to produce a distinct polypeptide chain. (Dobson, 1999; Hartl & Hayer-Hartl, 2002)

8.2. PROTEIN STRUCTURE

A protein's ability to interact with other molecules is determined by its shape, which is crucial to the protein's function. Since only recently have scientists been able to quickly and readily ascertain the atomic-level structure of entire proteins, due to the complexity of protein structures. To better understand the overall complexity of the structures, early structural biochemists conceptually divided protein structures into four "levels." We must comprehend the primary, secondary, tertiary, and quaternary levels of protein structure to ascertain how the protein acquires its final form or conformation.

8.2.1. Primary Structure

The distinct arrangement of amino acids in each polypeptide chain that constitutes a protein is its main structural component. This isn't really a structure; it's just a list of the amino acids that are present in a polypeptide chain in the order that they appear. However, this sequence is what ultimately determines the final protein structure, which is why it was dubbed the polypeptide chain's primary structure. For instance, the polypeptide chains A and B are present in the pancreatic hormone insulin (Figure 8.12).

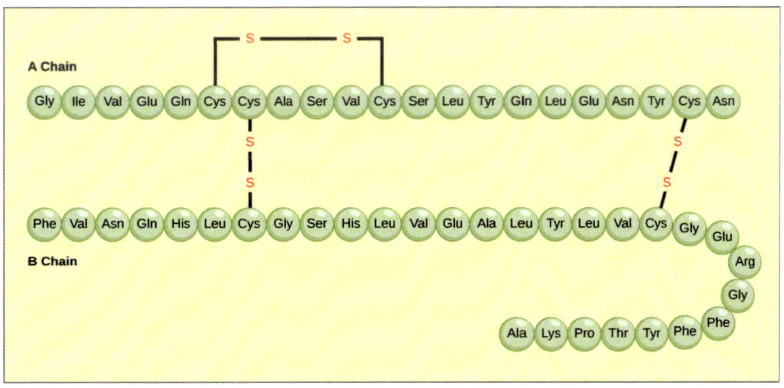

Figure 8.12. *Primary structure: The A chain of insulin is 21 amino acids long and the B chain is 30 amino acids long, and each sequence is unique to the insulin protein.*

Source: https://nigerianscholars.com/assets/uploads/2017/04/insulin-protein-1.jpg.

The specific amino acid sequence found in each peptide chain is ultimately determined by the gene or DNA sequence. A different amino acid may be added to the expanding polypeptide chain as a result of a change in the nucleotide sequence of the gene's coding region, altering the structure and function of the resulting protein.

Four polypeptide chains make up the oxygen-transport protein hemoglobin: two identical α chains and two identical β chains. A single amino substitution in the hemoglobin β chain results in a change in the entire protein's structure in sickle cell anemia. A malfunctioning hemoglobin protein is produced when valine, which is supposed to replace glutamic acid in the β chain, causes the polypeptide to fold into a slightly different shape. Thus, significant changes can result from changing just one amino acid. In low-oxygen environments, these malfunctioning hemoglobin proteins begin to associate with one another. This results in the formation of lengthy fibers from millions of aggregated hemoglobin molecules, which deform red blood cells into crescent or "sickle" shapes and clog arteries. The illness frequently causes headaches, dizziness, dyspnea, and stomach pain in its victims (Figure 8.13).

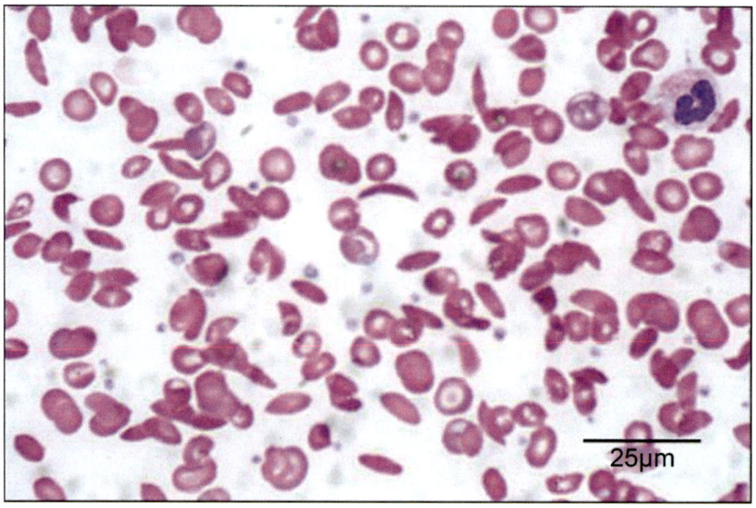

Figure 8.13. *Sickle cell disease: Sickle cells are crescent-shaped, while normal cells are disc-shaped.*

Source: https://cdn.lecturio.com/assets/Sickle-Cell-Anemia-Peripheral-blood-smear.jpg.

8.2.2. Secondary Structure

The regular structures that result from interactions between adjacent or nearby amino acids when the polypeptide begins to fold into its functional three-dimensional form are known as a protein's secondary structure. As H bonds between local groups of amino acids form in a region of the polypeptide chain, secondary structures are created. Seldom does a polypeptide chain contain a single secondary structure. Usually, it is limited to a certain chain link. The α-helix and β-pleated sheet structures are the most prevalent types of secondary structure, and they are crucial for the structural integrity of the majority of globular and fibrous proteins (Figure 8.14).

Within the α-helix, a hydrogen bond is formed between the oxygen atom in the carbonyl group of one amino acid and the hydrogen atom of the amino group of another amino acid, which is located four amino acids further down the chain. This maintains the amino acid stretch in a right-handed coil. An alpha helix has 3.6 amino acid residues in each helical turn. The polypeptide's R groups, or side chains, extend beyond the α-helix chain and do not participate in the H bonds that preserve the α-helix structure.

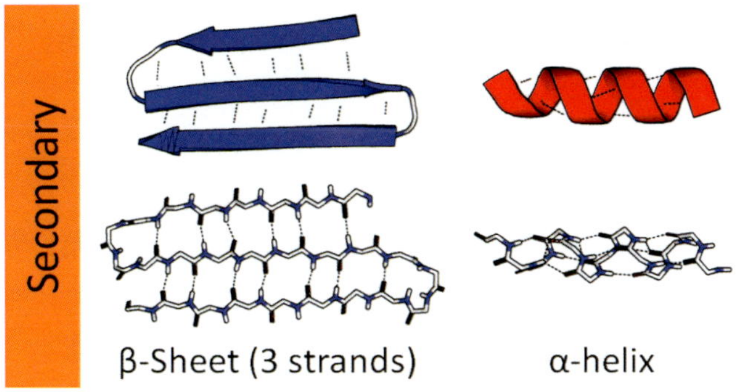

Figure 8.14. *Secondary structure: The α-helix and β-pleated sheet form because of hydrogen bonding between carbonyl and amino groups in the peptide backbone. Certain amino acids have a propensity to form an α-helix, while others have a propensity to form a β-pleated sheet.*

Source: https://en.wikipedia.org/wiki/File:Alpha_beta_structure_(full).png.

Stretches of amino acids are held in an almost fully-extended conformation that "pleats" or zigzags in β-pleated sheets because single C-C and C-N covalent bonds are non-linear. There are never just β-pleated sheets. Additional β-pleated sheets are required to secure them in place. Because of the formation of hydrogen bonds between the oxygen atom in one β-pleated sheet's carbonyl group and the hydrogen atom in another β-pleated sheet's amino group, the stretches of amino acids in β-pleated sheets are kept in their pleated sheet structure. The sheets that are held together by β-pleated sheets align either parallel or antiparallel to one another. The amino acid R groups in a β-pleated sheet are not responsible for preserving the β-pleated sheet structure; rather, they point out perpendicular to the hydrogen bonds holding the β-pleated sheets together.

8.2.3. Tertiary Structure

Once all of the secondary structure components have folded into one another, a polypeptide chain's overall three-dimensional shape is known as its tertiary structure. A protein's complex three-dimensional tertiary structure is the result of interactions between polar, nonpolar, acidic, and basic R groups within the polypeptide chain. The hydrophobic R groups of nonpolar amino acids are primarily found inside proteins during protein folding in the body's aqueous environment, whereas the hydrophilic R groups are primarily found outside. The only covalent bond that forms during protein folding is a disulfide linkage formed by cysteine side chains in the presence of oxygen. The final three-dimensional structure of the protein is determined by all of these interactions, strong and weak. A protein will cease to function if it loses its three-dimensional structure (Figure 8.15).

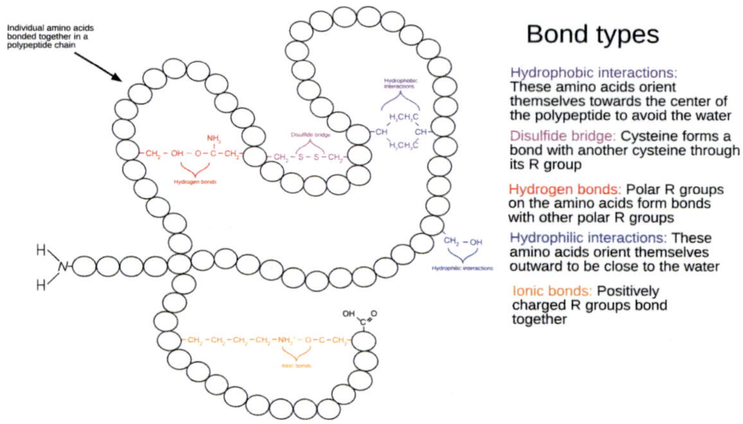

Figure 8.15. *Tertiary structure: The tertiary structure of proteins is determined by hydrophobic interactions, ionic bonding, hydrogen bonding, and disulfide linkages.*

Source: By WikiComTD – Own work, CC BY-SA 4.0, https://commons.wikimedia.org/w/index.php?curid=79148318.

8.2.4. Quaternary Structure

The orientation and arrangement of a protein's subunits with respect to one another is known as its quaternary structure.

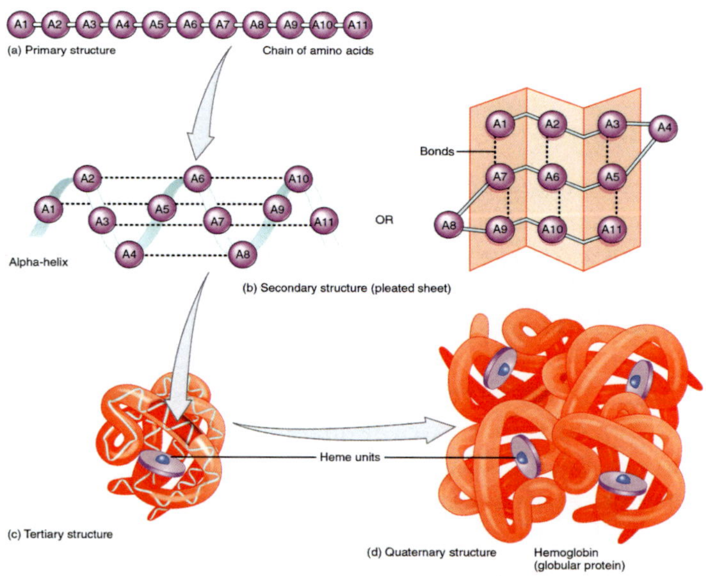

Figure 8.16. *Four levels of protein structure: The four levels of protein structure can be observed in these illustrations.*

Source: https://en.m.wikipedia.org/wiki/File:225_Peptide_Bond-01.jpg.

The quaternary structure therefore only pertains to multi-subunit proteins, or proteins composed of multiple polypeptide chains. A single polypeptide cannot form a protein with a quaternary structure. Weak interactions between subunits in proteins with multiple subunits contribute to the overall stability of the structure. Enzymes are frequently essential for joining subunits to create the finished, functional protein. For instance, the two polypeptide chains of insulin, a globular protein with a ball-like shape, are joined by hydrogen bonds as well as disulfide bonds (Dobson, 1999; Hartl & Hayer-Hartl, 2002).

A fibrous protein called silk is created when various β-pleated chains form hydrogen bonds with one another (Figure 8.16).

8.3. PROTEIN FOLDING AND PROCESSING

The flow of genetic information inside the cell is completed by translation. Now, the sequence of amino acids in a polypeptide chain has replaced the sequence of nucleotides in DNA. However, the synthesis of a polypeptide does not mean that a functional protein is produced. Polypeptides need to fold into specific three-dimensional conformations to be useful, and in many situations, multiple polypeptide chains need to come together to form a functional complex. Many proteins also go through additional changes, such as cleavage and the covalent attachment of lipids and carbohydrates, which are essential for the proper localization and operation of proteins within the cell.

8.3.1. Chaperones and Protein Folding

Proteins' three-dimensional conformations are the outcome of interactions between the side chains of the amino acids that make them up. The fundamental idea behind protein folding is that a protein's amino acid sequence contains all the information needed for it to adopt the proper three-dimensional conformation. Christian Anfinsen's research, which showed that denatured RNase can spontaneously refold *in vitro* to its active conformation, was the first to establish this. Thus, protein folding seemed to be a self-assembly process independent of other cellular factors. However, more recent research has demonstrated that this is insufficient to explain how proteins fold inside cells. Other protein functions mediate the correct folding of proteins within cells.

Molecular chaperones are proteins that aid in the folding of other proteins. Ron Laskey and his associates coined the term "chaperone" to refer to a protein called nucleoplasmin, which is necessary for the nucleosomes to form from DNA and histones. Although nucleoplasmin mediates the assembly of histones into nucleosomes by binding to them, nucleoplasmin is not integrated into the final structure of the nucleosome. Hence, without actually being a part of the assembled complex, chaperones serve as catalysts that aid in assembly. The idea has been expanded upon in later research to encompass proteins that mediate a range of additional assembly processes, most notably protein folding.

It is crucial to remember that chaperones do not transmit extra information needed for polypeptides to fold into their proper three-dimensional conformations; rather, a protein's folded conformation is exclusively dictated by its amino acid sequence. Instead, by aiding in the process of self-assembly, chaperones catalyze the folding of proteins. They seem to work by attaching to and stabilizing partially folded or unfolded polypeptides, which are the steps between the final correctly folded state and the intermediate state.

Polypeptide chains that are partially or unfolded would be unstable in the cell without chaperones, often folding incorrectly or aggregating into insoluble complexes. These unfolded polypeptides are stabilized by chaperone binding, which stops improper folding or aggregation and permits the polypeptide chain to fold into its proper conformation. Chaperones are a good example of this. They bind to polypeptide chains that are still in the process of being translated on ribosomes, preventing the incorrect folding or aggregation of the polypeptide's amino-terminal portion before the chain's synthesis is complete (Figure 8.17). This interaction is probably especially significant for proteins where proper folding of the amino terminus depends on the carboxy terminus, which is the last to synthesize. Chaperone binding maintains the amino-terminal segment in an unfolded conformation in these situations so that the finished protein can fold correctly while the remaining polypeptide chain is synthesized. Additionally, chaperones stabilize unfolded polypeptide chains during their passage into subcellular organelles, such as when proteins are transferred from the cytosol into mitochondria (Figure 8.18). Chaperones in the cytosol stabilize the partially unfolded conformations of proteins as they are transported across the mitochondrial membrane. The polypeptide chain is then transferred across the membrane and folded inside the organelle with the help of chaperones found within the mitochondrion. Furthermore, chaperones play a role in the assembly of proteins made up of several polypeptide chains, macromolecular structures (e.g., nucleoplasmin), as well as in controlling the breakdown of proteins.

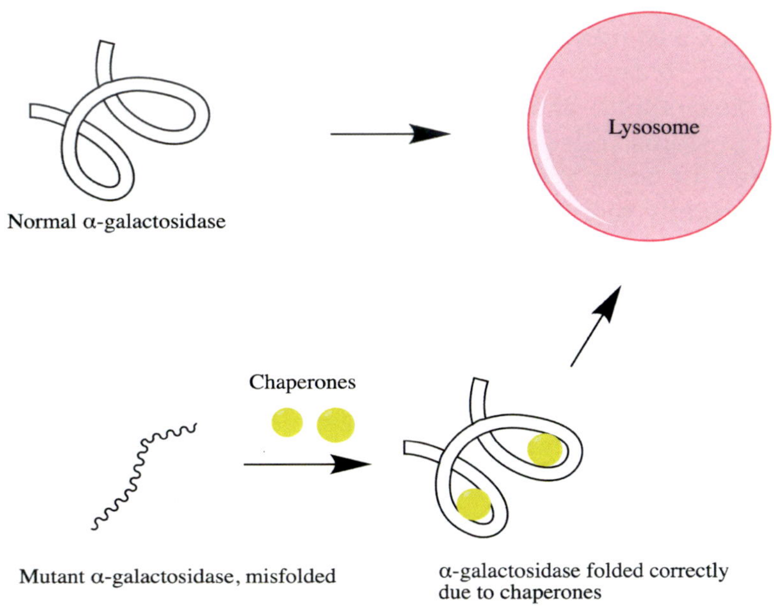

Figure 8.17. *Chaperones bind to the amino (N) terminus of the growing polypeptide chain, stabilizing it in an unfolded configuration until synthesis of the polypeptide is completed. The completed protein is then released from the ribosome and is able to fold into its correct three-dimensional conformation.*

Source: https://en.m.wikipedia.org/wiki/File:Chaperones.png.

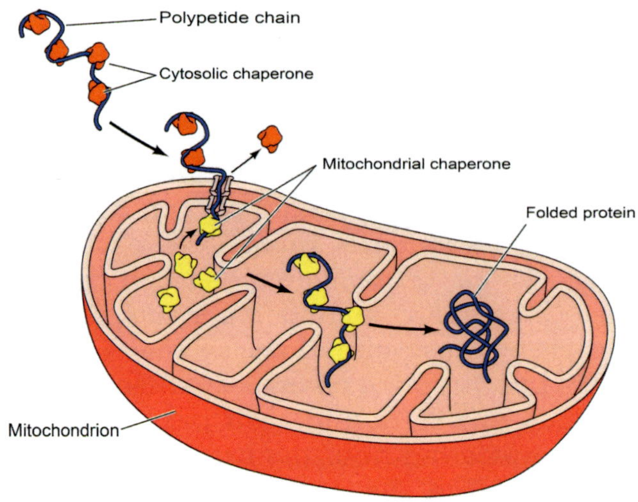

Figure 8.18. *A partially unfolded polypeptide is transported from the cytosol to a mitochondrion. Cytosolic chaperones stabilize the unfolded configuration. Mitochondrial chaperones facilitate transport and subsequent folding of the polypeptide chain within the organelle.*

Source: https://o.quizlet.com/FpAoSEqBaHF4-H1AVsrvSA.png.

A large number of the proteins that are currently understood to serve as molecular chaperones were first discovered to be heat-shock proteins, a class of proteins expressed in cells exposed to high temperatures or other types of environmental stress. It is believed that the heat-shock proteins, or Hsp, stabilize and aid in the refolding of partially denatured proteins that have been exposed to high temperatures. Hsp members are highly conserved in both prokaryotic and eukaryotic cells. Under normal growth conditions, however, many members of the heat-shock protein family are expressed and have vital cellular functions. These proteins function as molecular chaperones, which are essential for polypeptide transport and folding in both healthy cells and stressed cells.

The heat-shock protein families Hsp70 and Hsp60 seem to play a special role in the general pathways of protein folding in both prokaryotic and eukaryotic cells. The two family proteins work by attaching themselves to polypeptide chain segments that have unraveled. Hsp70 family members help to stabilize unfolded polypeptide chains during translation (Figure 8.17, for example) and when polypeptides are being transported into different subcellular compartments, including the endoplasmic reticulum and mitochondria. These proteins maintain the unfolded configuration of the polypeptide chain and inhibit aggregation by binding to short segments (seven or eight amino acid residues) of unfolded polypeptides.

Chaperonins, or members of the Hsp60 family, aid in the folding of proteins into their natural conformations. Chaperonins are made up of 14 subunits, each weighing about 60 kilodaltons (kd). These subunits are stacked in two rings to create the appearance of a "double doughnut" (Figure 8.19). Unfolded polypeptide chains are contained inside the chaperonin cylinder's central cavity, providing protection from the cytosol. Protein folding can occur in this isolated environment, and the chaperonin-bound unfolded polypeptide chain segments stop the polypeptide chain from aggregating. A reversible process linked to the hydrolysis of ATP as an energy source is the binding of unfolded polypeptides to the chaperonin. Thus, unfolded regions of the polypeptide chain are released and rebound to the chaperonin multiple times as a result of ATP hydrolysis, which enables the polypeptide to gradually fold into the correct conformation.

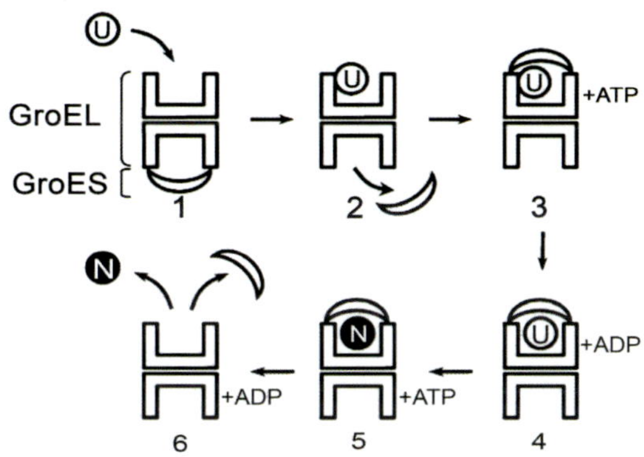

Figure 8.19. *Structure of a chaperonin. GroEL, a member of the Hsp60 family, is a porous cylinder composed of two stacked rings. Each ring consists of seven subunits.*

Source: Marchenkov VV, Semisotnov GV. GroEL-Assisted Protein Folding: Does It Occur Within the Chaperonin Inner Cavity? International Journal of Molecular Sciences. 2009; 10(5):2066–2083. https://doi.org/10.3390/ijms10052066.

It has been observed that members of the Hsp70 and Hsp60 families occasionally behave sequentially. For instance, in the folding of freshly synthesized proteins in *E. coli*, Hsp70 and Hsp60 family members function in succession during the transport of proteins into mitochondria. Before the completion of protein synthesis, an Hsp70 chaperone stabilizes developing polypeptide chains. Following the transfer of the unfolded polypeptide chain to an Hsp60 chaperonin, protein folding occurs, resulting in the production of a correctly folded protein in its functional three-dimensional conformation. Hsp60 and Hsp70 family members can be found in subcellular organelles (e.g., mitochondria) and the cytosol, respectively of eukaryotic cells and bacteria (refer to Table 8.1), suggesting that the sequential activity of Hsp70 and Hsp60 represents a general protein folding pathway. Although the exact role of Hsp90 is still unknown, there may be a different route for the folding of some proteins in the cytosol and endoplasmic reticulum that involves the sequential actions of Hsp70 and Hsp90 family members (Figure 8.20).

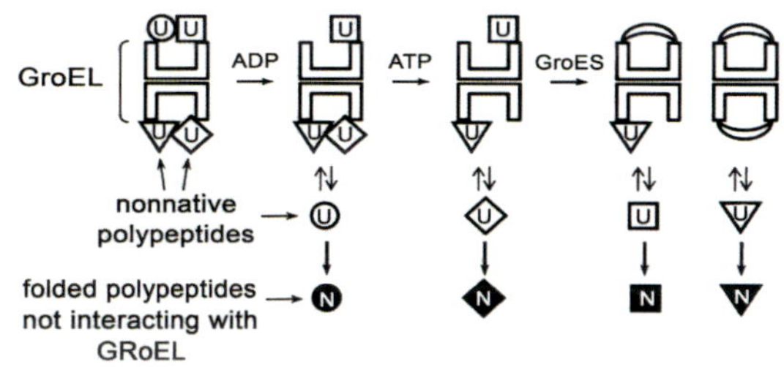

Figure 8.20. *Chaperones of the Hsp70 family bind to and stabilize unfolded polypeptide chains during translation. The unfolded polypeptide is then transferred to chaperones of the Hsp60 family, within which protein folding takes place. ATP hydrolysis is required for release of the unfolded polypeptide from Hsp70 as well as for folding within Hsp60.*

Source: Marchenkov VV, Semisotnov GV. GroEL-Assisted Protein Folding: Does It Occur Within the Chaperonin Inner Cavity? International Journal of Molecular Sciences. 2009; 10(5):2066–2083. https://doi.org/10.3390/ijms10052066.

8.3.2. Enzymes and Protein Folding

Cells contain at least two types of enzymes that catalyze protein folding by breaking and re-forming covalent bonds, in addition to chaperones, which aid in the process by binding to and stabilizing partially folded intermediates. The stabilization of many folded structures depends on the creation of disulfide bonds between cysteine residues. The enzyme that breaks and re-forms these bonds is called protein disulfide isomerase, and it was identified by Christian Anfinsen in 1963 (Figure 8.21). Protein disulfide isomerase (PDI), which facilitates quick exchanges between paired disulfides, is crucial for proteins with many cysteine residues because it helps the protein achieve the disulfide bond pattern that is consistent with its stably folded conformation. Because reducing agents found in the cytosol keep cysteine residues in their reduced (—SH) form, disulfide (S—S) linkages cannot form, and disulfide bonds are typically limited to secreted proteins and certain membrane proteins. Disulfide bonds arise in the endoplasmic reticulum of eukaryotic cells, where an oxidizing environment is preserved. The activity of PDI in the endoplasmic reticulum is correlated with the amount of protein secreted in various cell types, which is consistent with the role of disulfide bonds in stabilizing secreted proteins.

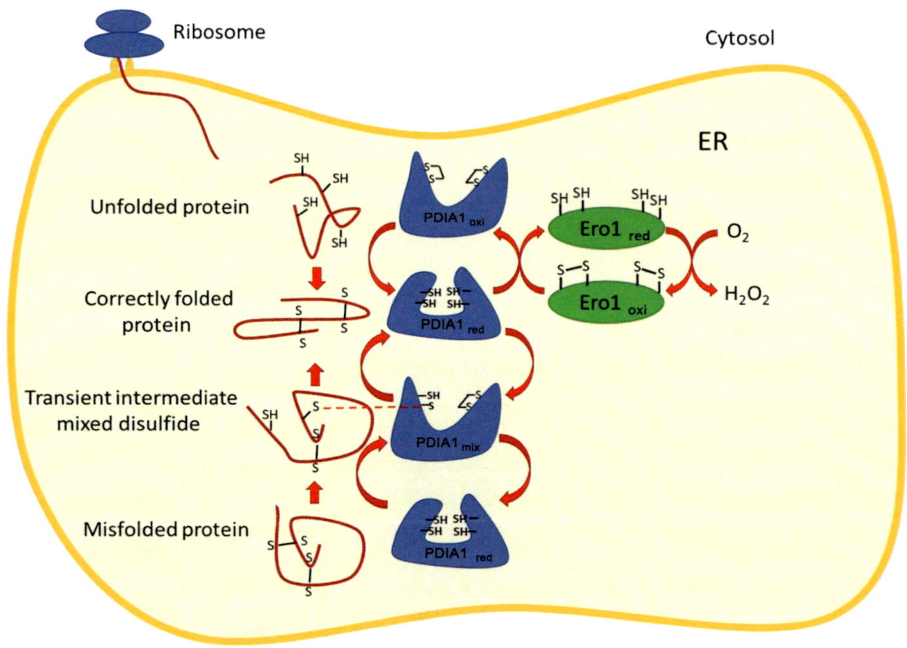

Figure 8.21. *The Protein disulfide isomerase (PDI) catalyzes the breakage and rejoining of disulfide bonds, resulting in exchanges between paired disulfides in a polypeptide chain. The enzyme forms a disulfide bond with a cysteine residue of the polypeptide and then exchanges its paired disulfide with another cysteine residue. In this example, PDI catalyzes the conversion of two incorrect disulfide bonds (1–2 and 3–4) to the correct pairing (1–3 and 2–4).*

Source: By Tiphany C. DE BESSA – Own work, CC BY-SA 4.0, https://commons.wikimedia.org/w/index.php?curid=68429099.

The isomerization of peptide bonds containing proline residues is catalyzed by the second enzyme involved in protein folding (Figure 8.22). Proline is a unique amino acid because peptide bonds that come before proline residues have an equilibrium between their cis and trans conformations that is tipped slightly in favor of the trans-form. Peptide bonds relating to other amino acids, on the other hand, are nearly always in the trans-form. The enzyme peptidyl prolyl isomerase catalyzes the isomerization between the cis and trans configurations of prolyl peptide bonds, which could otherwise represent a rate-limiting step in protein folding. This enzyme can catalyze the refolding of at least some proteins and is found in both prokaryotic and eukaryotic cells. Its physiologically significant substrates and function within cells are still unknown.

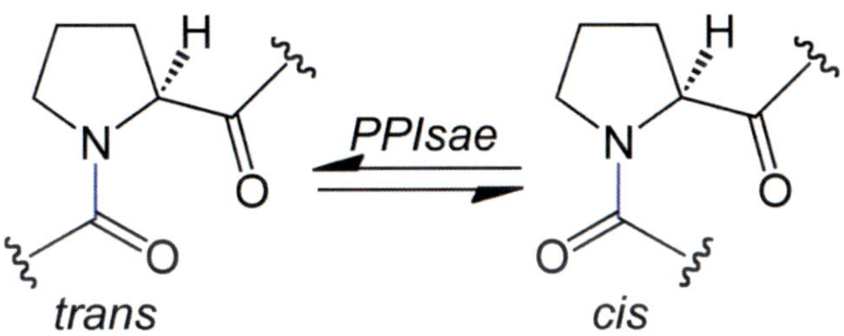

Figure 8.22. *The action of peptidyl prolyl isomerase. Peptidyl prolyl isomerase catalyzes the isomerization of peptide bonds that involve proline between the cis and trans conformations.*

Source: https://en.wikipedia.org/wiki/File:Prolyl_isomerization_General_Formulae_V.2.png.

8.3.3. Protein Cleavage

A crucial stage in the maturation of many proteins is the cleavage of the polypeptide chain or proteolysis. A straightforward illustration is the elimination of the initiator methionine from the amino terminus of numerous polypeptides, which happens shortly after the growing polypeptide chain's amino terminus exits the ribosome. The amino-terminal residues are then frequently supplemented with additional chemical groups, such as acetyl groups or fatty acid chains. Many proteins, including those that are secreted by bacteria and eukaryotes and those that are meant to be incorporated into the plasma membrane, lysosomes, mitochondria, and chloroplasts of eukaryotic cells, are also translocated across membranes as a result of proteolytic modifications of the amino terminus. As the protein crosses the membrane, proteolytic cleavage removes amino-terminal sequences that are responsible for targeting these proteins for transport to their destinations. For instance, while translation is still occurring, amino-terminal signal sequences, which are typically 20 amino acids long, direct secreted proteins to the endoplasmic reticulum of eukaryotic cells or the plasma membrane of bacteria (Figure 8.23). As it exits the ribosome, the signal sequence—which is primarily made up of hydrophobic amino acids—is inserted into the membrane. As translation continues, the remaining polypeptide chain travels through a membrane channel. The mature protein is then released after a particular membrane protease (signal peptidase) cleaves the signal sequence. When eukaryotic cells target proteins for secretion, plasma membrane incorporation, or lysosomal incorporation, the first step is the translocation of growing polypeptide chains into the endoplasmic reticulum the methods by which proteins are transported to these locations and the part played by additional targeting sequences in controlling the import of proteins into mitochondria and chloroplasts.

Other significant proteolytic processing scenarios involve the cleavage of larger precursors to form active enzymes or hormones. An excellent example is insulin, which is produced from a longer precursor polypeptide. Insulin is formed through two cleavages. An amino-terminal signal sequence found in the first precursor, preproinsulin, directs the polypeptide chain to the endoplasmic reticulum (Figure 8.24). Proinsulin is the second precursor that is produced when the signal sequence is eliminated during transfer to the endoplasmic reticulum. The internal peptide in this precursor is then removed by protease to create insulin, which is composed of two chains joined by disulfide bonds. Similar cleavage processes also activate other proteins, such as those involved in blood clotting and digestion enzymes.

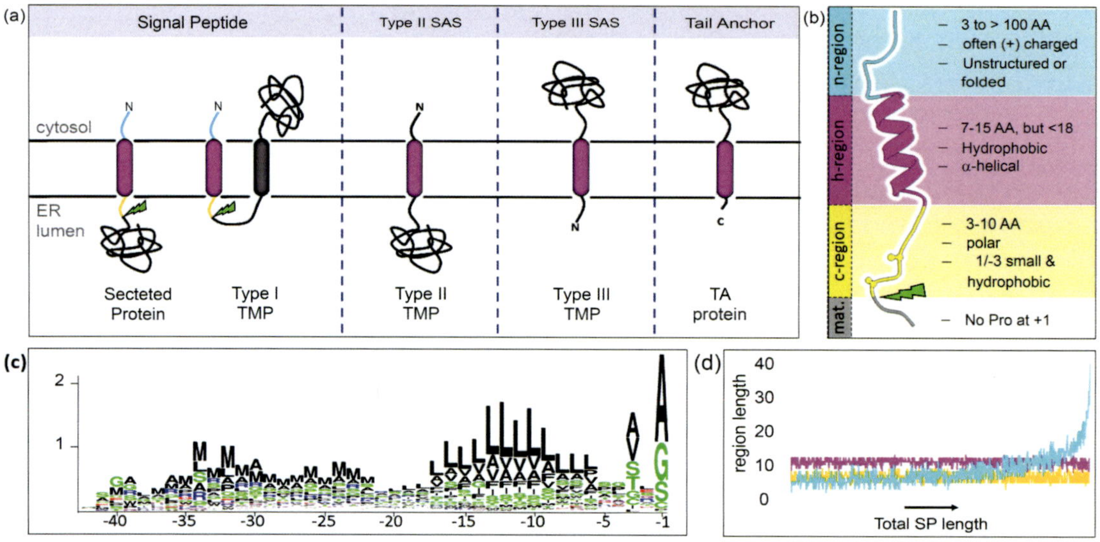

Figure 8.23. *The role of signal sequences in membrane translocation.*

Source: Liaci AM, Förster F. Take Me Home, Protein Roads: Structural Insights into Signal Peptide Interactions during ER Translocation. International Journal of Molecular Sciences. 2021; 22(21):11871. https://doi.org/10.3390/ijms222111871.

Interestingly, many animal viruses have proteins that are produced by the cleavage of larger precursors. HIV provides a particularly significant illustration of the function of proteolysis in viral replication. A protease encoded in the virus cleaves precursor polypeptides to create the structural proteins that the virus uses for replication. Along with reverse transcriptase, the HIV protease plays a crucial role in the replication of the virus, making it a target for drug development aimed at treating AIDS. In fact, among the most potent medicines on the market today for treating this illness are protease inhibitors of this kind.

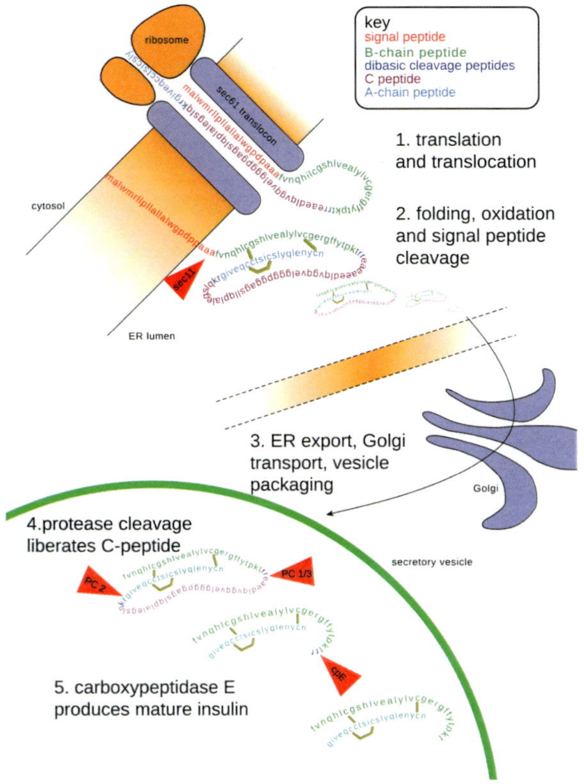

Figure 8.24. *Proteolytic processing of insulin.*

Source: By Uploaded by Fred the Oyster, CC BY-SA 4.0, https://commons.wikimedia.org/w/index.php?curid=35286441.

8.3.4. Glycosylation

Glycosylation is the process of adding carbohydrates to many proteins, especially in eukaryotic cells. While certain nuclear and cytosolic proteins are also glycosylated, glycoproteins—proteins to which carbohydrate chains have been added—are primarily secreted or localized to the cell surface. The glycoprotein carbohydrate moieties are crucial for the function of the endoplasmic reticulum in folding proteins, directing proteins to the right intracellular compartments for delivery, and serving as recognition sites in interactions between cells.

Glycoproteins are categorized as either O-linked or N-linked based on where the carbohydrate side chain attaches (Figure 8.25). The carbohydrate in N-linked glycoproteins is bound to the asparagine side chain's nitrogen atom. The site of carbohydrate attachment in O-linked glycoproteins is the side chain of serine or threonine's oxygen atom. These positions are typically directly attached to N-acetylglucosamine or N-acetylgalactosamine, respectively.

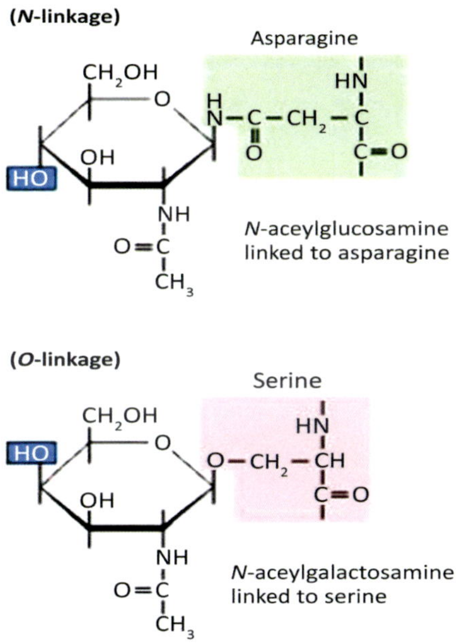

Figure 8.25. *The carbohydrate chains of N-linked glycoproteins are attached to asparagine; those of O-linked glycoproteins are attached to either serine (shown) or threonine. The sugars joined to the amino acids are usually either N-acetylglucosamine (N-linked) or N-acetylgalactosamine (O-linked).*

Source: https://qph.cf2.quoracdn.net/main-qimg-6adfb2202367b094b41b1ab31c59dba9.

In eukaryotic cells, the majority of glycoproteins are either integrated into the plasma membrane or secreted. These proteins are typically transferred while their translation is still ongoing into the endoplasmic reticulum (by cleaving a signal sequence). Before translation is finished, the endoplasmic reticulum also starts the process of glycosylation. A common oligosaccharide with 14 sugar residues—2 N-acetylglucosamine, 3 glucose, and 9 mannose—is transferred to an asparagine residue of the developing polypeptide chain as the first step (Figure 8.26). Within the endoplasmic reticulum, the oligosaccharide is assembled on a lipid carrier called dicholphosphate. After that, it is moved intact to an asparagine (Asn) residue acceptor in the Asn-X-Ser or Asn-X-Thr sequence (where X can be any amino acid other than proline).

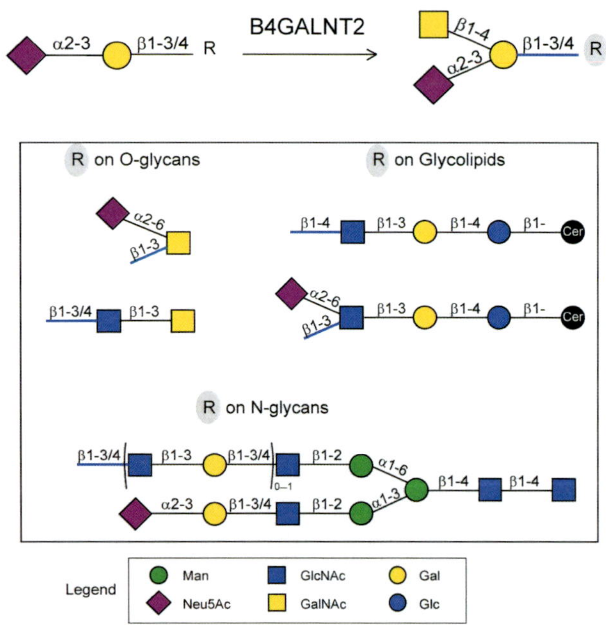

Figure 8.26. *Synthesis of N-linked glycoproteins.*

Source: Cogez V, Vicogne D, Schulz C, Portier L, Venturi G, de Ruyck J, Decloquement M, Lensink MF, Brysbaert G, Dall'Olio F, et al. N-Glycan on the Non-Consensus N-X-C Glycosylation Site Impacts Activity, Stability, and Localization of the Sda Synthase B4GALNT2. International Journal of Molecular Sciences. 2023; 24(4):4139. https://doi.org/10.3390/ijms24044139.

Subsequent processing results in modifications to the common N-linked oligosaccharide. While the glycoprotein is in the endoplasmic reticulum, three glucose residues and one mannose are eliminated. After that, the oligosaccharide undergoes additional modification in the Golgi apparatus, where glycoproteins from the endoplasmic reticulum are transported. As the glycoprotein is moved through the Golgi's compartments, these modifications include the addition and removal of carbohydrates residues (Figure 8.27). Different N-linked oligosaccharides undergo varying degrees of processing, based on the enzymes found in each cell as well as the oligosaccharide accessibility to the enzymes that catalyze its modification. In the Golgi apparatus, glycoproteins that contain inaccessible oligosaccharides do not acquire additional sugars. Because of the high proportion of mannose residues in these glycoproteins, akin to the common oligosaccharide that was first added in the endoplasmic reticulum, they are referred to as high-mannose oligosaccharides. On the other hand, more processing is done on glycoproteins that have accessible oligosaccharides, which leads to the creation of a range of complex oligosaccharides.

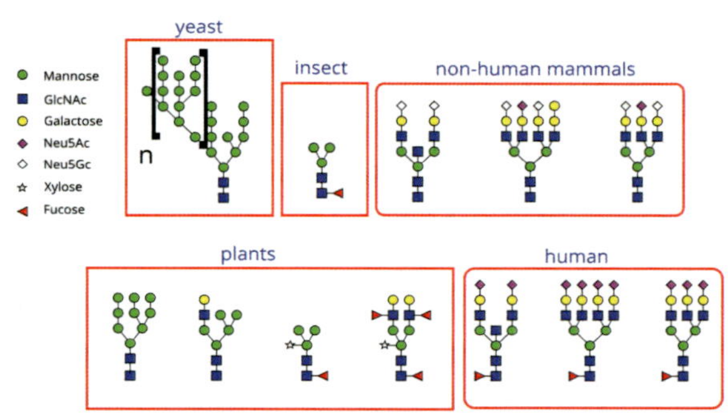

Figure 8.27. *Examples of N-linked oligosaccharides.*

Source: By DNA 621 – Own work, CC BY-SA 3.0, https://commons.wikimedia.org/w/index.php?curid=31825161.

The Golgi apparatus additionally contains O-linked oligosaccharides. O-linked oligosaccharides, in contrast to N-linked oligosaccharides, are created by adding one sugar at a time and typically have a small number of residues (Figure 8.28). Single O-linked N-acetylglucosamine residues are also added to many nuclear and cytoplasmic proteins, including a range of transcription factors, and this process is catalyzed by a distinct enzyme system. Nevertheless, it is currently unknown how carbohydrates affect the ability of these nuclear and cytoplasmic glycoproteins to function.

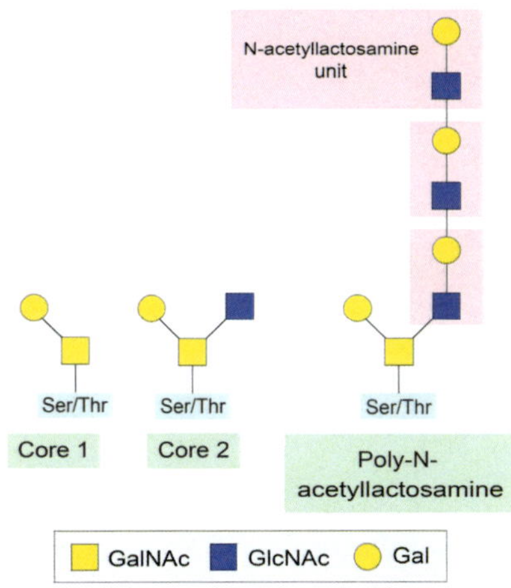

Figure 8.28. *Examples of O-linked oligosaccharides. O-linked oligosaccharides usually consist of only a few carbohydrate residues, which are added one sugar at a time.*

Source: By Waffeln – Own work, CC BY-SA 4.0, https://commons.wikimedia.org/w/index.php?curid=77471121.

8.3.5. Attachment of Lipids

In eukaryotic cells, lipids are attached to the polypeptide chain to modify certain proteins. These alterations usually aim to attach these proteins to the plasma membrane so that the hydrophobic lipid can interact with them. N-myristoylation, prenylation, and palmitoylation are the three general lipid additions that are frequently seen in eukaryotic proteins connected to the cytosolic face of the plasma membrane. The addition of glycolipids, a fourth type of modification, is crucial in anchoring certain cell surface proteins to the plasma membrane's extracellular face. During translation, a fatty acid is sometimes added to the amino terminus of the expanding polypeptide chain in proteins. This 14-carbon fatty acid process, known as N-myristoylation, involves attaching myristic acid to an N-terminal glycine residue (Figure 8.29). Typically, glycine is added to the polypeptide chain after the initiator methionine has been eliminated through proteolysis and before any fatty acids are added. Analysis of mutant proteins whose N-terminal glycine is changed to an alanine has clearly shown the role of the fatty acid in this association. Many proteins that are modified by N-myristoylation are associated with the inner face of the plasma membrane. By obstructing the mutant protein ability to associate with membranes, this substitution stops myristoylation and hinders their ability to function.

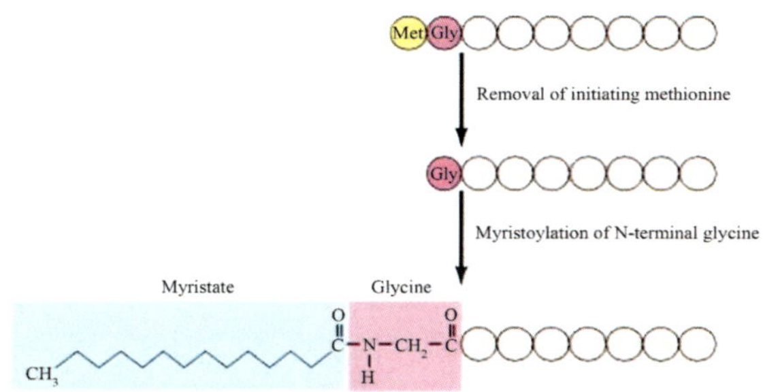

Figure 8.29. *Addition of a fatty acid by N-myristoylation. The initiating methionine is removed, leaving glycine at the N terminus of the polypeptide chain. Myristic acid (a 14-carbon fatty acid) is then added.*

Source: https://slideplayer.com/slide/10605175/36/images/129/Figure+9.35+Addition+of+a+fatty+acid+by+N-myristoylation.jpg.

Additionally, side chains of cysteine, serine, and threonine residues can have lipids attached to them. Prenylation is a significant example of this kind of modification, whereby particular lipid types (prenyl groups) are joined to the sulfur atoms in the side chains of cysteine residues that are close to the polypeptide chain's C terminus (Figure 8.30). This results in the modification of numerous plasma membrane-associated proteins that regulate cell growth and differentiation, such as the Ras oncogene proteins that cause the unchecked development of many malignancies in humans. These proteins go through three steps of prenylation. The prenyl group is first added to a cysteine that is three amino acids away from the polypeptide chain's carboxy terminus. Either farnesyl (15 carbons, as shown in Figure 8.30) or geranylgeranyl (20 carbons) are the prenyl groups added in this reaction. The cysteine residue is then followed by amino acids, which are eliminated, leaving cysteine at the carboxy terminus. Lastly, the carboxyl group of the C-terminal cysteine residue receives an addition of a methyl group.

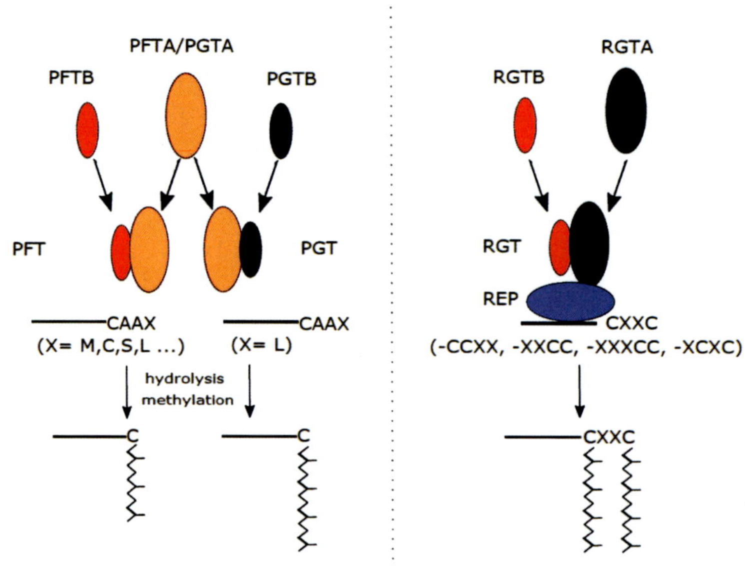

Figure 8.30. *Prenylation of a C-terminal cysteine residue.*

Source: Hála M, Žárský V. Protein Prenylation in Plant Stress Responses. Molecules. 2019; 24(21):3906. https://doi.org/10.3390/molecules24213906.

The fact that mutations of the critical cysteine prevent the membrane association and function of Ras proteins suggests the biological significance of prenylation. The possibility that inhibitors of the key enzyme, farnesyl transferase, may prove useful as drugs for the treatment of cancers involving Ras proteins has sparked interest in this reaction because farnesylation is a relatively uncommon modification of cellular proteins. These farnesylation inhibitors have been shown to impede the growth of cancer cells in model organisms and are currently being tested in humans to see if they work well against malignancies. Sulfur atoms of the side chains of internal cysteine residues are joined to palmitic acid, a 16-carbon fatty acid, in the third type of fatty acid modification known as palmitoylation (Figure 8.31). Similar to N-myristoylation and prenylation, palmitoylation is crucial for certain proteins to bind to the cytosolic face of the plasma membrane.

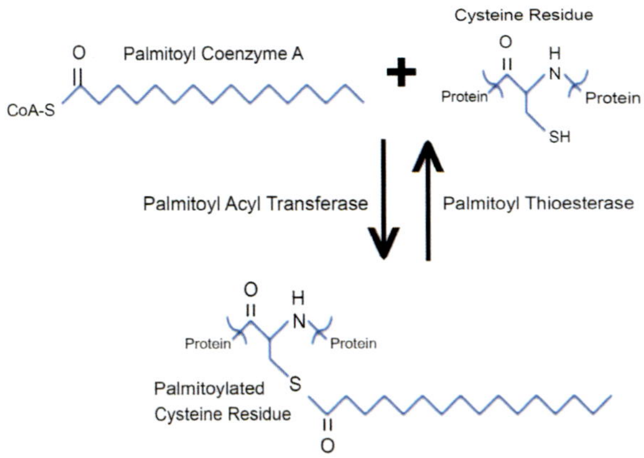

Figure 8.31. *Palmitoylation. Palmitate (a 16-carbon fatty acid) is added to the side chain of an internal cysteine residue.*

Source: By Sivart13 – Own work, CC BY-SA 4.0, https://commons.wikimedia.org/w/index.php?curid=41737823.

The C-terminal carboxyl groups of some proteins receive the addition of glycolipids, which are lipids linked to oligosaccharides. These glycolipids act as anchors to secure the proteins to the plasma membrane's exterior. The glycolipids that are linked to these proteins are commonly referred to as glycosylphosphatidylinositol, or GPI, anchors since they contain phosphatidylinositol (Figure 8.32). The terminal carboxyl group of polypeptide chains is where the oligosaccharide portions of GPI anchors are attached. The carbohydrate acts as a link between the protein and the fatty acid chains of phospholipid, with the inositol head group of phosphatidylinositol being attached to the oligosaccharide in turn. Within the endoplasmic reticulum, proteins are supplemented with preassembled GPI anchors that are synthesized. The polypeptide chain's C terminus contains a peptide that is roughly 20 amino acids long, and it cleaves as they are added. The fatty acid chains of the GPI anchor mediate the modified protein's attachment to the plasma membrane once it reaches the cell surface (Dobson, 1999; Hartl & Hayer-Hartl, 2002).

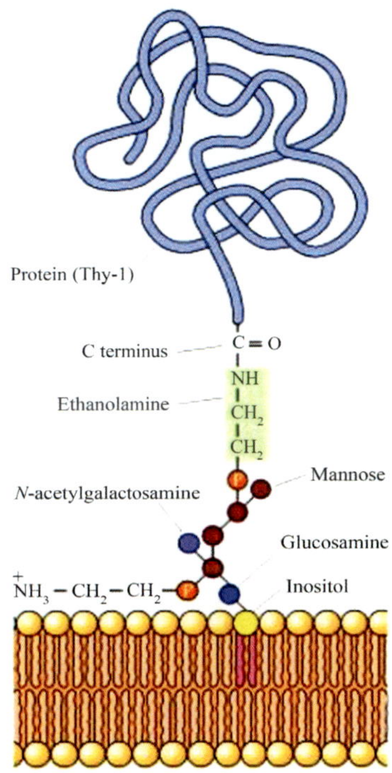

Figure 8.32. *Structure of a GPI anchor.*

Source: https://www.ncbi.nlm.nih.gov/books/NBK9843/bin/ch7f32.jpg.

8.4. CELL-FREE PROTEIN SYNTHESIS

In molecular biology, cell-free protein synthesis is one of the most popular techniques. While producing proteins in cells usually takes days or even weeks, producing proteins using cell-free protein synthesis usually only takes a few hours. As a matter of fact, even inexperienced users can frequently start using a commercial system the same day they receive freshly synthesized proteins. However, users should be aware of the mechanisms and processes that underlie cell-free protein synthesis, even though many will opt to use commercial systems rather than create them in their own labs.

8.4.1. Early History of Cell-Free Protein Synthesis

Several labs employed cell-free protein synthesis in the early days of molecular biology to investigate the process of amino acid incorporation into proteins. Scientists used extracts from rat liver cells, bacteria, human and rabbit reticulocytes, ascites cells, and wheat germ to perform cell-free protein synthesis. They discovered that the microsome fraction—which needs GTP and ATP—is where protein synthesis takes place. The first correlation between nucleotide triplets and the amino acids they encoded was discovered by Nirenberg and Matthaei in 1961, proving that polyphenylalanine could be synthesized cell-free from synthetic polyuridylic acid in the *E. coli* extract. The "central dogma of molecular biology" was established in the years that followed, and cell-free synthesis remained crucial in determining the amino acids encoded by the remaining triplets—that is, in deciphering the remaining characters in the genetic code.

8.4.2. Landmark Advances in the Development of Cell-Free Protein Synthesis Systems

Since the preparation processes did not remove messenger RNA (mRNA), the cell extracts that comprised early cell-free systems synthesized mostly native proteins from endogenous mRNA. These extracts were essentially supernatants obtained by high-speed centrifugation of lysates from disrupted cells for example, at 30,000g (the so-called S30 extract). To direct the protein synthesis machinery to translate exogenous mRNA, Nirenberg and Matthaei incubated the *E. coli* S30 extract in the presence of 20 amino acids and ATP regeneration components. Such 'incubated S30' improved the amount of the protein synthesized from exogenous mRNA, possibly by freeing ribosomes from attaching to endogenous mRNA (ribosome run-off) and allowing endogenous mRNA to be degraded by ribonucleases in the extract. Similarly, Pelham and Jackson added a Ca^{2+}-dependent micrococcal ribonuclease during the preparation of the rabbit reticulocyte extract (lysate) to degrade endogenous mRNA. The micrococcal ribonuclease can be subsequently inactivated by EGTA to prevent its digestion of exogenous mRNA. Interestingly, the wheat germ extract does not translate a significant amount of endogenous proteins, therefore, there is no need for incubation or nuclease treatment.

The next landmark improvement in cell-free protein synthesis was the introduction of coupled transcription-translation systems that used DNA as the template instead of mRNA. Zubay and colleagues used the *E. coli* RNA polymerase in the S30 extracts to transcribe DNA into mRNA for protein synthesis, mainly to study gene regulation in–vitro. To obtain higher amounts of mRNA, other researchers later used stronger phage promoters and the more efficient phage RNA polymerases (T7 or SP6) for the coupled transcription-translation from the DNA template. The coupled transcription-translation based on the phage promoters and RNA polymerases has since been adopted by most cell-free systems. In addition to using the more stable DNA template and avoiding a separate *in vitro* transcription step, coupled transcription and translation in the cell-free system often result in a higher protein synthesis yield, probably due to newly synthesized mRNA is immediately translated into protein, which may minimize the adverse effects of mRNA degradation or inhibitory structures.

Protein synthesis without cells is typically carried out in a batch format. Nevertheless, the processes involved in protein synthesis frequently come to an end quickly—in many cases, in less than an hour—because of the depletion of ATP and GTP, the breakdown of mRNA, and the build-up of byproducts to inhibitory levels. As a result, the protein produced by batch cell-free protein synthesis typically yields $\mu g\ ml^{-1}$. Continuous cell-free protein synthesis was first proposed by Spirin and colleagues in both *E. coli* and cell-free eukaryotic systems. By using a flow chamber or dialysis membrane to maintain the levels of small molecules energy sources and amino acids while also eliminating inhibitory by-products (like phosphate), continuous cell-free protein synthesis avoids some of the drawbacks of the batch format. Because of this, a continuous cell-free protein synthesis reaction can frequently produce mg μg^{-1} of protein and last longer than 10 hours.

Due to continuous improvement by academic and industrial researchers, the cell extracts in cell free systems have improved significantly. From the 'incubated S30' first described by Nirenberg and Matthaei, Zubay, Pratt and colleagues developed robust protocols for making the *E. coli* extract that have become the standards. Other important improvements included the use of *E. coli* strains that lacked RNase I (RNA) (MRE600) and Exonuclease V (A19) to stabilize RNA and linear DNA templates, respectively. Recently, Swartz and colleagues pioneered the use of genome engineered *E. coli* strains to stabilize amino acids during *in vitro* reactions, *in vitro* activation of endogenous metabolic pathways to sustain high-yield protein synthesis, and systematic optimization of *E. coli* growth conditions and extraction procedures to reduce the production cost (Figure 8.33).

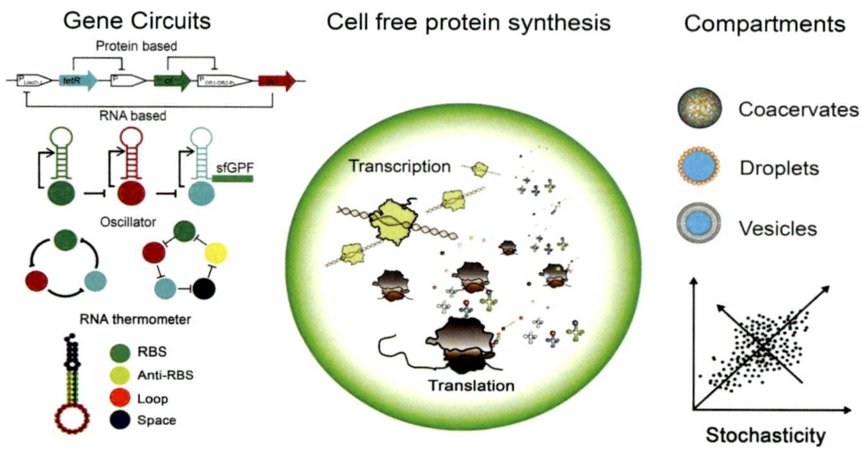

Figure 8.33. *Cell-free protein synthesis (CFPS0 for producing proteins).*

Source: Yue K, Zhu Y, Kai L. Cell-Free Protein Synthesis: Chassis toward the Minimal Cell. Cells. 2019; 8(4):315. https://doi.org/10.3390/cells8040315.

Since their initial development, the protocols for making the 'classic' eukaryotic extracts, rabbit reticulocyte lysate and wheat germ, remain largely unchanged, and the protein synthesis yields from these eukaryotic extracts are generally lower than that from the *E. coli* extract. Researchers developed improved protocols for making highly active extracts from wheat germ and mammalian cells. Endo and colleagues discovered that the remnants of endosperm on the surface of wheat germ were the sources of ribosome-inactivating proteins and other translation inhibitors, responsible for the poor stability and low yield of the conventional wheat germ cell-free system. By extensively washing isolated wheat germ, the Endo group developed a high-yield long-lasting wheat germ cell-free system. Until recently, with the exception of the rabbit reticulocyte lysate, the extracts of mammalian cells have not been widely used for cell-free protein synthesis mainly due to low yields. Imataka and colleagues at RIKEN found that the lack of functional translation initiation factors in the HeLa cell extract was a major limiting factor. By supplementing with exogenous translation initiation factors, the Imataka group developed significantly improved mammalian cell-free systems, effectively taking over the rabbit reticulocyte lysate system.

Finally, the Ueda group reconstituted the minimal protein translational machinery from *E. coli*, thereby completely circumventing the issues of inhibitory factors in cell extracts. The Ueda group developed the PURE system (Protein synthesis Using Recombinant Elements), which is made up of fully purified components with very few nucleases, proteases, or other enzymatic activity that is either unrelated to or detrimental to protein translation. The PURE system was the first reliable cell-free system with a full complement of purified translation factors and aminoacyl-tRNA synthetases, despite attempts by other labs to reconstitute the protein synthesis using purified components. Using the same methodology, another lab demonstrated cell-free protein synthesis at temperatures as high as 65°C and recreated the essential protein translational machinery from

Thermus thermophilus. To make cell-free protein synthesis a protein factory rather than just an analytic tool, Swartz and colleagues pioneered chemical engineering approaches to making the *E. coli* extract. Their goal was to manufacture therapeutic proteins in the cell-free system at liter or any scale. The entire processes of cell growth, extract preparation, protein synthesis and downstream purification were subject to cost analyzes and process engineering. The 'black-box' nature of the *E. coli* extract became sets of biochemical reactions amenable to pathway engineering. One of the key advances was the creation of the 'Cytomim' system that can activate central metabolism and oxidative phosphorylation in the *E. coli* extract, allowing not only the use of inexpensive energy substrates, such as glutamate or glucose but also continuous protein synthesis for more than 6 hr in the batch format. The Cytomim system was the basis for further improvements that led to the cell-free synthesis of recombinant human granulocyte-macrophage colony-stimulating factor in a 100-liter industrial stir tank.

8.4.3. Use of Commercially Available Cell-Free Protein Synthesis Systems

The quickest and easiest method for producing proteins *in vitro* is to use a cell-free system that is readily available for purchase. The majority of commercial cell-free systems offer DNA expression vectors that allow researchers to clone their genes and produce proteins using the cloned gene. The promoter and other sequences in these expression vectors are frequently tailored for cell-free protein synthesis. Coupled transcription and translation have been implemented in nearly all commercial cell-free systems, enabling the use of PCR-generated linear DNA templates for protein synthesis. To choose an appropriate commercial cell-free system, one should consider protein size and origin, protein folding, post-translational modification, downstream assays, applications and costs. In general, the prokaryotic systems (for example, *E. coli* extracts) produce more proteins, (particularly tiny proteins mass<50 kDa) than the eukaryotic systems, whereas the latter is more suited for expressing large multidomain eukaryotic proteins. The continuous cell-free synthesis format generally produces more proteins (often > 1 mgμl^{-1}) in comparison to the batch format. While reconstituted systems offer mRNA and protein stability, systems based on cell extracts typically offer superior folding environments and post-translational modifications. It is feasible to perform downstream characterization right after protein synthesis without purification because the reconstituted systems contain few macromolecules and enzymatic activities outside of those required for protein translation.

8.4.4. Application of Cell-Free Proteins

8.4.4.1. High-Throughout Cell-Free Protein Synthesis for Functional and Structural Analyzes

The use of cell-free protein synthesis has made the most impact on functional and structural genomics. For the first time, researchers have been able to express and purify a large number of proteins in a short period of time for subsequent high-throughput functional and structural analyzes. For instance, many laboratories have used low-cost *E. coli* extract and wheat germ cell-free systems in high throughput automated format to produce stable isotope-labeled proteins for nuclear magnetic resonance analyzes. The advantage is that only newly synthesized proteins are labeled during cell-free protein synthesis and can be analyzed without extensive purification.

Protein arrays have been used for high throughput analyzes of protein functions, similar to DNA microarrays for gene expression. Initially, each protein had to be separately purified by researchers and then identified on solid surfaces. Surface-bound proteins may become inactive during storage, and creating these protein arrays is a laborious task. By enabling proteins to be produced and captured close to immobilized DNA templates, cell-free protein synthesis avoids these issues and successfully creates just-in-time protein arrays from more stable DNA arrays. Functional proteomics studies of protein-protein interactions have made use of two such protein array technologies: protein in situ array (PISA) and nucleic acid programmable protein array (NAPPA).

8.4.4.2. Incorporation of Unnatural Amino Acids into Proteins

The translational machinery, including aminoacyl-tRNA synthetases, can tolerate a wide range of unnatural amino acids, allowing the genetic code to be expanded or even reassigned. One of the earliest applications of cell-free protein synthesis was the incorporation of unnatural amino acids into a specific position of an *in vitro* synthesized protein. Schultz and colleagues first demonstrated the incorporation of unnatural amino acids into an amber stop codon that substituted a sense codon in a protein in an *E. coli* extract. It took more than a decade for the Schultz laboratory to engineer living cells to allow such expansion of the genetic code *in vivo*. Still, compared to cell-based approaches, cell-free systems are easier to manipulate for the incorporation of unnatural amino acids. The reconstituted cell-free systems are especially suited for these applications. Using the reconstituted cell-free system, researchers can simply remove the competing factors, such as release factors, tRNAs or aminoacyl-tRNA synthetases, or replace them with an tRNA-synthetase pair from another organism, or an engineered aminoacyl-tRNA synthetase. Moreover, tRNAs can be pre-acylated with unnatural amino acids using natural enzymes or engineered ribozymes (so-called 'Flexizymes'). The combined use of pre-charged tRNAs and the reconstituted cell-free system has allowed unnatural peptides or peptidomimetics containing mostly unnatural amino acids to be synthesized for potential applications in therapeutics.

8.4.4.3. Directed Evolution of Proteins

Directed protein evolution is the process of repeatedly selecting and mutating proteins to improve their properties, including binding affinity, thermostability, and catalytic activity. Throughout the selection process, a stable physical bond between the protein (phenotype) and nucleic acid (genotype) is essential for directed protein evolution. The reason behind the maintenance of genotype-phenotype linkage in cell-based approaches like phage display and yeast surface display is that the distinct nucleic acid sequences are limited to distinct phage particles or cells. Phage packaging, cellular transformation efficiencies, and growth biases may be able to restrict the application of these cell-based techniques (since phages expressing distinct protein library members may amplify at varying rates). In contrast, protein libraries with greater complexity (more genes to choose from) and less amplification bias may be incorporated into *in vitro* directed evolution based on cell-free protein synthesis. These *in vitro* methods include the encapsulation of DNA, mRNA, and protein in water-in-oil emulsion droplets by in-vitro compartmentalization, as well as ribosome display and mRNA display, which both establish a physical connection between mRNA and protein. Researchers have demonstrated the benefits of reconstituted cell-free systems, despite the fact that extract-based cell-free systems have been successfully used for mRNA and ribosome display. The reconstituted cell-free systems exhibit minimal nuclease and protease activities due to their purified components, resulting in a significant stabilization of the mRNA-protein linkage. When ribosome display is carried out using a reconstituted cell-free system devoid of release factor 1, the efficiency of ribosome stalling at the stop codon and displaying the entire protein is increased. Since only the components required for invitro synthesis are included in the reconstituted cell-free systems, they become the only option when the incorporation of multiple unnatural amino acids is required for in-vitro evolution and selection of drug-like molecules.

8.4.4.4. Cell-free Synthetic Biology Towards Artificial Cells

For any attempt to create an artificial cell, whether for practical applications or to understand the origin of life, cell-free protein synthesis is probably a crucial first step. A significant fraction of the 200 or so essential genes that are thought to be the bare minimum for a free-living cell are directly related to protein synthesis. DNA/RNA replication and cell-free transcription regulation have been linked to protein synthesis. In parallel, water-in-oil droplets, phospholipid vesicles, and liposomes have all been used for cell-free protein synthesis. By doing this, the reconstituted cell-free systems may prove to be a valuable instrument in the process of reconstructing the minimal cell from the bottom up.

8.4.5. Comparison of Protein Synthesis in Prokaryotes and Eukaryotes

The process by which cells assemble proteins is known as protein synthesis. Mostly referring to a multistep process in which cells follow a very systematic procedure that first transcribes DNA into mRNA and then translates the mRNA into chains of amino acids, the term is sometimes used to refer only to protein translation. After that, the chain of amino acids folds to create useful proteins.

Through the decoding of the genetic instructions carried by the mRNAs, translation plays a role in both prokaryotic and eukaryotic protein synthesis. Codons, which are triplets of nucleotides on the mRNA, are translated into an amino acid sequence during translation. The fundamental strategy for both prokaryotic and eukaryotic translation is the same. There are, nevertheless, many distinctions between these translation procedures. The primary distinction between translation in prokaryotes and eukaryotes is that the former happens simultaneously with transcription, while the latter happens asynchronously.

8.4.5.1. Prokaryotic Translation

Translation is the process by which transcription and protein synthesis occur simultaneously in prokaryotes. Translation starts immediately after the gene's 5′ end is converted to mRNA. Three steps make up prokaryotic translation: initiation, elongation, and termination. The two subunits, 30S and 50S, are assembled to start the translation. The initiation complex is put together with assistance from the three initiation factors, IF1, IF2, and IF3. The first amino acid that is added during translation is N-formylmethionine. Peptide bonds are formed between the incoming and residual nucleotides using GTP as the energy source. EF-P is the translation initiation factor. The ribosome's binding to the Shine--Dalgarno sequence aids in the selection of the start codon. Upstream of the AUG start codon lies the purine-rich Shine-Dalgarno sequence. This sequence complements the pyrimidine-rich region of 16S rRNA. The 30S subunit contains the 16S rRNA. By pairing these two complementary nucleotides, double-stranded RNA is formed. The initiation codon enters the ribosome's P-site as a result of this pairing. The P site is bound by the initial amino acid. Three active sites make up a ribosome: the A, P, and E sites. Other than the first aminoacyl tRNA, incoming aminoacyl tRNAs bind to the A site. At the P site, peptide bonds are formed. The E site is the exit site for the uncharged tRNA.

The two elongation factors are EF-G and EF-Tu. The translation elongates until the ribosome reaches one of the three stop codons: UAA, UGA, and UAG. Release factors other than the tRNAs, recognize the stop codon. The mRNA with the termination codon at the A site is referred to as the termination complex. Three release factors can be identified: RF1, RF2 and the RF3. RF1 and RF2 recognize the UAA/UAG and UAA/UGA and hydrolyze the ester bond in peptidyl-tRNA to release the nascent polypeptide chain. RF3 catalyzes the releasing of RF1 and RF2. Once the new protein releases, the ribosome undergoes recycling. The Ribosome Recycling Factor and EF-G, are involved in releasing mRNA and tRNAs from the ribosome and dissociation of 70S ribosome into 30S and 50S subunits. IF3 releases the mRNA by replacing the deacylated tRNA. When bacteria enter the stationary phase, the translation is downregulated by the dimerization of ribosomes. Ribosome dimerization is facilitated by RMF, HPF and YfiA. The ribosome -dissociation factors are RsfA and HflX.

8.4.5.2. Eukaryotic Translation

The second stage of eukaryotic gene expression, translation occurs independently of eukaryotic transcription. In eukaryotes, transcription and translation take place in two distinct compartments. As a result, the two processes cannot take place at the same time. Eukaryotic mRNAs are monocistronic, and before being released into the cytoplasm, they undergo processing in the nucleus that includes the addition of a 5′ cap, poly-A tail, and splicing out of introns. By co-translationally folding the newly synthesized polypeptide chain on the ribosome, ribosomal pausing also influences translation. This procedure causes a delay in the translation, allowing for more time. Eukaryotic mRNAs have a poly A tail and a 5′ cap. Thus, there are two distinct ways that translation can begin: cap-dependent initiation and cap-independent initiation. The initiation factors bind to the 5′ end of the mRNA during cap-dependent initiation. The mRNA is held in the ribosome's small subunit by these initiation factors. Internal ribosome entry sites enable ribosome trafficking to the start site by direct binding during cap-independent initiation. Methionine is the first amino acid that binds in eukaryotes. The 40S and 60S subunits combine to form the 80S ribosome. eEF-1 and eEF-2 are two elongation factors that are involved in eukaryotic translation. The process of elongation is comparable to that of prokaryotes. The translation process ends in the same way as it does in the prokaryotic system. Nevertheless, all three stop codons can be recognized by the universal release factor eRF1. To assist eRF1 in releasing the polypeptide chain, eRF3 is a release factor. (Dobson, 1999; Hartl & Hayer-Hartl, 2002)

CASE STUDY

UNDERSTANDING PROTEIN SYNTHESIS IN HEALTH AND DISEASE

Background:

Protein synthesis is a complex biological process crucial for cellular function and organismal survival. This case study focuses on how research into protein synthesis has contributed to understanding diseases and developing therapeutic interventions.

Challenges:

- **Regulation of Gene Expression:** Coordinating transcription and translation processes to ensure accurate synthesis of functional proteins.
- **Genetic Mutations and Disorders:** Identifying genetic mutations or environmental factors that disrupt protein synthesis, leading to diseases.
- **Therapeutic Target Identification:** Discovering molecular targets within the protein synthesis pathway for drug development.

Solutions Implemented

- Mechanisms of Protein Synthesis:
  - – **Action:** Researchers investigate the molecular machinery involved in protein synthesis, including ribosomes, transfer RNAs (tRNAs), messenger RNAs (mRNAs), and initiation factors.
  - **Rationale:** Understanding the mechanisms ensures efficient and accurate translation of genetic information into proteins.
  - **Implementation:** Utilizing techniques such as ribosome profiling, cryo-electron microscopy, and biochemical assays to elucidate protein synthesis dynamics at molecular levels.
- Impact of Mutations on Protein Synthesis:
  - **Action:** Genetic studies and clinical research focus on mutations affecting components of the translation machinery or regulatory elements.
  - **Rationale:** Mutations can impair protein synthesis, leading to diseases such as cystic fibrosis (CF), Duchenne muscular dystrophy (DMD), and various cancers.
  - **Implementation:** Characterizing disease-associated mutations, developing

animal models, and conducting functional assays to link molecular defects to clinical phenotypes.

- Therapeutic Interventions:
  - **Action:** Pharmaceutical companies and academic researchers explore targeting protein synthesis pathways for therapeutic development.
  - **Rationale:** Modulating translation rates or correcting translation errors could offer potential treatments for genetic disorders and cancers.
  - **Implementation:** Screening small molecules or RNA-based therapies that enhance translation fidelity, suppress aberrant translation, or target specific disease-causing mutations.

Results and Future Considerations

- **Scientific Insights:** Research into protein synthesis has uncovered mechanisms underlying genetic diseases and provided insights into cellular adaptations under stress conditions.
- **Clinical Applications:** Therapeutic strategies targeting protein synthesis pathways show promise in preclinical and clinical studies, offering new avenues for treating previously incurable conditions.
- **Challenges Ahead:** Addressing specificity, delivery methods, and potential off-target effects remain critical for translating protein synthesis-based therapies into clinical practice.

Conclusion

The study of protein synthesis exemplifies how basic research into fundamental cellular processes can drive medical advances and therapeutic innovations. By unraveling the intricacies of translation mechanisms and their dysregulation in diseases, researchers are paving the way for precision medicine and personalized treatments tailored to patients' genetic profiles.

This case study underscores the significance of protein synthesis in both health and disease contexts, highlighting its role as a cornerstone of cellular biology and a frontier for therapeutic breakthroughs.

CLASS ACTIVITY

OBJECTIVE: TO DEMONSTRATE THE PROCESS OF PROTEIN SYNTHESIS FROM DNA TO FUNCTIONAL PROTEIN.

Materials Needed:

Paper strips or index cards (representing mRNA)

- Different colored beads or paper cutouts (representing amino acids)
- Scissors and glue/tape
- Diagrams of protein synthesis (DNA to mRNA to protein)
- Models of ribosomes, tRNA, and amino acids

Assessment:

Observation: Monitor student participation and understanding during the activity.

Completion of Protein: Evaluate if students accurately constructed the protein sequence from the mRNA provided.

Discussion: Assess student engagement and understanding during the discussion and review phase.

SUMMARY

- Protein synthesis is a fundamental process within living organisms, essential for the growth, development, and maintenance of cellular structures and functions. This intricate molecular mechanism is governed by a series of tightly regulated steps that ensure precise control over the production of proteins. Protein synthesis involves the transcription of genetic information encoded in DNA into messenger RNA (mRNA), followed by the translation of this mRNA into a sequence of amino acids, forming a polypeptide chain—the building blocks of proteins.
- Primary Structure refers to the linear sequence of amino acids in the protein chain, determined by the sequence of nucleotides in the gene encoding it. The primary structure is fundamental because it dictates how the protein will fold into its three-dimensional shape and ultimately determines its function.
- Secondary structure refers to the local folding patterns within a protein chain, primarily driven by hydrogen bonding between amino acids. The two most common types of secondary structures are alpha helices and beta sheets. Alpha helices are coiled structures stabilized by hydrogen bonds between nearby amino acids, while beta sheets are formed when segments of the protein chain align side by side, forming a sheet-like structure.
- Tertiary structure refers to the overall three-dimensional shape of a single protein molecule. It is stabilized by interactions between amino acid side chains (such as hydrogen bonds, disulfide bonds, hydrophobic interactions, and ionic bonds), which fold the protein into a compact and functional form. The tertiary structure is critical because it determines the specific functional properties of the protein, such as enzyme activity, binding specificity, and stability.
- Protein folding is the process by which a protein chain assumes its functional three-dimensional structure, known as its native conformation, from a linear sequence of amino acids. This process is crucial because a protein's function is intimately linked to its specific shape.
- Cell-free protein synthesis (CFPS) is a laboratory technique that allows for the production of proteins outside of living cells. It operates by utilizing the cellular machinery necessary for protein synthesis—such as ribosomes, tRNAs, amino acids, and enzymes—extracted from cells and placed into a test tube or other suitable container.

REVIEW QUESTIONS

1. What is the role of mRNA in protein synthesis?
2. Describe the process of translation, including its key stages and molecular components.
3. Discuss the four levels of protein structure, providing examples of each.

4. How do primary and secondary protein structures influence tertiary and quaternary structures?
5. Describe the process of protein folding, including the factors that influence it.
6. Explain the role of post-translational modifications in protein processing and function.
7. What is cell-free protein synthesis, and how does it differ from traditional *in vivo* protein synthesis?

REFERENCES

1. Aartsma-Rus, A., & den Dunnen, J. T. (2019). Phenotype predictions for exon deletions/duplications: A user guide for professionals and clinicians using Becker and Duchenne muscular dystrophy as examples. *Human Mutation, 40*(10), 1630–1633.
2. Austin, D. J., Crabtree, G. R., & Schreiber, S. L. (1994). Proximity versus allostery: The role of regulated protein dimerization in biology. *Chemistry & Biology, 1*(3), 131–136.
3. Azar, S., & Wong, T. E. (2017). Sickle cell disease: A brief update. *Medical Clinics of North America, 101*(2), 375–393.
4. Baker-Fulco, C. J. (1995). Overview of dietary intakes during military exercises. In B. M. Marriott (Ed.), *Not Eating Enough, Overcoming Underconsumption of Military Operational Rations* (pp. 121–149). Washington, D.C.: National Academy Press.
5. Balagopal, P., Rooyackers, O. E., Adey, D. B., Ades, P. A., & Nair, K. S. (1997). Effects of aging on *in vivo* synthesis of skeletal muscle myosin heavy-chain and sarcoplasmic protein in humans. *American Journal of Physiology, 273*(3), E790–E800.
6. Barbul, A., Lazarou, S. A., Efron, D. T., Wasserkrug, H. L., & Efron, G. (1990). Arginine enhances wound healing and lymphocyte immune responses in humans. *Surgery, 108*(2), 331–336.
7. Burack, W. R., & Shaw, A. S. (2000). Signal transduction: Hanging on a scaffold. *Current Opinion in Cell Biology, 12*(2), 211–216.
8. Chee, N. T., Lohse, I., & Brothers, S. P. (2019). mRNA-to-protein translation in hypoxia. *Molecular Cancer, 18*(1), 49.
9. Cohn, S. H., Vartsky, D., Yasumura, S., Sawitsky, A., Zanzi, I., Vaswani, A., & Ellis, K. J. (1980). Compartmental body composition based on total-body nitrogen, potassium, and calcium. *American Journal of Physiology, 239*(6), E524–E530.
10. Coleman, G. L., Barthold, W., Osbaldiston, G. W., Foster, S. J., & Jonas, A. M. (1977). Pathological changes during aging in barrier-reared Fischer 344 male rats. *Journal of Gerontology, 32*(3), 258–278.
11. Coleman, W., & DuBois, E. F. (1915). Clinical calorimetry. VII. Calorimetric observations on the metabolism of typhoid patients with and without food. *Annals of Internal Medicine, 15*, 887–938.
12. Collins, D. M., Rezzo, C. T., Kopp, J. B., Ruiz, P., Coffman, T. M., & Klotman, P. E. (1990). Chronic high protein feeding does not produce glomerulonephritis or renal

insufficiency in the normal rat. *Journal of the American Society of Nephrology, 1*(5), 624.

13. Collipp, P. J., Thomas, J., Curti, V., Sharma, R. K., Maddaiah, V. T., & Cohn, S. E. (1973). Body composition changes in children receiving human growth hormone. *Metabolism, 22*(5), 589–595.
14. Conlay, L. A., Wurtman, R. J., Blusztajan, J. K., Coviella, I. L. G., Maher, T. J., & Evoniuk, G. E. (1986). Marathon running decreases plasma choline concentrations. *New England Journal of Medicine, 315*(14), 892.
15. Cox, S., Radzio-Andzelm, E., & Taylor, S. S. (1994). Domain movements in protein kinases. *Current Opinion in Structural Biology, 4*(6), 893–901.
16. Dobson, C. M. (1999). Protein misfolding, evolution, and disease. *Trends in Biochemical Sciences, 24*(9), 329–332.
17. Gao, H., Zhou, Z., Rawat, U., Huang, C., Bouakaz, L., Wang, C., Cheng, Z., Liu, Y., Zavialov, A., Gursky, R., Sanyal, S., Ehrenberg, M., Frank, J., & Song, H. (2007). RF3 induces ribosomal conformational changes responsible for dissociation of class I release factors. *Cell, 129*(5), 929–941.
18. Hammer III, J. A., & Wu, X. S. (2002). Rabs grab motors: Defining the connections between Rab GTPases and motor proteins. *Current Opinion in Cell Biology, 14*(1), 69–75.
19. Harish, A., & Caetano-Anollés, G. (2012). Ribosomal history reveals origins of modern protein synthesis. *PLoS ONE, 7*(3), e32776.
20. Hartl, F. U., & Hayer-Hartl, M. (2002). Molecular chaperones in the cytosol: From nascent chain to folded protein. *Science, 295*(5561), 1852–1858.
21. Horovitz, A., Fridmann, Y., Kafri, G., & Yifrach, O. (2001). Allostery in chaperonins. *Journal of Structural Biology, 135*(1), 104–114.
22. Kawasaki, H., Nakayama, S., & Kretsinger, R. H. (1998). Classification and evolution of EF-hand proteins. *Biometals, 11*(3), 277–295.
23. Kornitzer, D., & Ciechanover, A. (2000). Modes of regulation of ubiquitin-mediated protein degradation. *Journal of Cellular Physiology, 182*(1), 1–11.
24. Lim, W. A. (2002). The modular logic of signaling proteins: Building allosteric switches from simple binding domains. *Current Opinion in Structural Biology, 12*(1), 61–68.
25. Lin, R. C., & Scheller, R. H. (2000). Mechanisms of synaptic vesicle exocytosis. *Annual Review of Cell and Developmental Biology, 16*, 19–49.
26. Medzihradszky, K. F., & Chalkley, R. J. (2015). Lessons in de novo peptide sequencing by tandem mass spectrometry. *Mass Spectrometry Reviews, 34*(1), 43–63.
27. Nissen, P., Hansen, J., Ban, N., Moore, P. B., & Steitz, T. A. (2000). The structural basis of ribosome activity in peptide bond synthesis. *Science, 289*(5481), 920–930.
28. Parker, N., Schneegurt, M., Thi Tu, A. H., Lister, P., & Forster, B. M. (2016). Microbiology [online]. Figure 11.15 Translation in bacteria begins with the formation of the initiation complex. In *Microbiology* (Section 11-4). OpenStax.

29. Ptashne, M., & Gann, A. (1998). Imposing specificity by localization: Mechanism and evolvability. *Current Biology, 8*, R812–R822.

30. Rochet, J. C., & Landsbury, P. T. (2000). Amyloid fibrillogenesis: Themes and variations. *Current Opinion in Structural Biology, 10*, 60–68.

31. Saibil, H. R., Horwich, A. L., & Fenton, W. A. (2001). Allostery and protein substrate conformational change during GroEL/GroES-mediated protein folding. *Advances in Protein Chemistry, 59*, 45–72.

32. Schmeing, T. M., & Ramakrishnan, V. (2009). What recent ribosome structures have revealed about the mechanism of translation. *Nature, 461*(7268), 1234–1242.

33. Segev, N. (2001). Ypt and Rab GTPases: Insight into functions through novel interactions. *Current Opinion in Cell Biology, 13*, 500–511.

34. Simonovi⊠, M., & Steitz, T. A. (2009). A structural view on the mechanism of the ribosome-catalyzed peptide bond formation. *Biochimica et Biophysica Acta (BBA) – General Subjects, 1789*(9–10), 612–623.

35. Spudich, J. A. (2001). The myosin swinging cross-bridge model. *Nature Reviews Molecular Cell Biology, 2*, 387–392.

CHAPTER

9

Heterologous Protein Production in Eukaryotic Cells

LEARNING OBJECTIVES

After studying this chapter, you will be able to:

- Define posttranslational modification of eukaryotic proteins;
- Describe general features of eukaryotic expression systems;
- Understand the fungus-based expression systems;
- Explain baculovirus–insect cell expression systems.

KEY TERMS FROM THIS CHAPTER

Chaperones	Cytoplasm
Endoplasmic reticulum	Eukaryotic expression system
Expression vector	Glycosylation
Host cell	Insect cells
Mammalian cells	Preproinsulin
Promoter	Protein folding
Secretion signal	Transfection

INTRODUCTION

Prokaryotic expression systems have been used to successfully produce heterologous (recombinant) proteins from cloned DNA originating from a wide variety of organisms. The eukaryotic proteins generated in bacteria may not always have the appropriate biological activity or stability, despite the fact that it is theoretically feasible to express any gene from any source organism in a prokaryotic host. Furthermore, even with rigorous purification protocols, the finished product may be contaminated by toxic or animal-heating-producing bacteria (pyrogens). For the purpose of producing uncontaminated therapeutic agents for humans or animals, as well as large amounts of stable, biologically active proteins for biochemical, biophysical, and structural studies and proteins for industrial processes, researchers have developed eukaryotic expression systems in fungal, insect, and mammalian cells. Furthermore, all the properties of a human protein meant for medicinal application must match those of the natural protein. For the most part, incorrect posttranslational protein processing—including incorrect protein cleavage and folding—as well as the lack of suitable mechanisms that add chemical groups to particular amino acid acceptor sites account for prokaryotic organisms' incapacity to produce authentic versions of eukaryotic proteins.

9.1. POSTTRANSLATIONAL MODIFICATION OF EUKARYOTIC PROTEINS

Since the steps involved in protein synthesis are not segregated in prokaryotes, translation of messenger RNA (mRNA) happens simultaneously with transcription. The ribosome can access the developing transcript as soon as it emerges from RNA polymerase and can start translation. Proteins are folded into their correct three-dimensional configuration during synthesis with the help of folding proteins called chaperones, which bind to polypeptides as they are being synthesized. Eukaryotes, on the other hand, move mRNA from the nucleus to ribosomes on the endoplasmic reticulum or in the cytoplasm, where translation takes place. During synthesis, proteins made on ribosomes connected to the endoplasmic reticulum are either secreted into the endoplasmic reticulum's lumen where they are processed further or inserted into the endoplasmic reticulum's membrane (Barnes, and Dickson. 2006).

Many proteins go through posttranslational processing, which is necessary for protein stability and activity. This includes the majority of proteins that are of interest as therapeutic agents for the treatment of diseases in humans or animals. Proteases must cleave certain precursor polypeptides of certain proteins at specific sites to convert them into the active form. For example, the small peptide hormone insulin is produced in animal pancreatic cells as a single polypeptide, preproinsulin, that is cleaved to produce two shorter peptides that are joined by disulfide bonds (Figure 9.1). When cleaved mature insulin is secreted, it can act on other cells, but production of inactive preproinsulin guarantees that the peptide is not active in the pancreatic cells that produce it. Similar to this, to prevent the breakdown of components of the producing cell, the digestive enzyme trypsinogen, which breaks down proteins, is produced as an inactive polypeptide. Trypsinogen is broken down into the active enzyme trypsin by an enteropeptidase once it is secreted into the small intestine. Chaperones are necessary for the correct folding of proteins in eukaryotic cells, just like in prokaryotes. The chaperones BiP and calnexin in the endoplasmic reticulum bind developing polypeptides, while protein disulfide isomerases catalyze the creation of disulfide bonds between neighboring cysteine residues. Not only is proper folding necessary for the protein to reach its ideal configuration for activity, but it also shields portions of the protein from recognition by proteases that would otherwise break the protein down. Only correctly folded proteins are released from the endoplasmic reticulum and delivered in vesicles to the Golgi apparatus for additional processing, thanks to quality control systems. Proteins that are supposed to be secreted from the cell are then brought to the membrane by particular transport vesicles and released through the process of exocytosis.

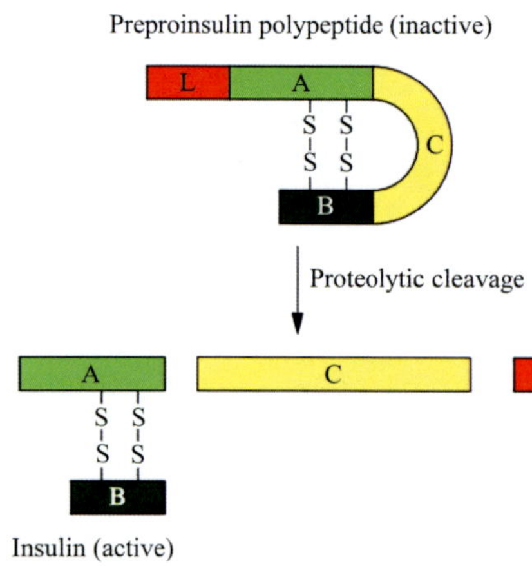

Figure 9.1. *Cleavage of inactive preproinsulin to yield active mature insulin. Proteases remove the leader peptide (L) and an internal peptide (C), yielding a peptide that consists of chains A and B.*

Source: Bernard R. Glick, Jack J. Pasternak, and Cheryl L. Patten.; Molecular biotechnology: principles and applications of recombinant DNA; ISBN 978-1-55581-498-4.

Glycosylation, the addition of particular sugars to specific amino acids, is a significant alteration that gives proteins their stability and unique binding characteristics. To target a protein to a specific location, for example, through interaction with a receptor molecule, or to increase protein stability by shielding it from proteases, proper protein glycosylation is crucial. It also influences protein folding, which influences conformation. Certain enzymes called glycosylases and glycosyltransferases bind oligosaccharides to freshly synthesized proteins in the endoplasmic reticulum and the Golgi apparatus of the cell. The same protein may be glycosylated differently in different tissues, which increases the heterogeneity of the protein. Protein engineering has the potential to improve a protein's effectiveness or change its activity because different sugar modifications can change a protein's properties. Glycosylation is necessary for the activity of several therapeutic proteins, such as blood factors, antibodies, certain interferons, and hormones.

O-linked glycosylation, which is the most prevalent type of glycosylation, involves attaching particular sugars to the hydroxyl group of either serine or threonine (Figure 9.2) as well as to the asparagine amide group (N-linked glycosylation) (Figure 9.3). The glycosylated portion of all human proteins makes up about 50%.

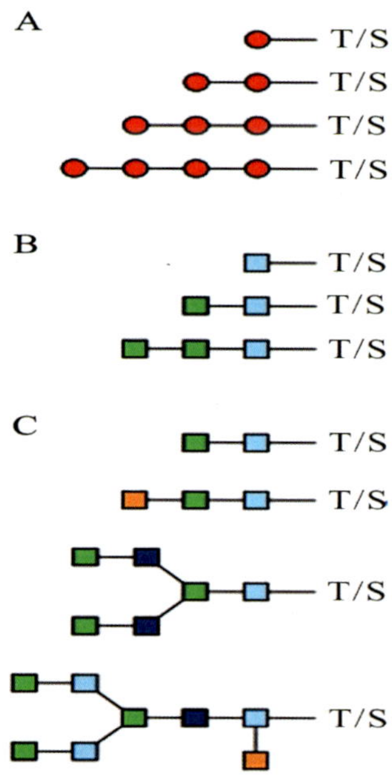

Figure 9.2. *Examples of some O-linked oligosaccharides in yeasts (A), insects (B), and mammals (C). O-linked oligosaccharides have a number of arrangements with different combinations of sugars. Some of the more prevalent forms are shown here. S, serine; T, threonine; red circles, mannose; darkblue squares, N-acetylglucosamine; light-blue squares, N-acetylgalactosamine; green squares, galactose; orange squares, sialic acid.*

Source: Bernard R. Glick, Jack J. Pasternak, and Cheryl L. Patten.; Molecular biotechnology: principles and applications of recombinant DNA; ISBN 978-1-55581-498-4.

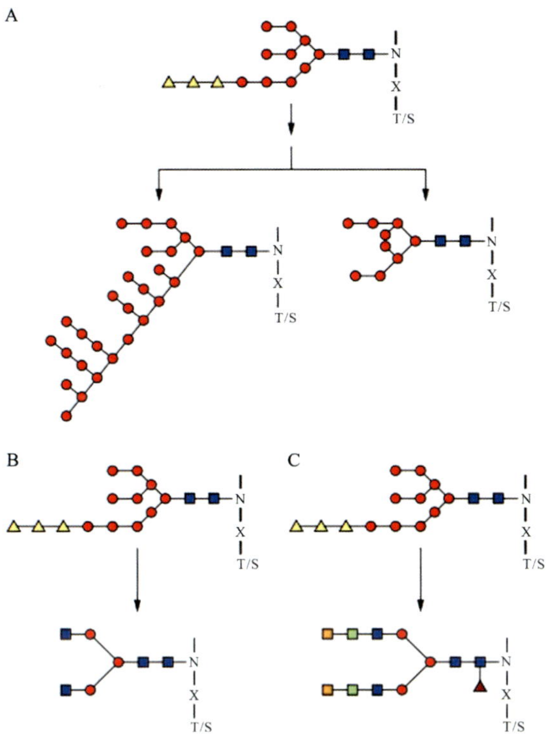

Figure 9.3. *Examples of some N-linked oligosaccharides in yeasts (A), insects (B), and mammals (C). All N-linked glycosylations in eukaryotes start with the same initial group, which is subsequently trimmed and then elaborated in diverse ways within and among species. Some yeast sites have 15 or fewer mannose units (core series), and others have more (outer-chain family). In S. cerevisiae, the chains frequently have 50 or more mannose units. An asparagine (N) residue next to any amino acid (X) followed by either threonine (T) or serine (S) can be targeted for glycosylation. Red circles, mannose; dark blue squares, N-acetylglucosamine; yellow triangles, glucose; green squares, galactose; orange squares, sialic acid; maroon triangle, fucose.*

Source: Bernard R. Glick, Jack J. Pasternak, and Cheryl L. Patten.; Molecular biotechnology: principles and applications of recombinant DNA; ISBN 978-1-55581-498-4.

Although the subsequent elaborations among yeasts, insects, and mammals are quite diverse, particularly for N-linked glycosylation, the initial core sugar groups added to these amino acid acceptor sites tend to be similar among eukaryotes. Additional changes to amino acids include acylation, γ-carboxylation, sulfation, phosphorylation, acylation, and the inclusion of C_{14} and C_{16} fatty acids, i.e., palmitoylation (also called palmitylation) and myristoylation (also called myristylation), in that order (Condreay and. Kost. 2007.)

Regretfully, there isn't a single eukaryotic host cell that can accurately modify every protein in all situations. A host cell may occasionally add strange sugars to real or fictitious amino acid sites, which can result in the creation of a highly antigenic protein or perhaps a protein that is malfunctioning. Recombinant proteins may still be valuable for scientific or commercial applications, even if they don't meet the strict requirements needed to qualify as a therapeutic agent.

To find out which eukaryotic expression system produces the most functional recombinant protein, it is necessary to test various systems. The quality of the recombinant protein that is generated is the main factor in selecting an expression system; other factors to take into account include product yield, ease of use, and production and purification costs.

9.2. GENERAL FEATURES OF EUKARYOTIC EXPRESSION SYSTEMS

Similar prerequisites apply to prokaryotes and eukaryotic hosts for the expression of a target protein. The vectors used to clone the target gene and transfer it to the host cell include specialized plasmids like the yeast 2μm plasmid, viruses specific to the host like the insect baculovirus, or artificial chromosomes like the yeast artificial chromosome (YAC). A sequence that permits polyadenylation of the mRNA, a selectable eukaryotic marker gene, eukaryotic transcriptional and translational stop signals, and a eukaryotic promoter that drives the transcription of the cloned gene of interest are all required for the vector (Figure 9.4). Since eukaryotic cells are technically challenging to work with for recombinant DNA procedures, the majority of eukaryotic vectors are shuttle vectors with two replication origins and selectable marker genes. Two sets of functions are found in the bacterium *Escherichia coli* and the eukaryotic host cell, respectively. If a plasmid is to be created using a eukaryotic expression vector, i.e., since it is extrachromosomal replicating DNA, its replication must also have a eukaryotic origin. Alternatively, to facilitate insertion into a chromosomal site, the vector must have a sequence complementary to a segment of the host chromosomal DNA if it is intended for stable integration into the host chromosomal DNA.

Transformation is the process of introducing DNA into cells, such as yeast and bacteria. The phrase in these systems refers to an inherited alteration brought about by the influx of exogenous (foreign) DNA. On the other hand, transformation in animal cells describes alterations in the characteristics of the growth of cells in culture once they develop cancer. To prevent misunderstandings, the term "transfection" has been used to refer to hereditary modifications in animal cells that result from the insertion of foreign DNA.

Yeasts can be transformed via three common methods: cell wall removal (protoplast formation), lithium acetate treatment, and electroporation. Animal cells that have been cultured can be transfected by electroporation, incubation with DNA that has been coprecipitated with either calcium phosphate or diethylaminoethyl (DEAE)-dextran. By exposing cells to brief electric current pulses, a process known as electroporation, DNA can enter cells temporarily through pores that are created. In addition, viruses, lipid-DNA complexes, and protein-DNA aggregates can be employed to introduce foreign DNA into an animal cell that is the recipient.

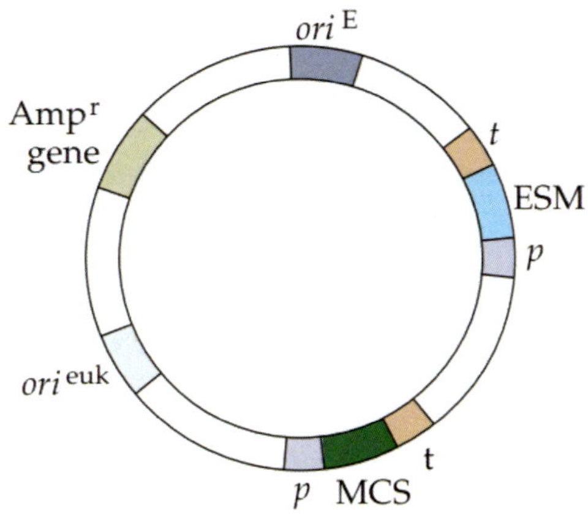

Figure 9.4. *Generalized eukaryotic expression vector. The major features of a eukaryotic expression vector are a eukaryotic transcription unit with a promoter (p), a multiple cloning site (MCS) for a gene of interest, and a DNA segment with termination and polyadenylation signals (t); a eukaryotic selectable marker (ESM) gene system; an origin of replication that functions in the eukaryotic cell (orieuk); an origin of replication that functions in E. coli (oriE); and an E. coli selectable marker (Ampr) gene.*

Source: Bernard R. Glick, Jack J. Pasternak, and Cheryl L. Patten.; Molecular biotechnology: principles and applications of recombinant DNA; ISBN 978-1-55581-498-4.

9.3. FUNGUS-BASED EXPRESSION SYSTEMS

Fungi are a good option for heterologous protein production because they have many of the molecular, genetic, and biochemical characteristics of other, "higher" eukaryotes. They can process eukaryotic proteins correctly, secrete large amounts of heterologous proteins, grow quickly in inexpensive media, and typically do not require the addition of growth factors to the growth medium. These growth advantages are similar to those of prokaryotes. Originally, cloned eukaryotic genes were expressed in large quantities using Saccharomyces cerevisiae as the host cell. It has been a longstanding application in conventional biotechnologies within the baking and brewing sectors. Fungal expression systems have been optimized for the expression of recombinant proteins and are now readily available in a variety of forms. Because it is frequently necessary to determine the ideal host for producing a specific target protein experimentally in a variety of systems, versatile expression vectors with broad host ranges have been created.

9.3.1. Saccharomyces cerevisiae Expression Systems

S. cerevisiae has been used to produce recombinant proteins at high levels. Using this one-celled yeast has a number of benefits. First, the biochemistry, genetics, and cell biology of the fungus are well understood. *S. cerevisiae,* whose genome sequence was finished in 1996, is frequently used as a model organism for cell function in research. It can be quickly grown in both small culture vessels and large-scale bioreactors to high cell densities on comparatively basic media. Furthermore, an endogenous yeast expression vector system can incorporate the 2μm plasmid, a naturally occurring plasmid, and several potent promoters that have been isolated and characterized from the yeast. Third, *S. cerevisiae* has a wide range of posttranslational modification capabilities. Fourth, the yeast typically secretes so little protein that purification of the result is simple when it is modified to release a recombinant protein extracellularly. Fifth, because of its years of use in the baking and brewing industries, *S. cerevisiae* has been listed by the U.S. Food and Drug Administration as a "generally recognized as safe" organism. It does not harbor human pathogens or produce fever-stimulating pyrogens. Therefore, the use of the organism for the production of human therapeutic agents (drugs or pharmaceuticals) does not require the same extensive experimentation demanded for unapproved host cells. Nowadays, a variety of *S. cerevisiae*-produced proteins are sold commercially as medications, vaccines, and diagnostic tools (Table 9.1). For instance, *S. cerevisiae* currently produces more than half of the insulin used worldwide. Major producers of a hepatitis B vaccine are also engineered *S. cerevisiae* strains (Hamilton and Gerngross. 2007).

Table 9.1. *Recombinant Proteins Produced by S. cerevisiae Expression Systems*

Vaccines	
	Hepatitis B virus surface antigen
	Malaria circumsporozoite protein
	HIV-1 envelope protein
Diagnostics	
	Hepatitis C virus protein
	HIV-1 antigens
Human therapeutic agents	
	Epidermal growth factor
	Insulin
	Insulin-like growth factor
	Platelet-derived growth factor
	Proinsulin
	Fibroblast growth factor
	Granulocyte-macrophage colony-stimulating factor
	α_1 -Antitrypsin
	Blood coagulation factor XIIIa
	Hirudin
	Human growth factor
	Human serum albumin

Source: Bernard R. Glick, Jack J. Pasternak, and Cheryl L. Patten.; Molecular biotechnology: principles and applications of recombinant DNA; ISBN 978-1-55581-498-4.

9.3.1.1. S. cerevisiae Vectors

S. cerevisiae expression vectors fall into three main categories: YACs, integrating vectors (yeast integrating plasmids [YIps]), and episomal, or plasmid, vectors (yeast episomal plasmids [YEps]). The most widely used of these to produce extracellular or intracellular heterologous proteins is the episomal vector. The vectors usually have characteristics that enable them to work with both *S. cerevisiae* and bacteria. An *E. coli* since the vector typically contains genes for the replication origin and bacterial antibiotic resistance, all manipulations can be carried out in *E. coli* before *S. cerevisiae* receives the vector to be expressed in.

The 2μm plasmid, a small circular plasmid with a high copy number that is present in most natural strains of *S. cerevisiae,* is the basis for the YEp vectors. The vector is maintained in more than 30 copies per cell and replicates independently of the host chromosome through a single replication origin (autonomous replicating sequence [ARS]/STB loci). Mutant host strains that need a specific amino acid (histidine, tryptophan, or leucine) or nucleotide (uracil) for growth are the foundation of many *S. cerevisiae* selection strategies. Because a particular nutrient needs to be added to the minimal growth medium, these strains are referred to as auxotrophic. In actuality, a functional (wild-type) copy of a gene that balances the mutated gene in the host strain is carried by the vector. For instance, only cells carrying the plasmid will grow when a YEp with a wild-type LEU2 gene is transformed into a mutant leu2 host cell and plated onto a leucine-free medium.

For the purpose of efficiently engineering the transcription of heterologous genes in yeast vectors, a variety of promoters derived from *S. cerevisiae* genes are available (Table 9.2). For the purpose of producing large amounts of recombinant protein at a specific time during large-scale growth, tightly regulatable, inducible promoters are generally preferred. In this case, the addition of galactose causes the galactose-regulated promoters to respond quickly by increasing transcription by a factor of 1,000. There are also constitutive, repressible, and hybrid promoters that combine the qualities of several promoters. Effective transcriptional termination is essential for maximum expression. In many YEp vectors, the promoter and terminator sequences come from the same gene.

A DNA coding sequence known as a signal sequence, signal peptide, or leader sequence is attached to a large number of heterologous genes. This sequence helps the recombinant protein pass through cell membranes and be released into the surrounding environment. This change was primarily motivated by the fact that purifying a secreted protein is far simpler than purifying one from a cell lysate. The mating factor α gene yields the *S. cerevisiae* signal sequence that is most frequently used. To increase the amount of protein secreted, synthetic leader sequences have also been developed. It is possible to fuse additional sequences onto the heterologous gene's coding sequence to stabilize the recombinant protein, shield it from proteolytic degradation, and provide an affinity tag—a particular amino acid sequence—that is used for selective purification. These additional amino acid sequences have a protease cleavage site attached to them, allowing the recombinant protein to be purified without containing them.

Table 9.2. *Promoters for S. cerevisiae Expression Vectors*

Promoter	Expression conditions	Status
Acid phosphatase (PH05)	Phosphate-deficient medium	Inducible medium
Alcohol dehydrogenase I (ADHI)	2–5% Glucose	Constitutive
Alcohol dehydrogenase II (ADHII)	0.1–0.2%	Glucose Inducible
Cytochrome c_1 (CYC1)	Glucose	Repressible
Gal-1-P Glc-1-P uridyltransferase	Galactose	Inducible
Galactokinase (GAL1)	Galactose	Inducible
Glyceraldehyde-3-phosphate		
dehydrogenase (GAPD, GAPDH)	2–5% Glucose	Constitutive
Metallothionein (CUP1)	0.03–0.1 mM	Copper Inducible
Phosphoglycerate kinase (PGK)	2–5% Glucose	Constitutive
Triosephosphate isomerase (TPI)	2–5% Glucose	Constitutive
UDP-galactose epimerase (GAL10)	Galactose	Inducible

Source: Bernard R. Glick, Jack J. Pasternak, and Cheryl L. Patten.; Molecular biotechnology: principles and applications of recombinant DNA; ISBN 978-1-55581-498-4.

Even in the presence of selection pressure, plasmid-based yeast expression systems are frequently unstable in large-scale (≥10 liters) growth conditions. A heterologous gene is inserted into the host genome to address this issue and create a more dependable production system. Many strategies have been developed for integrating a selectable marker gene and cloned gene into a *S. cerevisiae* chromosome. In summary, two DNA segments derived from the ends of a nonessential yeast gene are inserted between a heterologous gene containing transcription and translation control sequences specific to yeast and a functional selectable marker gene. In this case, the plasmid typically lacks a replication origin that is functional in yeast cells. After the plasmid is broken, *S. cerevisiae* is created from the linear fragment. The fragment of DNA containing both the target and marker genes is inserted into a particular chromosome location through a double recombination event between homologous sequences on the linearized plasmid and a chromosome in the host (Figure 9.5). Because linearized DNA is more likely to recombine with chromosome DNA than circular DNA, plasmid DNA has undergone linearization. Any non-integrated DNA is lost through subsequent cell divisions. This strategy's main flaw is the low yield of recombinant protein produced from a single gene copy.

One way to target an integrated heterologous gene to increase its copy number and, consequently, the total yield of the recombinant protein is to target repetitive DNA sequences found in yeast, such as ribosomal DNA (rDNA), which codes for ribosomal RNA and is present in about 200 copies. As an alternative, the heterologous gene can be incorporated into multiple copies of the 400 δ sequences found in the genome of *S. cerevisiae.* The segments of non-essential DNA elements originating from retrotransposons comprise the δ sequences.

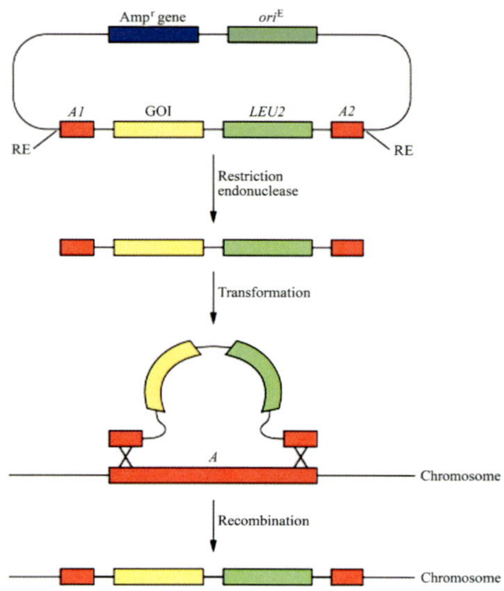

Figure 9.5. *Schematic representation of integration of DNA with a YIp vector. A selectable marker gene (LEU2) and a gene of interest (GOI) with transcription and translation control elements (not shown) are inserted into a YIp vector between two segments from the ends of a nonessential yeast gene (A1 and A2). The ampicillin resistance (Ampr) gene and the origin of replication (oriE) function in E. coli. A leucine-requiring (leu2) yeast strain is transformed with restriction endonuclease digested (RE) vector DNA because chromosomal DNA is more likely to recombine with linearized DNA than with circular DNA. The restriction endonuclease sites flank the segments from the nonessential gene. The DNA sequences at the ends of nonessential gene A undergo recombination (×) that leads to the incorporation of both the gene of interest and the LEU2 gene into the corresponding chromosome site. Transformants grow on medium that is not supplemented with leucine. Nonrecombined DNA is degraded.*

Source: Bernard R. Glick, Jack J. Pasternak, and Cheryl L. Patten.; Molecular biotechnology: principles and applications of recombinant DNA; ISBN 978-1-55581-498-4.

In short, chromosomal DNA sequences that are transcribed but not translated into proteins are known as retrotransposons. The reverse transcriptase enzyme uses the RNA molecules as templates to synthesize DNA, and the duplicated DNA sequences integrate into the chromosome at various locations. In one study, δ sequences were inserted with 10 copies of a heterologous gene, which resulted in a sizable amount of recombinant protein.

A YAC is made to clone a sizable DNA segment (100 kilobase pairs [kb]), which the host yeast cell keeps as a distinct chromosome. Because of its great stability, the YAC system has been used to physically map human genomic DNA, analyze large transcription units, and create genomic libraries made up of DNA from individual human chromosomes. Because it contains sequences that function as origins of DNA replication (ARS), yeast centromeres to guarantee that each daughter cell receives a copy of the YAC after cell division, and telomere sequences that are present at both ends after the YAC DNA has linearized for stability, a YAC vector resembles a chromosome (Figure 9.6). Sometimes a yeast marker gene is disrupted by cloning the input DNA into a different location. When recipient cells are grown on a specialized medium, a colorimetric response is seen in the absence of the marker gene's product. Alternatively, a selectable marker gene that is unrelated to the cloning site can be found in certain YAC vectors. Although they have the capacity to generate large quantities of either a single protein from multiple copies of the same gene or a heterologous protein with different subunits, YACs have not yet been employed as expression systems for the commercial production of heterologous proteins (Kost, Condreay, and Jarvis, 2005)

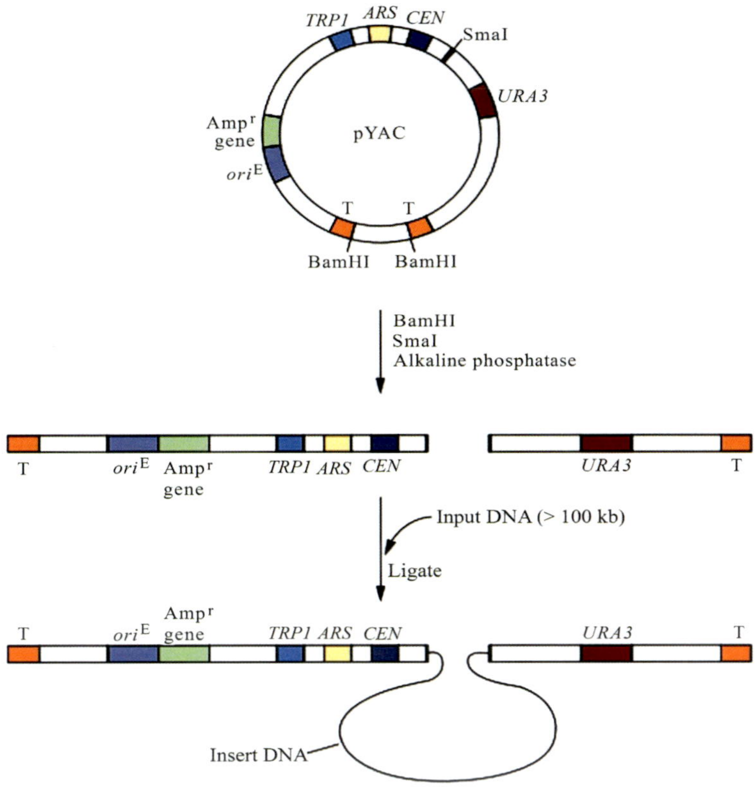

Figure 9.6. *YAC cloning system. The YAC plasmid (pYAC) has an E. coli selectable marker (Ampr) gene; an origin of replication that functions in E. coli (oriE); and yeast DNA sequences, including URA3, CEN, TRP1, and ARS. CEN provides centromere function, ARS is a yeast autonomous replicating sequence that is equivalent to a yeast origin of replication, URA3 is a functional gene of the uracil biosynthesis pathway, and TRP1 is a functional gene of the tryptophan biosynthesis pathway. The T regions are yeast chromosome telomeric sequences. The SmaI site is the cloning insertion site. pYAC is first treated with SmaI, BamHI, and alkaline phosphatase and then ligated with size-fractionated (100-kb) input DNA. The final construct carries cloned DNA and can be stably maintained in double-mutant ura3 and trp1 cells.*

Source: Bernard R. Glick, Jack J. Pasternak, and Cheryl L. Patten.; Molecular biotechnology: principles and applications of recombinant DNA; ISBN 978-1-55581-498-4.

9.3.1.2. Intracellular Production of Heterologous Proteins in S. cerevisiae

The fundamental characteristics of most *S. cerevisiae* intracellular expression systems are similar. Here, the process will be demonstrated by the synthesis of the human enzyme superoxide dismutase (SOD). An byproduct of aerobic organisms using oxygen is superoxide anion. This anion aids in both stimulating phagocytes' inflammatory response and guiding leukocytes to the infection site in humans. On the other hand, excessive amounts of the molecule and its derivatives can harm cells. The naturally occurring cytoplasmic enzyme Cu/Zn-SOD scavenges the superoxide radical and combines it with a hydrogen ion to form hydrogen peroxide, which is then broken down to water and oxygen by catalase or peroxidase, thereby reducing these potentially cytotoxic effects. Superoxide anion is also generated during reperfusion, the process of blood returning to an organ following blood loss during surgery. According to clinician conjecture, an organ receiving a reperfusion could be given Cu/Zn-SOD to prevent this source of superoxide anion damage. Furthermore, Cu/Zn-SOD may have therapeutic effects on inflammatory illnesses like rheumatoid arthritis, ankylosing spondylitis, osteoarthritis, and scleroderma. To prevent any negative immunological reactions that could arise from using an enzyme from a different species, an authentic human form of Cu/Zn-SOD is recommended for both of these uses.

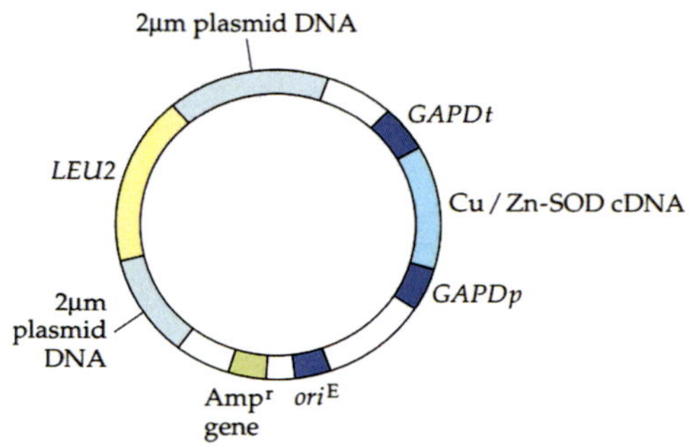

Figure 9.7. *S. cerevisiae expression vector. The cDNA for human Cu/Zn-SOD was cloned between the promoter (GAPDp) and termination–polyadenylation sequence (GAPDt) of the S. cerevisiae glyceraldehyde phosphate dehydrogenase gene. The LEU2 gene that was cloned between segments of the yeast 2µm plasmid DNA encodes a functional enzyme of the leucine biosynthesis pathway. The yeast origin of replication is included in the 2µm plasmid DNA. The ampicillin resistance (Ampr) gene and the E. coli origin of replication (oriE) are derived from plasmid pBR322.*

Source: Bernard R. Glick, Jack J. Pasternak, and Cheryl L. Patten.; Molecular biotechnology: principles and applications of recombinant DNA; ISBN 978-1-55581-498-4.

Initially, a complementary DNA (cDNA) for human Cu/Zn-SOD was cloned into an *E. coli* expression system. As expected, the *E. coli* host cells removed the initiator N-terminal methionine from the Cu/Zn-SOD pro- 4e 1st Proof 2nd Proof 3rd Proof 4rd Proof Fin Tein. However, the next amino acid (alanine) was not acetylated, as it is in human cells. To produce a fully functional protein, the human Cu/Zn-SOD cDNA was cloned into a YEp vector (Figure 9.7). This YEp vector contained (1) a 2µm plasmid origin of DNA replication, (2) a yeast gene for leucine biosynthesis (LEU2) for selection in yeast, (3) an *E. coli* origin of replication, (4) the ampicillin resistance (Amp^r) gene for selection in *E. coli*, and (5) the human Cu/Zn-SOD cDNA inserted between the promoter region of the yeast glyceraldehyde phosphate dehydrogenase gene (GAPDp) and a sequence containing the signals for transcription termination and polyadenylation of mRNA from the same gene (GAPDt). A leucine-defective (leu2) yeast strain was transformed with the vector, and the cells were plated onto medium that lacked leucine. Only cells with the functional LEU2 gene, which was supplied by the vector, could grow under these conditions. The GAPD promoter is transcribed continuously (constitutively) during cell growth. In this experiment, the yeast cells produced high levels of intracellular Cu/Zn-SOD that, like the authentic protein from human cells, had an acetylated N-terminal alanine residue (Barnes and Dickson, 2006).

9.3.1.3. Secretion of Heterologous Proteins by S. cerevisiae

Every glycosylated protein produced by *S. cerevisiae* is secreted, and in order for them to all get past the secretory system, they all need a leader sequence. As such, a leader sequence needs to be incorporated into the coding sequences of recombinant proteins whose biological activity depends on O- or N-linked sugars. Typically, the cDNA of the target gene is inserted right in front of (upstream) the leader sequence derived from the yeast mating type α-factor gene, known as prepro-α-factor. An active recombinant protein is secreted under these circumstances, and proper disulfide bond formation, proteolytic removal of the leader sequence, and suitable posttranslational modifications frequently take place. An endoprotease that detects the dipeptide Lys-Arg eliminates the leader peptide during this process. In order for the recombinant protein to have the correct amino acid residue at its N terminus after the leader peptide is removed, the Lys-Arg codons need to be placed close to the cDNA sequence. For instance, an *S. cerevisiae* strain carrying a YEp vector with the prepro-α-factor sequence added to the hirudin coding sequence synthesized and secreted an appropriately processed and active form of the protein hirudin. The leech *Hirudo medicinalis* is the source of the hirudin gene. This protein is a potent blood thinner that does not cause allergic reactions in people.

The amount of heterologous protein that can be produced in a liter of yeast culture has increased 100-fold in the last 10 years (from approximately 0.02 to 2 g/liter). The main cause of this increase is the increased ability to grow cultured cells to high densities; the amount of protein produced per cell has not changed significantly. The ability to increase a target gene's number of copies in a host cell and its expression levels has greatly advanced, but overexpressed proteins typically form intracellular aggregates—often linked to molecular chaperones—instead of being secreted into the medium, which makes purification easier. The improper folding of the polypeptide, the activation of cellular mechanisms to deal with the stress of protein overproduction, and the aberrant processing and release of the target protein from the endoplasmic reticulum are the main issues that need to be resolved to increase heterologous protein secretion in yeast cells.

To guarantee that the protein is correctly processed after synthesis, one of the main justifications for creating recombinant proteins in yeast as opposed to bacteria for use in human treatments. In eukaryotes, correct folding of proteins takes place in the endoplasmic reticulum and is supported by various proteins such as enzymes that form disulfide bonds, signal transduction proteins that keep track of the demand and availability of the machinery responsible for folding proteins, and proteases that break up improperly folded or aggregated proteins (Figure 9.8). The proper formation of disulfide bonds within proteins is facilitated by the eukaryotic enzyme protein disulfide isomerase. Protein configuration is altered by aberrant disulfide bond formation, which eliminates protein activity and results in instability. When a cell's ability to correctly fold and secrete proteins is exceeded, poor yields of overexpressed proteins frequently result. As a result, a number of techniques have been used to improve the host cell's ability to process proteins at levels above normal. For instance, overexpression of protein disulfide isomerases and molecular chaperones may lead to a higher yield of recombinant proteins, particularly those containing disulfide bonds. To investigate this theory, a transcription terminator sequence and the constitutive glyceraldehyde phosphate dehydrogenase promoter were cloned in a YIp vector, allowing the yeast protein disulfide isomerase gene to be integrated into a chromosomal location. When compared to the wild-type strain, the modified strain produced 16 times more protein disulfide isomerase. A 10 fold increase in recombinant protein secretion was observed when human platelet-derived growth factor B gene-carrying YEp vector was used to transform protein disulfide isomerase-overproducing cells. This was in comparison to transformed cells that had normal levels of protein disulfide isomerase. The secretion of proteins with disulfide bonds is specifically increased by overproduction of protein disulfide isomerase. In *S. cerevisiae* cells that overexpress the chaperone BiP, higher levels of secreted products are also obtained for the recombinant proteins leech hirudin, bovine prochymosin, and human erythropoietin.

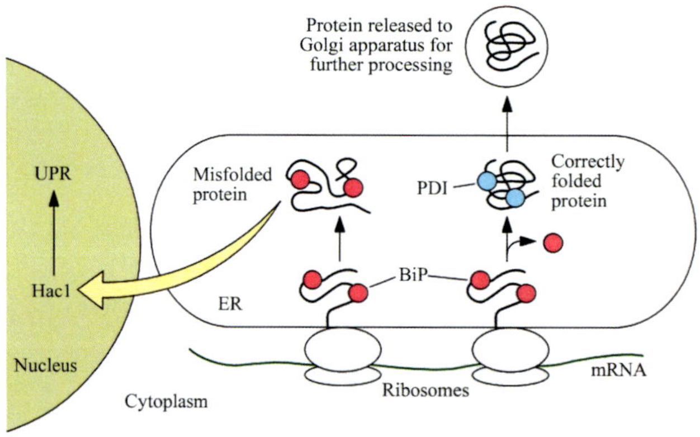

Figure 9.8. *Summary of protein folding in the endoplasmic reticulum of yeast cells. During synthesis on ribosomes associated with the endoplasmic reticulum (ER), nascent proteins are bound by the chaperones BiP and calnexin, which aid in the correct folding of the protein. Protein disulfide isomerases (PDI) catalyze the formation of disulfide bonds between cysteine amino acids that are nearby in the folded protein. Quality control systems ensure that only correctly folded proteins are released from the ER. Proteins released from the ER are transported to the Golgi apparatus for further processing. Prolonged binding of BiP to misfolded proteins leads to activation of the S. cerevisiae transcription factor Hac1, which controls the expression of several proteins that mediate the unfolded-protein response (UPR).*

Source: Bernard R. Glick, Jack J. Pasternak, and Cheryl L. Patten.; Molecular biotechnology: principles and applications of recombinant DNA; ISBN 978-1-55581-498-4.

The secretion of some heterologous proteins was enhanced by overexpressing the molecular chaperone BiP or protein disulfide isomerase; however, overexpressing a single chaperone might not have the desired effect and, in certain cases, might even accelerate the degradation of the target protein. This is due to the fact that correct protein folding necessitates the concerted efforts of numerous interdependent components (Figure 9.8). The levels of cochaperones or cofactors may be limiting, even in cases where levels of one chaperone are sufficient. Yeast cells' unfolded-protein response orchestrates the expression of multiple chaperones and cochaperones. The unfolded-protein response raises the expression of chaperones, protein disulfide isomerase, and other proteins involved in protein secretion when the demand for protein folding exceeds the endoplasmic reticulum's folding capacity. By preserving the proper ratios of all necessary components, engineering the proteins of the unfolded-protein response might be a more effective strategy to boost the cell's overall ability to fold proteins in a coordinated manner. The expression of Hac1 was targeted for genetic manipulation because the accumulation of unfolded proteins in the endoplasmic reticulum triggers the *S. cerevisiae* transcription factor Hac1, which in turn activates the expression of proteins of the unfolded-protein response. Hac1 overexpression in *S. cerevisiae* enhanced the secretion of α-amylase, a crucial industrial enzyme used for starch hydrolysis in various processes like oil drilling, paper recycling, and alcohol production.

9.3.2. Pichia Pastoris Expression Systems

Cloned genes from various sources have been successfully used to produce recombinant proteins in *S. cerevisiae*. However, protein yields are often low and expression levels are low. The tendency of *S. cerevisiae* to hyperglycosylate heterologous proteins by adding 50–150 mannose residues to N-linked oligosaccharide side chains is one of the main disadvantages of using this yeast, as it frequently modifies protein function. While the process of adding glycan chains to proteins in the endoplasmic reticulum lumen is similar in yeast and humans, there are notable differences in the subsequent processing that occurs once the protein is transferred to the Golgi apparatus. As a result, a hypermannosylated protein is produced in yeast and a sialylated protein in humans. The heterologous protein can become antigenic by forming α-1,3 bonds between its sugar residues (Figure 9.3). Additionally, proteins that are intended for secretion are frequently held in the periplasmic space, which lengthens the purification process and raises costs. Last but not least, *S. cerevisiae* reduces the amount of secreted protein by producing ethanol at high cell densities, which is harmful to the cells (the Crabtree effect). For these reasons, scientists have looked into eukaryotic cells and other yeast species that might serve as useful hosts for the synthesis of recombinant proteins.

P. pastoris is a methylotrophic yeast that is able to utilize methanol as a source of energy and carbon. It is an attractive host for recombinant protein production because glycosylation occurs to a lesser extent and the linkages between sugar residues are of the α-1,2 type, which are not allergenic to humans. With these natural characteristics as a starting point, a *P. pastoris* strain was extensively engineered with the aim of developing a "humanized" strain that glycosylates proteins in a manner identical to that of human cells. The same tiny (10-residue), branched oligosaccharide is added by both yeast and human cells to developing proteins in the endoplasmic reticulum (Figure 9.9). But after the protein is delivered to the Golgi apparatus, additional processing occurs differently, making this the final common precursor between the two cell types. Yeast cells add an α-1,6 mannose residue to the oligosaccharide in the Golgi apparatus, which causes hypermannosylation. In contrast, mammalian cells trim the precursor of some mannose residues and then add particular sugars one after the other until they produce a glycoprotein with an oligosaccharide that ends in sialic acid. To create a "humanized" strain, the enzyme responsible for addition of the α-1,6 mannose was first eliminated from *P. pastoris* to prevent hypermannosylation. Subsequently, the yeast genome was altered to contain the gene encoding a mannosidase, an enzyme that trims mannose to a precursor similar to that of humans, from the filamentous fungus *Trichoderma reesei*. Additionally, genes encoding the enzymes responsible for adding sugar residues one after the other to break the oligosaccharide chains in galactose were added. It should be highlighted that a secretion signal for the localization of the encoded protein to the Golgi apparatus was present in the coding sequences of every engineered gene. Finally, several genes for proteins that catalyze the synthesis, transport to the Golgi apparatus, and addition of sialic acid to the terminal galactose on the protein precursor were inserted into the *P. pastoris* genome. Properly sialylated recombinant proteins that can be used as human therapeutic agents have been produced by the "humanized" *P. pastoris*, including erythropoietin and antibodies.

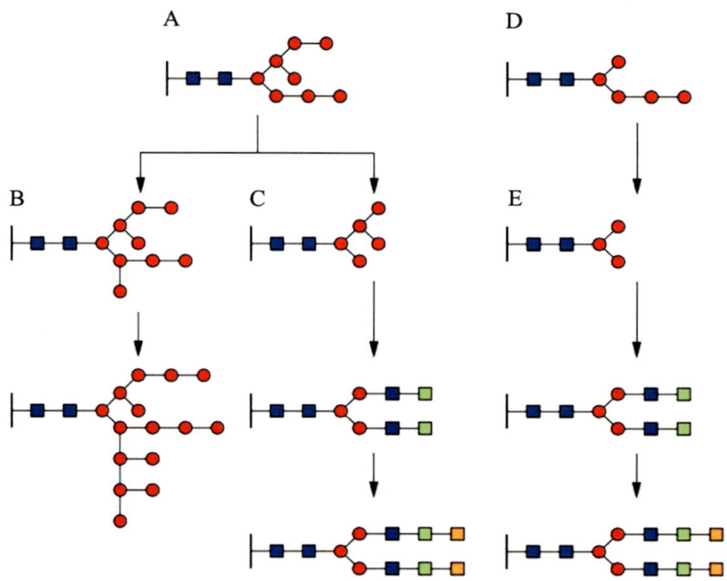

Figure 9.9. *Differential processing of glycoproteins in P. pastoris, humans, and "humanized" P. pastoris. Initial additions of sugar residues to glycoproteins in the endoplasmic reticulum are similar in human and P. pastoris cells.*

Source: Hamilton and Gerngross, Curr. Opin. Biotechnol. 18:387–392, 2007.

Note: (A). However, further N glycosylation in the Golgi apparatus differs significantly between the two cell types. N-glycans are hypermannosylated in P. pastoris (B), while in humans, mannose residues are trimmed and specific sugars are added, leading to termination of the oligosaccharide in sialic acid (C). P. pastoris cells have been engineered to produce enzymes that process glycoproteins in a manner similar to that of human cells. In "humanized" P. pastoris, a recombinant glycoprotein produced in the endoplasmic reticulum (D) is transported to the Golgi apparatus, where it is further processed to yield a properly sialylated glycoprotein (E). Blue squares, N-acetylglucosamine; red circles, mannose; green squares, galactose; orange squares, sialic acid.

Enzymes needed for the catabolism of methanol are expressed at very high levels when the cell is growing on it. The first enzyme in the methanol utilization pathway, alcohol oxidase, is encoded by the gene AOX1, and it can make up as much as 30% of the protein in the cell. The tightly controlled transcription of AOX1 results in a completely off-regulated gene that reacts quickly when methanol is added to the medium. Thus, under controlled conditions, the AOX1 promoter is a great option for producing large amounts of recombinant protein. Furthermore, during large-scale fermentations, the cloned gene's induction can be timed to maximize the production of recombinant proteins. In contrast to *S. cerevisiae*, *P. pastoris* does not synthesize ethanol, which can limit cell yields; therefore, very high cell densities of *P. pastoris* are attained, with the concomitant secretion of large quantities of protein. *P. pastoris* normally secretes very few proteins, thus simplifying the purification of secreted recombinant proteins.

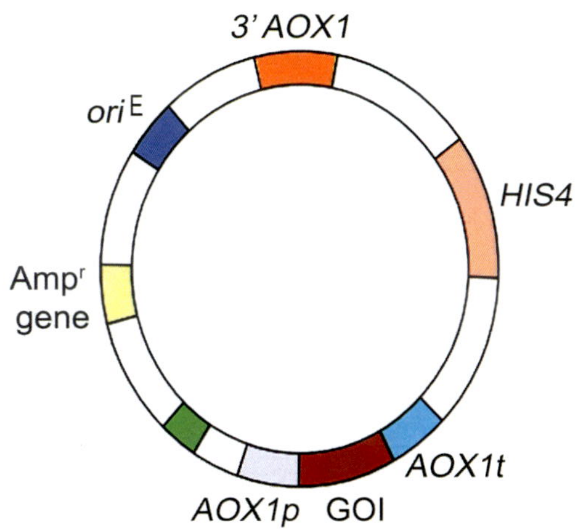

Figure 9.10. *P. pastoris integrating expression vector. The gene of interest (GOI) is cloned between the promoter (AOX1p) and termination-polyadenylation sequence (AOX1t) of the P. pastoris alcohol oxidase 1 gene. The HIS4 gene encodes a functional histidinol dehydrogenase of the histidine biosynthesis pathway. The ampicillin resistance (Ampr) gene and an origin of replication (oriE) function in E. coli. The segment marked 3′ AOX1 is a piece of DNA from the 3′ end of the alcohol oxidase 1 gene of P. pastoris. A double recombination event between the AOX1p and 3′ AOX1 regions of the vector and the homologous segments of chromosome DNA results in the insertion of the DNA carrying the gene of interest and the HIS4 gene.*

Source: Bernard R. Glick, Jack J. Pasternak, and Cheryl L. Patten.; Molecular biotechnology: principles and applications of recombinant DNA; ISBN 978-1-55581-498-4.

Many *P. pastoris* expression vectors have been devised, each one having more or less the same format. The basic features include a gene of interest under the control of promoter and transcription termination sequences from *P. pastoris* AOX1 gene, an *E. coli* origin of replication and selectable marker gene, and a yeast selectable marker gene

(Figure 9.10). The addition of a signal sequence from either the *P. pastoris* phosphatase PHO1 gene or another yeast gene facilitates the secretion of a recombinant protein. To prevent plasmid instability issues during prolonged growth, the majority of *P. pastoris*. The goal of *P. pastoris* vectors is to integrate themselves into the host genome, typically through the rDNA, HIS4 gene for histidine biosynthesis, or AOX1 gene. A single (Figure 9.11A) or multiple insertions of the engineered gene of interest and a yeast selectable marker gene are made into a particular chromosome site. (Figure 9.11B) or a double. Recombination event. The *P. pastoris* expression system has been used to produce more than 100 different biologically active proteins from bacteria, fungi, invertebrates, plants, and mammals, including humans. Many of these proteins, such as the hepatitis B virus surface antigen, human serum albumin, and bovine lysozyme, are identical to the native proteins and thus authentic. (Barnes and Dickson, 2006)

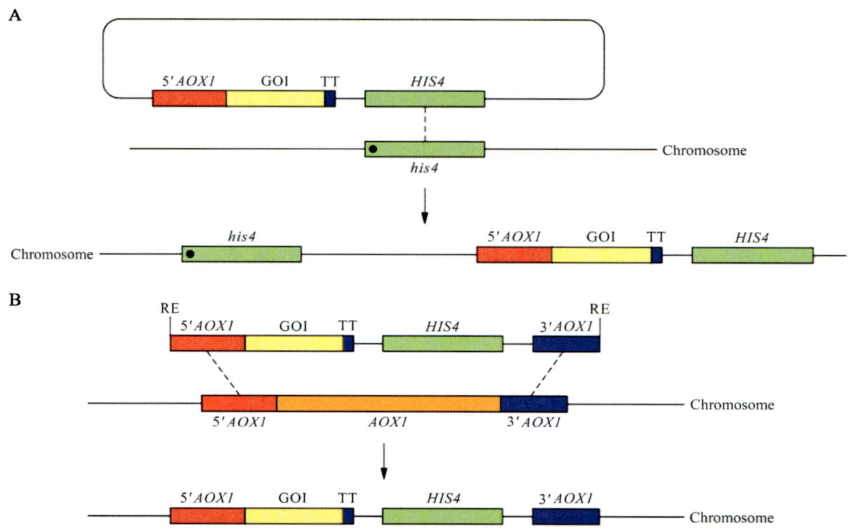

Figure 9.11. *Integration of DNA into a specific P. pastoris chromosome site by single (A) or double (B) recombination.*

Source: Bernard R. Glick, Jack J. Pasternak, and Cheryl L. Patten.; Molecular biotechnology: principles and applications of recombinant DNA; ISBN 978-1-55581-498-4.

9.4. BACULOVIRUS–INSECT CELL EXPRESSION SYSTEMS

A vast and varied class of viruses known as baculoviruses only infect arthropods, which includes a wide variety of insect species. They do not spread to other animals. Two distinct baculovirus forms are generated throughout the infection cycle (Figure 9.12). The virus in its occluded form is what starts the infection. In this form, the viral virions, or nucleocapsids, are grouped together in a matrix composed of the polyhedrin protein. The polyhedron-shaped occluded virions enclosed in this protein matrix are shielded from environmental agents' inactivation. After the virus enters the insect's midgut—usually through the consumption of contaminated plant material—the alkaline gut environment causes the polyhedrin matrix to dissolve, allowing the virions to enter midgut cells and start the infection cycle in the nucleus. As viral particles, or single nucleocapsids, bud off from an infected cell and infect other midgut cells, the infection can spread from cell to cell within the insect midgut. Although it can infect cultured insect cells, this virus form, referred to as the budding form, is not embedded in a polyhedrin matrix and is not contagious to other individual insect hosts. The morphology of plaques produced by the budding form of baculovirus in insect cell cultures differs from that of the occluded form. Massive amounts of polyhedrin protein are produced in the last stages of the infection cycle in the insect host, approximately 36 to 48 hours after infection. This process lasts for 4 to 5 days, or until the infected cells burst and the host organism dies. Released occluded virions have the ability to infect fresh hosts.

There is a very potent promoter for the polyhedrin (polyh) gene, and transcription from this promoter can produce up to 25% of the mRNA in virus-infected cells. In addition, the production of viruses does not require the polyhedrin protein. Therefore, the theory went that substituting a heterologous protein's coding sequence for the polyhedrin gene and then infecting grown insect cells would produce a significant amount of the heterologous protein. Furthermore, it was expected that the recombinant protein would closely, if not exactly, mimic the original protein's authentic form due to the similarities between insects and mammals' posttranslational modification systems. These assumptions led to the development of a baculovirus–insect cell expression vector system. When introducing target genes for the production of heterologous proteins in insect cells, baculoviruses have proven to be an extremely effective delivery system. With the help of this system, over a 1,000 distinct proteins, including enzymes, transport proteins, receptors, and secreted proteins, have been created for a range of uses (Table 9.3) (Ansari and Emery, 1998).

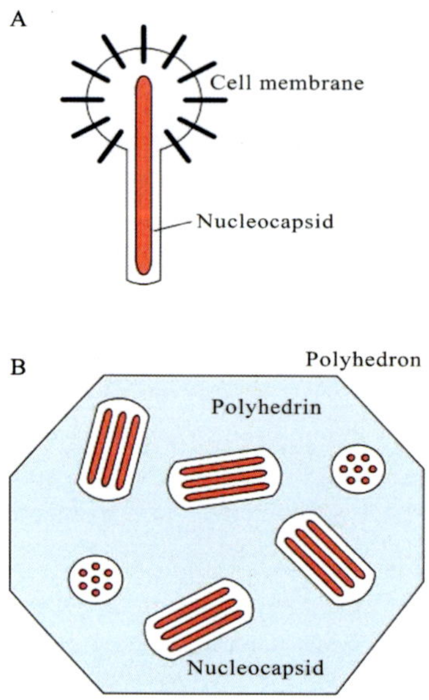

Figure 9.12. *Budded (A) and occluded (B) forms of AcMNPV. During budding, a nucleocapsid becomes enveloped by the membrane of an infected cell. A polyhedron consists of clusters of nucleocapsids (occluded virions) embedded in various orientations in a polyhedrin matrix.*

Source: Bernard R. Glick, Jack J. Pasternak, and Cheryl L. Patten.; Molecular biotechnology: principles and applications of recombinant DNA; ISBN 978-1-55581-498-4.

Table 9.3. *Some of the Recombinant Proteins that have been Produced by the Baculovirus Expression Vector System*

α-Interferon Adenosine deaminase Anthrax antigen β-Amyloid precursor protein β-Interferon Bovine rhodopsin Cystic fibrosis transmembrane conductance regulator Dengue virus type 1 antigen Erythropoietin	G-protein-coupled receptors HIV-1 envelope protein HSV capsid proteins Human alkaline phosphatase Human DNA polymerase α Human pancreatic lipase Influenza virus hemagglutinin Interleukin-2 Lassa virus protein	Malaria proteins Mouse monoclonal antibodies Multidrug transporter protein Poliovirus proteins Pseudorabies virus glycoprotein 50 Rabies virus glycoprotein Respiratory syncytial virus antigen Simian rotavirus capsid antigen Tissue plasminogen activator

Source: Bernard R. Glick, Jack J. Pasternak, and Cheryl L. Patten.; Molecular biotechnology: principles and applications of recombinant DNA; ISBN 978-1-55581-498-4.

The particular baculovirus known as *Autographa californica* multiple nuclear polyhedrosis virus (AcMNPV) has been utilized widely as an expression vector. A. AcMNPV infects over 30 other insect species, including *A. californica* (the alfalfa looper). Moreover, this virus thrives on a variety of insect cell lines. The most widely utilized cell line for genetically modified AcMNPV comes from the *Spodoptera frugiperda* fall armyworm. The polyhedrin promoter is very active in these cells, and high amounts of polyhedrin are synthesized during infections with wild-type baculovirus. The Bombyx mori nuclear polyhedrosis virus has been utilized to express heterologous proteins in silkworm larvae, albeit to a lesser degree (Ansari and Emery, 1998).

9.4.1. Baculovirus Expression Vectors

The creation of a transfer vector is the initial stage in the recombinant AcMNPV production process, which will be utilized to introduce the desired gene into the insect host cell. An serves as the transfer vector *E. coli*-based plasmid that contains an AcMNPV DNA segment containing a multiple cloning site, the polyhedrin termination and polyadenylation signal regions, the polyhedrin promoter region, and an adjacent section of upstream AcMNPV DNA (Figure 9.13). In this DNA block, the polyhedrin gene's coding region has been removed. Regions for homologous recombination with AcMNPV are provided by the transfer vector's upstream and downstream AcMNPV DNA segments. The transfer vector is propagated in *E. coli* after a desired gene is inserted into the multiple cloning site between the termination sequences and the polyhedrin promoter (Figure 9.13).

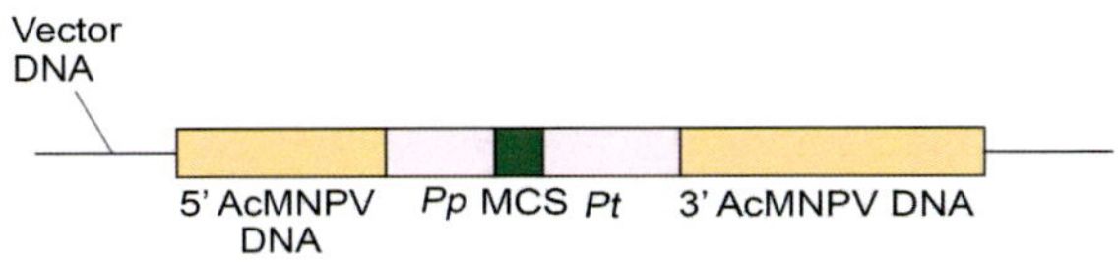

Figure 9.13. *Organization of the expression unit of a baculovirus (AcMNPV) transfer vector. The gene of interest is inserted into the multiple cloning site (MCS) that lies between the polyhedrin gene promoter (Pp) and polyhedrin gene transcription termination (Pt) sequences. The AcMNPV DNA upstream from the polyhedrin promoter (5′ AcMNPV DNA) and downstream from the polyhedrin transcription termination sequence (3′ AcMNPV DNA) provides sequences for integration of the expression unit by homologous recombination into an AcMNPV genome.*

Source: Bernard R. Glick, Jack J. Pasternak, and Cheryl L. Patten.; Molecular biotechnology: principles and applications of recombinant DNA; ISBN 978-1-55581-498-4.

Subsequently, the cloned gene transfer vector and AcMNPV DNA are cotransfected into insect cells grown in culture. A double-crossover recombination event takes place at homologous polyhedrin gene sequences on the transfer vector and in the AcMNPV genome in some of the doubly transfected cells, resulting in the integration of the cloned gene containing polyhedrin promoter and termination regions into the AcMNPV DNA (Figure 9.14), along with the simultaneous deletion of the polyhedrin gene. Recombinant baculovirus is isolated from unique zones of cell lysis (occlusion-negative plaques) caused by virions devoid of the polyhedrin gene.

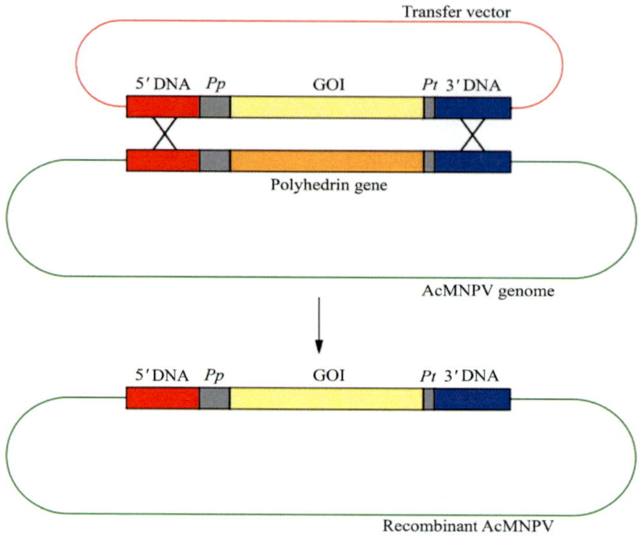

Figure 9.14. *Replacement of the AcMNPV polyhedrin gene with an expression unit from a transfer vector. A double crossover event (×) between homologous DNA segments of the transfer vector and the AcMNPV genome results in the integration of the expression unit into the AcMNPV genome. GOI, gene of interest; Pp, polyhedrin promoter; Pt, polyhedrin polyadenylation and termination sequence.*

Source: Bernard R. Glick, Jack J. Pasternak, and Cheryl L. Patten.; Molecular biotechnology: principles and applications of recombinant DNA; ISBN 978-1-55581-498-4.

DNA hybridization or a polymerase chain reaction (PCR) assay can be used to confirm the presence of recombinant baculovirus. To facilitate the detection of recombinant plaques, the *E. coli* lacZ gene, which encodes β-galactosidase, is put under the control of a baculovirus promoter that is turned on during the early to late stages of the infection cycle, and this construct is made part of the DNA unit that is incorporated into the AcMNPV genome. Recombinant plaques turn blue when a chromogenic substrate for β-galactosidase is added to the medium. Additional rounds of infection of insect cells with viruses from occlusion-negative plaques increase the concentration of recombinant viruses. Heterologous protein is harvested 4 to 5 days after host insect cells are infected with a high-titer recombinant baculovirus stock.

9.4.2. Increasing the Yield of Recombinant Baculovirus

Improvements have been made to the original procedure to increase the frequency of recombinant viruses because the identification of occlusion-negative plaques is subjective and the low frequency of recombination (~0.1. percent) between the AcMNPV DNA and the transfer plasmid makes purification of recombinant baculovirus laborious. To enhance the frequency of recombinant plaques to approximately 30 percent, a straightforward yet efficient method involves linearizing the AcMNPV genome before transfection into insect cells. Reduced infectivity of linearized baculovirus genomes results in fewer nonrecombinant plaques, and an infective closed circular DNA molecule is established through a double-crossover event between a linearized AcMNPV DNA molecule and a circular transfer vector.

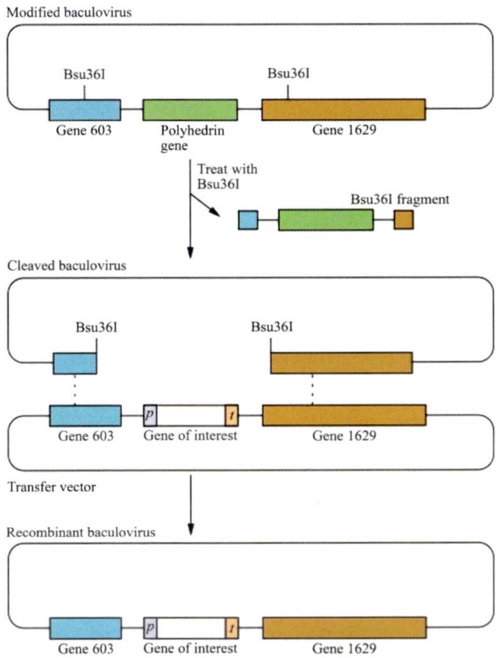

Figure 9.15. *Production of recombinant baculovirus. Single Bsu36I sites are engineered into gene 603 and a gene (1629) that is essential for AcMNPV replication. These genes flank the polyhedrin gene in the AcMNPV genome. After a baculovirus with two engineered Bsu36I sites is treated with Bsu36I, the segment between the Bsu36I sites is deleted. Insect cells are cotransfected with a Bsu36I-treated baculovirus DNA and a transfer vector with a gene of interest under the control of the promoter (p) and terminator (t) elements of the polyhedrin gene and the complete sequences of both genes 603 and 1629. A double crossover event (dashed lines) generates a recombinant baculovirus with a functional gene 1629. With this system, almost all of the progeny baculoviruses are recombinant.*

Source: Bernard R. Glick, Jack J. Pasternak, and Cheryl L. Patten.; Molecular biotechnology: principles and applications of recombinant DNA; ISBN 978-1-55581-498-4.

To create a system that yields recombinant baculovirus plaques at an even higher frequency, the linearization technique was refined. Two Bsu36I sites were inserted on either side of the polyhedrin gene to modify the AcMNPV genome (Figure 9.15). One is located in the open reading frame 603 [ORF603] gene, while the other is located in the ORF1629 gene, which is necessary for the replication of the virus. This modified baculovirus's DNA does not replicate when it is transfected into insect cells after being treated with Bsu36I because a portion of the critical gene (ORF1629) is absent. A transfer vector containing the desired gene is created as part of this system, sandwiched between intact copies of genes 603 and 1,629. Insect cells that were previously transfected with linearized, replication-defective AcMNPV DNA lacking the segment between the two Bsu36I sites are then exposed to this transfer vector. A double-crossover event integrates the cloned gene into the AcMNPV genome and restores a functional copy of ORF1629 (Figure 9.15). More than 90% of the baculovirus plaques in this system are recombinant (Ansari and Emery, 1998).

9.4.3. Integration of Target Genes into Baculovirus by Site-Specific Recombination

To eliminate the need to use plaque assays to identify and purify recombinant viruses, several methods have been developed that introduce the target gene into the baculovirus genome at a specific nucleotide sequence by recombination, either in an intermediate bacterial host, such as *E. coli*, or *in vitro* outside of a host cell. Transfection of insect cells is required only for the production of the heterologous protein. AcMNPV DNA can be maintained in *E. coli* as a plasmid known as a bacmid, which is a baculovirus– plasmid hybrid molecule. In addition to AcMNPV genes, the bacmid contains an origin of replication for propagation in *E. coli*, a kanamycin resistance gene for selection of the bacmid, and an integration site (attachment site) that is inserted into the lacZ′ gene without impairing its function (Figure 9.16A). Another component of this system is the transfer vector that carries the gene of interest cloned between the polyhedrin promoter and a terminator sequence. In the transfer vector, the target gene expression unit (expression cassette) and a gentamicin resistance gene are flanked by DNA attachment sequences that can bind to the attachment site in the bacmid (Figure 9.16B). An ampicillin resistance gene lies outside the expression cassette for selection of the transfer vector.

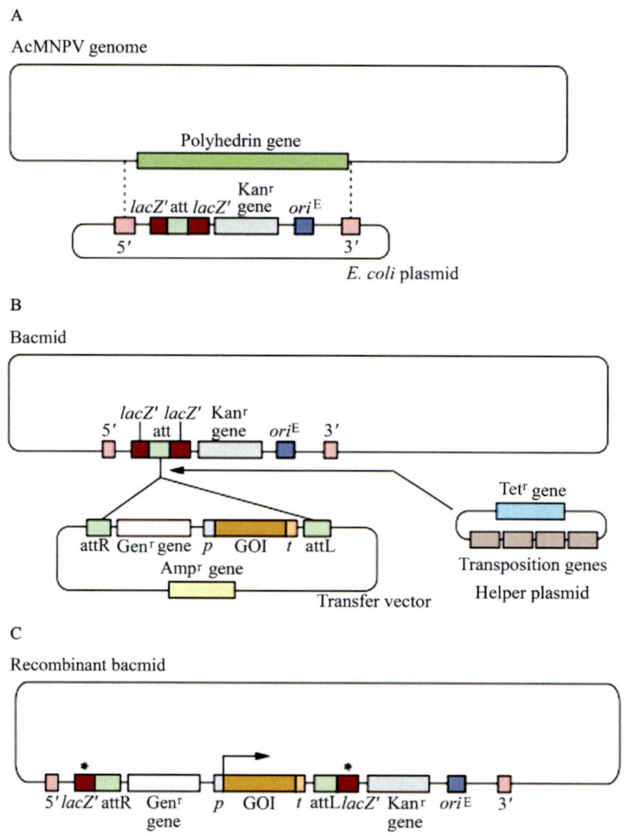

Figure 9.16. *Construction of a recombinant bacmid.*

Source: Bernard R. Glick, Jack J. Pasternak, and Cheryl L. Patten.; Molecular biotechnology: principles and applications of recombinant DNA; ISBN 978-1-55581-498-4.

Note: (A) An E. coli plasmid is incorporated into the AcMNPV genome by a double crossover event (dashed lines) between DNA segments (5′ and 3′) that flank the polyhedrin gene to create a shuttle vector (bacmid) that replicates in both E. coli and insect cells. The gene for resistance to kanamycin (Kanr), an attachment site (att) that is inserted in frame in the lacZ′ sequence, and an E. coli origin of replication (oriE) are introduced as part of the plasmid DNA. (B) The transposition proteins encoded by genes of the helper plasmid facilitate the integration (transposition) of the DNA segment of the transfer vector that is bounded by two attachment sequences (attR and attL).

The gene for resistance to gentamicin (Genr) and a gene of interest (GOI) that is under the control of the promoter (p) and transcription terminator (t) elements of the polyhedrin gene are inserted into the attachment site (att) of the bacmid. The helper plasmid and transfer vector carry the genes for resistance to tetracycline (Tetr) and ampicillin (Ampr), respectively. (C) The recombinant bacmid has a disrupted lacZ′ gene (*). The right-angled arrow denotes the site of initiation of transcription of the cloned gene after transfection of the recombinant bacmid into an insect cell. Cells that are transfected with a recombinant bacmid are not able to produce functional β-galactosidase.

To mediate recombination between the attachment sites on the transfer vector and the bacmid and to carry a tetracycline resistance gene, bacterial cells harboring a bacmid are cotransformed with the transfer vector and a helper plasmid. Following recombination, the lacZ′ gene's reading frame is destroyed when the DNA segment bounded by the two attachment sites on the transfer vector—the expression cassette carrying the target gene—is transposed into the attachment site on the bacmid (Figure 9.16). As a result, in the presence of 5-bromo-4-chloro-3-indolyl-β-d-galactopyranoside (X-Gal) and isopropyl-β-d-thiogalactopyranoside (IPTG), bacteria with recombinant bacmids form white colonies. Furthermore, only a recombinant bacmid and no transfer or helper plasmids are present in white colonies that are sensitive to ampicillin and tetracycline but resistant to gentamicin and kanamycin. Following all of these modifications, PCR can be used to verify the cloned gene's integrity. Ultimately, the cloned gene can be transcribed and the heterologous protein can be produced by transfecting insect cells with recombinant bacmid DNA.

A different technique uses the bacteriophage λ integration system to recombine λ DNA into the genome of a host bacterium at specific sites. The baculovirus expression vector is constructed through *in vitro* genetic manipulations, and the recombinant virus is then transferred into the insect host for protein production and selection. First, the tk gene, which codes for the herpes simplex virus enzyme thymidine kinase, was flanked by two distinct integration sequences (attachment sites attR1 and attR2) derived from bacteriophage λ that was engineered into the AcMNPV genome (Figure 9.17A). Thymidine kinase is a synthetic nucleotide analog that is toxic to insect cells. This product, ganciclovir, is part of the engineered AcMNPV construct that will be selected against nonrecombinant viruses in the future. The polyhedrin gene promoter was positioned downstream of the attR1-tk-attR2 DNA segment (cassette). Following that, the target gene was placed between two attachment sites (attL1 and attL2) on a donor plasmid, which matched the attachment sites in the engineered AcMNPV genome (Figure 9.17B). Recombination between the attachment sites flanking the target gene on the plasmid and flanking the tk gene in the AcMNPV genome integrates the target gene into the AcMNPV genome. The *in vitro* reaction mixture is supplemented with integration enzymes (integrase, excisionase, and integration host factor) derived from bacteriophage λ to facilitate site-specific recombination between the attachment sequences (Figure 9.17C). The recombinant baculovirus is then transfected into insect cells for selection and replication. The baculovirus genome's thymidine kinase gene is excised through recombination, so in the presence of ganciclovir, only cells transfected with viruses containing the target gene will endure (Ansari and Emery, 1998).

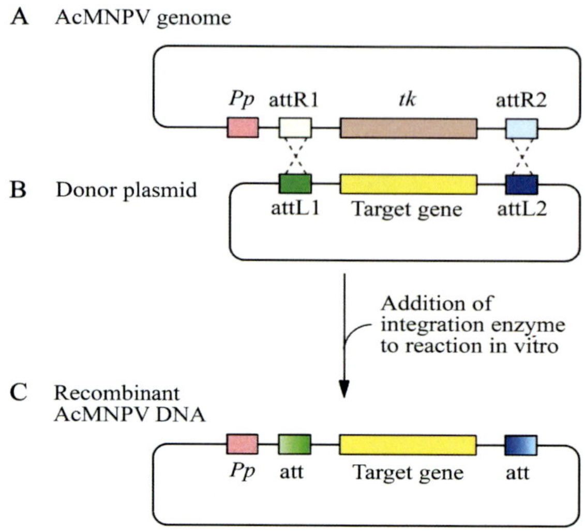

Figure 9.17. *Construction of a recombinant baculovirus expression vector in vitro.*

Note: (A) The AcMNPV genome was engineered to carry the thymidine kinase gene (tk) between attachment sequences attR1 and attR2 and downstream of the baculovirus polyhedrin gene promoter (Pp). (B) The target gene is cloned into a donor plasmid between attachment sequences attL1 and attL2. (C) The AcMNPV DNA carrying the tk gene and the donor plasmid are combined in a test tube, and integration enzymes are added to catalyze the recombination between corresponding attR and attL sites. Recombination results in the replacement of the tk gene with the target gene in the AcMNPV DNA so that, following transfection with the recombinant AcMNPV DNA, the insect cells are resistant to the toxic effects of ganciclovir and express the target gene from the polyhedrin gene promoter.

Source: Bernard R. Glick, Jack J. Pasternak, and Cheryl L. Patten.; Molecular biotechnology: principles and applications of recombinant DNA; ISBN 978-1-55581-498-4.

CASE STUDY

HETEROLOGOUS PROTEIN PRODUCTION IN EUKARYOTIC CELLS

Background: Biotech Innovations Ltd. specializes in developing biopharmaceuticals, including therapeutic proteins used in treating various diseases. They are focused on producing a complex protein, human insulin-like growth factor 1 (IGF-1), which is essential for growth and development.

Problem Statement: Producing sufficient quantities of functional IGF-1 is challenging due to its complex structure and post-translational modifications. Traditional bacterial expression systems are inadequate as they cannot perform eukaryotic-like modifications necessary for proper protein folding and activity. Biotech Innovations Ltd. seeks a robust method for heterologous protein production that ensures high yields of bioactive IGF-1.

Solution: To address these challenges, Biotech Innovations Ltd. adopts a eukaryotic expression system using mammalian cells. This approach allows for the production of IGF-1 with correct folding and post-translational modifications essential for biological activity.

Implementation

- Vector Design and Cell Line Selection:
  - **Expression Vector:** Biotech Innovations Ltd. constructs an expression vector containing the gene encoding human IGF-1, along with regulatory elements (promoters, enhancers) suitable for driving gene expression in mammalian cells.
  - **Host Cell Line:** They select a well-characterized mammalian cell line, such as Chinese hamster ovary (CHO) cells or HEK293 cells, known for their ability to efficiently produce complex proteins and perform eukaryotic-like post-translational modifications.
- Transfection and Cell Culture:
  - **Transfection:** The expression vector carrying the IGF-1 gene is transfected into the selected mammalian cell line using suitable transfection methods (e.g., lipid-mediated transfection, electroporation).
  - **Cell Culture:** Transfected cells are cultured in bioreactors under optimized conditions (temperature, pH, nutrient supply) to promote cell growth and protein production.
- Protein Expression and Purification:

  - **Expression Optimization:** Biotech Innovations Ltd. optimizes culture conditions and medium composition to enhance IGF-1 expression levels while maintaining cell viability and productivity.
  - **Purification:** They employ chromatographic techniques (e.g., affinity chromatography, ion exchange chromatography) to purify bioactive IGF-1 from cell culture supernatants or cell lysates, removing impurities and ensuring high product purity.
- Quality Control and Characterization:
  - **Bioactivity Assays:** The purified IGF-1 undergoes bioactivity assays to confirm its biological function, such as promoting cell growth or activating downstream signaling pathways.
  - **Analytical Methods:** Biotech Innovations Ltd. uses analytical techniques (e.g., SDS-PAGE, mass spectrometry) to verify the protein's molecular weight, identity, and post-translational modifications (e.g., glycosylation).
- Scale-Up and Commercialization:
  - Scale-Up: Once optimized, the production process is scaled up to industrial levels in bioreactors to meet commercial demand for IGF-1.
  - Regulatory Compliance: They adhere to regulatory guidelines (e.g., Good Manufacturing Practices, GMP) for biopharmaceutical production, ensuring product quality, safety, and consistency.

Results: Biotech Innovations Ltd. successfully produces bioactive IGF-1 in mammalian cells using a eukaryotic expression system. They achieve high yields of properly folded protein with authentic post-translational modifications, suitable for therapeutic applications.

Conclusion: This case study demonstrates the efficacy of using eukaryotic expression systems in heterologous protein production, particularly for complex proteins like IGF-1. By leveraging mammalian cell culture technology, Biotech Innovations Ltd. overcomes the limitations of bacterial systems and achieves reliable production of biopharmaceuticals with enhanced bioactivity and therapeutic potential.

SUMMARY

- Heterologous (recombinant) proteins from cloned DNA originating from a wide variety of organisms have been successfully produced using prokaryotic expression systems. Although the expression of any gene from any source organism in a prokaryotic host is theoretically possible, in practice, the eukaryotic proteins produced in bacteria do not always have the desired biological activity or stability.
- In prokaryotes, the steps in protein synthesis are not compartmentalized, and therefore, translation of messenger RNA (mRNA) occurs concurrently with transcription; as soon as the nascent transcript emerges from RNA polymerase, it is accessible to the ribosome to begin translation.
- All eukaryotic expression vectors have the same basic format. The gene of interest, which may be equipped with sequences that facilitate the secretion and purification of the heterologous protein, is under the control of eukaryotic promoter and polyadenylation and transcription terminator sequences. To simplify both maintenance and recombinant DNA manipulations, eukaryotic expression vectors are routinely maintained in *E. coli*.
- Fungi share many of the molecular, genetic, and biochemical features of other, "higher" eukaryotes and are therefore a good choice for heterologous protein production. They have growth advantages similar to those of prokaryotes, such as rapid growth in low-cost medium; generally, do not require growth factors to be added to the growth medium; can correctly process eukaryotic proteins; and can secrete large amounts of heterologous proteins.
- The YEp vectors are based on the high-copy-number 2 µm plasmid, a small, circular plasmid found in most natural strains of *S. cerevisiae.* The vector replicates independently of the host chromosome via a single origin of replication (autonomous replicating sequence [ARS]/STB loci) and is maintained in more than 30 copies per cell.
- The plasmid DNA is linearized because DNA in this form is more likely than circular DNA to recombine with chromosome DNA. The DNA that is not integrated is lost during successive cell divisions. The major drawback of this strategy is the low yield of recombinant protein from a single gene copy.
- The eukaryotic enzyme protein disulfide isomerase is instrumental in forming the correct disulfide bonds within a protein. Aberrant disulfide bond formation changes a protein's configuration, which abolishes protein activity and causes instability.
- Many *P. pastoris* expression vectors have been devised, each one having more or less the same format. The basic features include a gene of interest under the control of promoter and transcription termination sequences from the *P. pastoris* AOX1 gene, an *E. coli* origin of replication and selectable marker gene, and a yeast selectable marker gene.
- Baculoviruses are a large, diverse group of viruses that specifically infect arthropods, including many insect species, and are not infectious to other animals. During the infection cycle, two forms of baculovirus are produced.

REVIEW QUESTIONS

1. What are the major posttranslational modifications of eukaryotic proteins in the endoplasmic reticulum and Golgi apparatus?
2. In general, how do N glycosylation patterns differ among yeast, insects, and mammalian proteins?
3. Describe the features of a eukaryotic expression vector.
4. What are the advantages and disadvantages of the different classes of yeast vectors for producing a biotechnology product?
5. Describe some of the strategies that have been used to increase proper folding and secretion of recombinant proteins from yeast cells.
6. Discuss the salient features of a *P. pastoris* high-expression integrating vector system. How has *P. pastoris* been "humanized"?

REFERENCES

1. Ansari, A., & Emery, V. C. (1998). Baculoviruses. In R. Rabley & J. M. Walker (Eds.), *Molecular Bio Methods Handbook* (pp. 219–233). Humana Press Inc.
2. Barnes, L. M., & Dickson, A. J. (2006). Mammalian cell factories for efficient and stable protein expression. *Current Opinion in Biotechnology, 17*, 381–386.
3. Berger, I., Fitzgerald, D. J., & Richmond, T. J. (2004). Baculovirus expression system for heterologous multiprotein complexes. *Nature Biotechnology, 22*, 1583–1587.
4. Bowman, K. K., Clark, J., Yu, L., Mortara, K., Radika, K., Wang, J., & Zhan, H. (2000). Expression, purification, and characterization of deglycosylated human pro-prostate-specific antigen. *Protein Expression and Purification, 20*, 405–413.
5. Condreay, J. P., & Kost, T. A. (2007). Baculovirus vectors for insect and mammalian cells. *Current Drug Targets, 8*, 1126–1131.
6. Cregg, J. M. (1999). Expression in the methylotrophic yeast *Pichia pastoris*. In J. Fernandez & J. P. Hoeffler (Eds.), *Gene Expression Systems: Using Nature for the Art of Expression* (pp. 157–191). Academic Press.
7. Gasser, B., Saloheimo, M., Rinas, U., Dragosits, M., Rodríguez-Carmona, E., Baumann, K., Giuliani, M., Parrilli, E., Branduardi, P., Lang, C., Porro, D., Ferrer, P., Tutino, M. L., Mattanovich, D., & Villaverde, A. (2008). Protein folding and conformational stress in microbial cells producing recombinant proteins: A host comparative overview. *Microbial Cell Factories, 7*, 11–29.
8. Gellissen, G., Kunze, G., Gaillardin, C., Cregg, J. M., Berardi, E., Veenhuis, M., & van der Klei, I. J. (2005). New yeast expression platforms based on methylotrophic *Hansenula polymorpha* and *Pichia pastoris* and on dimorphic *Arxula adeninivorans* and *Yarrowia lipolytica*—A comparison. *FEMS Yeast Research, 5*, 1079–1096.
9. Hallewell, R. A., Mills, R., Tekamp-Olson, P., Blacher, R., Rosenberg, S., Otting, F., Masiarz, F. R., & Scandella, C. J. (1987). Amino terminal acetylation of authentic human Cu,Zn superoxide dismutase produced in yeast. *Bio/Technology, 5*, 363–366.

10. Hamilton, S. R., & Gerngross, T. U. (2007). Glycosylation engineering in yeast: The advent of fully humanized yeast. *Current Opinion in Biotechnology, 18*, 387–392.

11. Jarvis, D. L. (2003). Developing baculovirus-insect cell expression systems for humanized recombinant glycoprotein production. *Virology, 310*, 1–7.

12. Kim, M. D., Rhee, S. K., & Seo, J. H. (2001). Enhanced production of anticoagulant hirudin in recombinant *Saccharomyces cerevisiae* by chromosomal δ-integration. *Journal of Biotechnology, 85*, 41–48.

13. Kitts, P. A., & Possee, R. D. (1993). A method for producing recombinant baculovirus expression vectors at high frequency. *BioTechniques, 14*, 810–817.

14. Kjeldsen, T., Hach, M., Balschmidt, P., Havelund, S., Pettersson, A. F., & Markussen, J. (1998). Prepro-leaders lacking N-linked glycosylation for secretory expression in the yeast *Saccharomyces cerevisiae*. *Protein Expression and Purification, 14*, 309–316.

15. Kjeldsen, T., Pettersson, A. F., & Hach, M. (1999). Secretory expression and characterization of insulin in *Pichia pastoris*. *Biotechnology Applied Biochemistry, 29*, 79–86.

16. Kost, T. A., Condreay, J. P., & Jarvis, D. L. (2005). Baculovirus as versatile vectors for protein expression in insect and mammalian cells. *Nature Biotechnology, 23*, 567–575.

17. Kwaks, T. H. J., & Otte, A. P. (2006). Employing epigenetics to augment the expression of therapeutic proteins in mammalian cells. *Trends in Biotechnology, 24*, 137–142.

18. Kwaks, T. H. J., Sewalt, R. G. A. B., van Blokland, R., Siersma, T. J., Kasiem, M., Kelder, A., & Otte, A. P. (2005). Targeting of a histone acetyltransferase domain to a promoter enhances protein expression levels in mammalian cells. *Journal of Biotechnology, 115*, 35–46.

19. Liang, M., Dübel, S., Li, D., Queitsch, I., Li, W., & Bautz, E. K. (2001). Baculovirus expression cassette vectors for rapid production of complete human IgG from phage display selected antibody fragments. *Journal of Immunological Methods, 247*, 119–130.

20. Liu, X., Constantinescu, S. N., Sun, Y., Bogan, J. S., Hirsch, D., Weinberg, R. A., & Lodish, H. F. (2000). Generation of mammalian cells stably expressing multiple genes at predetermined levels. *Analytical Biochemistry, 280*, 20–28.

21. Lucas, B. K., Giere, L. M., DeMarco, R. A., Shen, A., Chisholm, V., & Crowley, C. W. (1996). High-level production of recombinant proteins in CHO cells using a dicistronic DHFR intron expression vector. *Nucleic Acids Research, 24*, 1774–1779.

22. Lucknow, V. A., Lee, S. C., Barry, G. F., & Olins, P. O. (1993). Efficient generation of infectious recombinant baculoviruses by site-specific transposon-mediated insertion of foreign genes into a baculovirus genome propagated in *Escherichia coli*. *Journal of Virology, 67*, 4566–4579.

23. Makrides, S. C. (1999). Components of vectors for gene transfer and expression in mammalian cells. *Protein Expression and Purification, 17*, 183–202.

CHAPTER 10

Biochemistry and Molecular Biology Techniques

LEARNING OBJECTIVES

After studying this chapter, you will be able to:

- Explain the mass spectroscopic techniques;
- Define electrophoretic techniques;
- Understand the chromatographic techniques;
- Define spectroscopic techniques.

KEY TERMS FROM THIS CHAPTER

DNA extraction
Electrospray ionization (ESI)
Gas chromatography (GC)
Immunoprecipitation
Mass spectrometry
Protein expression
Quantitative mass spectrometry
Electrophoresis
Flow cytometry
Gene expression analysis
Liquid chromatography (LC)
Nuclear magnetic resonance (NMR)
Protein purification
Tandem mass spectrometry

INTRODUCTION

A variety of methods are used in molecular biology and biochemistry to study the interactions, structure, and function of biomolecules. Comprehensive genome analysis has been made possible by DNA sequencing techniques like Sanger and next-generation sequencing. A common technique in gene expression analysis and genetic testing is polymerase chain reaction (PCR), which amplifies particular DNA sequences. Applications such as DNA fingerprinting and protein variant identification are made easier by gel electrophoresis, which divides DNA, RNA, and proteins according to size and charge. Western blotting uses antibodies to identify and measure particular proteins, revealing information about the expression and interactions of proteins. Biomolecule identification and characterization are aided by mass spectrometry, which examines the mass and charge of molecules. The three-dimensional structure of molecules, such as proteins and nucleic acids, can be ascertained through X-ray crystallography. NMR spectroscopy, or nuclear magnetic resonance, studies the dynamics, structure, and interactions of molecules in solution. These methods, along with others, are essential for expanding our knowledge of biological systems, providing light on underlying mechanisms and aiding in the creation of diagnostics and treatments.

10.1. MASS SPECTROSCOPIC TECHNIQUES

In biochemistry and molecular biology, mass spectrometry is a potent analytical method that is frequently used for the identification, description, and quantification of biomolecules. It offers data on the mass-to-charge ratio (m/z) of ions, making it possible to analyze molecular interactions, determine molecular weight, and clarify structures. Numerous methods are included in mass spectrometry, each with specific uses and advantages. Here, we will provide an overview of some commonly used mass spectroscopic techniques:

Electrospray Ionization (ESI): Proteins, peptides, and nucleic acids are examples of large biomolecules that are commonly analyzed using ESI, a soft ionization technique. It involves applying a high voltage to a liquid that contains the analyte, causing ions to form in the gas phase. Because ESI preserves the natural structure of biomolecules and permits gentle ionization, it is a valuable tool for researching post-translational modifications and non-covalent interactions.

Matrix-Assisted Laser Desorption/Ionization (MALDI): Another popular soft ionization method for the examination of big biomolecules is MALDI. The analyte is enmeshed in a matrix substance, which absorbs laser light and helps the analyte molecules move into the gas phase. When analyzing proteins, peptides, and other biomolecules, MALDI is especially helpful because it produces intact molecular ions with little fragmentation.

- **Tandem Mass Spectrometry (MS/MS):** The sequential examination of ions produced in a mass spectrometer is known as tandem mass spectrometry. By breaking apart precursor ions and examining the resulting fragments, it enables the structural elucidation of complex molecules. The sequence, modifications, and interactions of biomolecules can be learned a great deal about them by using MS/MS techniques like electron capture dissociation (ECD) and collision-induced dissociation (CID).
- **High-Resolution Mass Spectrometry (HRMS):** The capacity of HRMS to precisely and reliably measure the mass-to-charge ratio of ions is one of its distinguishing features. It improves the accuracy of determining molecular weights and permits the discrimination of closely related ions. Fourier transform ion cyclotron resonance (FT-ICR) and Orbitrap are two useful HRMS techniques for analyzing complex mixtures and identifying unknown compounds.
- **Quantitative Mass Spectrometry:** Biomolecules can also be quantitatively analyzed using mass spectrometry. Protein or peptide abundance can be

compared between samples using stable isotope labeling techniques like isobaric tags for relative and absolute quantitation (iTRAQ) or stable isotope labeling by amino acids in cell culture (SILAC). Targeted protein or peptide quantification is possible using parallel reaction monitoring (PRM) or selected reaction monitoring (SRM) techniques.

- **Imaging Mass Spectrometry:** Molecules within tissues or cells can be seen and mapped thanks to imaging mass spectrometry, which combines mass spectrometry with spatial data. It allows the identification of molecular signatures linked to particular locations or illnesses and sheds light on the spatial distribution of biomolecules like metabolites, lipids, and peptides.

10.1.1. Electrospray Ionization (ESI)

In mass spectrometry, electrospray ionization (ESI) is a popular ionization technique that makes it possible to analyze big biomolecules like proteins, peptides, nucleic acids, carbohydrates, and fats. Biochemistry and molecular biology benefit greatly from ESI because it permits the transfer of labile biomolecules into the gas phase while maintaining their original structure and noncovalent interactions.

In the ESI process, a nebulizing gas is exposed to a high voltage to produce ions from a liquid sample. Generally, a capillary emitter is used to introduce the sample into the mass spectrometer after it has been dissolved in a volatile solvent like methanol or acetonitrile. A high voltage is applied while the liquid passes through the emitter, causing a Taylor cone to form at the tip of the emitter.

Tiny droplets of liquid are produced by the high electric field at the Taylor cone, and these droplets go through a sequence of processes that ultimately lead to ionization. Charges build up on the droplets' surface as they continue to evaporate the solvent, reducing in size. Ions are eventually ejected into the gas phase when the droplets eventually become extremely charged and reach a critical point where coulombic repulsion overcomes surface tension.

The produced ions are subsequently brought into the mass spectrometer, where they are separated using a variety of mass analyzers, including quadrupoles, time-of-flight (TOF) analyzers, and ion traps, according to their mass-to-charge ratio (m/z). The molecular weight, charge state, and relative abundance of the ions are all revealed by the resulting mass spectrum. When it comes to the analysis of biomolecules, ESI has various benefits. First of all, it makes it possible to analyze big, complex molecules that are challenging to ionize and volatilize with conventional ionization techniques. Second, the preservation of noncovalent interactions through ESI makes it possible to study protein-ligand or protein-protein complexes and ascertain their binding affinities and stoichiometry. Furthermore, the analysis of post-translational modifications—such as phosphorylation, glycosylation, or acetylation—made easier by ESI offers new perspectives on the roles played by proteins and signaling pathways.

Numerous research fields, such as proteomics, metabolomics, lipidomics, and drug discovery, have found extensive uses for ESI. It has completely changed the field of mass spectrometry and made it possible to analyze biomolecules with extreme precision and sensitivity. Expanding the capabilities of ESI in the analysis of complex biological samples are new developments in ESI interface technologies and hybrid mass spectrometry platforms.

10.1.2. Matrix-Assisted Laser Desorption/Ionization (MALDI)

A potent ionization method used in mass spectrometry for the analysis of biomolecules, especially big proteins and peptides, is called matrix-assisted laser desorption/ionization, or MALDI. Through the use of a matrix material to aid in the desorption and ionization of the analyte molecules upon laser irradiation, MALDI enables the generation of intact molecular ions from a sample. During the MALDI process, the target analyte is combined with an organic compound, such as a polymer or small molecule, to form a solid-state matrix. A particular wavelength of laser light, usually in the ultraviolet (UV) or near-infrared (NIR) range, is absorbed by the matrix. When exposed to laser light, the matrix material absorbs the energy and gets excited, which transfers energy to nearby analyte molecules.

The analyte molecules desorb from the solid matrix and become ionized as a result of this energy transfer. After being desorbed and ionized, the analyte ions are accelerated into the mass spectrometer, where they are separated using a variety of mass analyzers, including ion traps and time-of-flight (TOF) analyzers, according to their mass-to-charge ratio (m/z). Information about the molecular weight and abundance of the analyte ions can be found in the resulting mass spectrum.

There are various benefits that MALDI provides when analyzing biomolecules. The capacity to analyze big biomolecules—such as proteins and peptides—while maintaining their structural integrity is a significant benefit. Analyte incorporation into a matrix material minimizes fragmentation and maintains noncovalent interactions within the molecule by allowing for mild ionization. This is especially significant for protein analysis, where understanding protein-protein interactions, post-translational modifications, and structural characterization depend on preserving the native conformation.

Many domains, such as proteomics, lipidomics, metabolomics, and clinical diagnostics, have made extensive use of MALDI. Its high sensitivity, mass accuracy, and capacity to analyze a broad range of biomolecules have completely changed the analysis of complex biological samples. Specifically, MALDI imaging mass spectrometry has made it possible to map and visualize biomolecules within tissues, giving researchers the ability to study spatial distributions and pinpoint molecular signatures connected to particular areas or illnesses.

Large biomolecule analysis in mass spectrometry uses a potent ionization technique called Matrix-Assisted Laser Desorption/Ionization (MALDI). The process entails integrating the analyte into a matrix substance to expedite the analyte's desorption and ionization upon exposure to laser radiation. With its ability to provide important insights into the structure, function, and interactions of proteins, peptides, and other biomolecules, MALDI has completely changed the analysis of these molecules. It has been widely applied in many different domains and is still advancing the study of molecular biology and biochemistry.

10.1.3. Tandem Mass Spectrometry

In analytical chemistry and proteomics, tandem mass spectrometry (MS/MS) is a potent method for obtaining comprehensive structural data about biomolecules. To gain insight into the structure, composition, and interactions of the molecule, it entails the sequential analysis of ions, in which a chosen precursor ion is isolated and fragmented. Three primary steps are usually involved in tandem mass spectrometry: fragmentation, product ion analysis, and precursor ion selection.

Initially, a target precursor ion is chosen from the mass spectrum according to its mass-to-charge ratio (m/z). A variety of mass analyzers, including quadrupoles and ion traps, which enable the isolation of a particular ion from the mixture, can be used for this.

Following its isolation, the precursor ion fragments to produce product ions. A variety of fragmentation methods, such as higher-energy collisional dissociation (HCD), electron transfer dissociation (ETD), electron capture dissociation (ECD), and collision-induced dissociation (CID), can be used. With the help of these methods, the precursor ion's chemical bonds are broken, forming fragments with particular m/z ratios.

To gather structural information, the generated product ions are finally examined. A mass analyzer that measures the fragments' m/z ratios, such as a quadrupole or a time-of-flight (TOF) analyzer, is usually used for this kind of analysis. The resulting tandem mass spectrum can be used to determine the molecular sequence, post-translational modifications, and other structural characteristics by giving information about the fragment ions and their relative abundance. Tandem mass spectrometry finds wide-ranging uses in molecular biology and biochemistry. It is frequently utilized in proteomics for peptide sequencing, which makes it possible to identify, characterize, and alter proteins. Proteins can be confidently identified by comparing the obtained tandem mass spectra with databases of known peptides. To determine the structures of small organic molecules and clarify metabolic pathways, tandem mass spectrometry is also essential for the study of metabolomics, lipidomics, and small organic molecule analysis.

Tandem mass spectrometry's capabilities can also be increased by combining it with other methods. For instance, the identification and quantification of complex mixtures of biomolecules is made possible by liquid chromatography-tandem mass spectrometry (LC-MS/MS), which combines the separation capability of liquid chromatography with the structural information offered by MS/MS. Tandem mass spectrometry is an adaptable method for obtaining precise structural data about biomolecules in molecular biology and biochemistry. Tandem mass spectrometry determines the molecular sequence, modifications, and other structural characteristics by successively isolating and fragmenting ions. Its uses, which include metabolite profiling and protein identification and characterization, greatly advance our knowledge of biological systems.

10.1.4. High-Resolution Mass Spectrometry (HRMS)

In analytical chemistry and biochemistry, High-Resolution Mass Spectrometry (HRMS) is a potent method for accurately determining the mass-to-charge ratio (m/z) of ions. With greater mass accuracy and resolution than traditional mass spectrometry methods, HRMS enables precise molecular mass measurement and the separation of closely related ions. HRMS makes use of sophisticated mass analyzers with high-resolution capabilities like Orbitrap or Fourier transform ion cyclotron resonance (FT-ICR). These analyzers enable precise ion mass determination by operating on the principles of ion motion in an electrostatic trap or a magnetic field.

The remarkable mass accuracy of HRMS is one of its key benefits. HRMS can distinguish between ions with a few atomic mass differences by obtaining mass accuracy within parts per million (ppm). This is especially helpful when analyzing complicated mixtures because different ions may have similar masses but distinct compositions. High mass accuracy lowers the number of false-positive and false-negative results while enhancing the dependability of compound identification. HRMS has the capacity to provide high resolution along with mass accuracy. The ability to separate ions with closely spaced m/z ratios is referred to as resolution. HRMS can differentiate between ions with a small mass increment difference with high resolution, giving more precise information about the sample. This is especially helpful in omics studies like proteomics, metabolomics, and others where separating and identifying multiple compounds is necessary for analyzing complex mixtures.

In the fields of molecular biology and biochemistry, HRMS is widely used. HRMS is used in proteomics to identify and characterize proteins, as well as to determine protein variants, post-translational modifications, and protein-protein interactions. HRMS makes it possible to identify and profile small molecules in metabolomics, such as lipids and metabolites, which sheds light on metabolic pathways and disease markers. In forensic science, environmental analysis, and drug development, HRMS is also utilized.

The capabilities of HRMS technology have been expanded, making it an essential tool in contemporary research. These advancements include increased mass resolving power, improved mass accuracy, and enhanced sensitivity. Its analytical capabilities can also be increased by combining it with other separation methods like gas chromatography (GC) or liquid chromatography (LC).

10.1.5. Quantitative Mass Spectrometry

The technique known as quantitative mass spectrometry is employed to ascertain the concentration or abundance of particular molecules within a sample. The mass-to-charge ratio (m/z) of analytes is quantitatively evaluated, and the relative or absolute concentrations of target molecules in intricate mixtures can be measured.

There are different approaches to quantitative mass spectrometry, each with its own advantages and limitations. Here are a few commonly used techniques:

- **Label-based Quantification:** Stable isotopic labels are inserted into the analyte molecules using this method either before or during sample preparation. Metabolic labeling techniques or isotopically labeled reagents can be used for isotope labeling. The target molecule can then be relatively quantified by differentiating between the labeled and unlabeled forms of the analyte according to their mass shift.

- **Label-free Quantification:** Instead of using isotopic labels, label-free quantification compares the strength or quantity of analyte signals across various samples. When comparing multiple samples, this method—which is frequently based on peak intensities or spectral counting—is helpful for relative quantification.
- **Selected Reaction Monitoring (SRM) or Multiple Reaction Monitoring (MRM):** Targeted quantification techniques like SRM/MRM are used to measure particular analytes in complex mixtures. For every analyte, it entails choosing a particular precursor ion and keeping an eye on the fragmentation pattern of a predetermined product ion. Because of its great sensitivity and selectivity, this method makes it possible to quantify target molecules exactly.
- **Absolute Quantification Using Stable Isotope-Labeled Internal Standards:** Using this method, the sample is supplemented with known concentrations of internal standards labeled with stable isotopes. The mass shift of the internal standards allows them to be identified from the analyte despite having the same chemical characteristics. The analyte's absolute concentration can be ascertained by contrasting its intensity or abundance with that of the internal standard.

Proteomics, metabolomics, pharmacokinetics, and environmental analysis are just a few of the many areas in which quantitative mass spectrometry finds use. Among many other uses, it helps scientists monitor drug concentrations, detect biomarkers, analyze changes in protein expression, and evaluate the effects of environmental contaminants. It is noteworthy that precise and dependable quantification in quantitative mass spectrometry necessitates meticulous experimental planning, data interpretation, and the application of suitable internal standards or reference materials. Accurate and repeatable quantitative results require consideration of factors like sample preparation, calibration of the instrument, and data normalization methods.

10.1.6. Imaging Mass Spectrometry

Combining the molecular information from mass spectrometry with the spatial information from microscopy creates a potent analytical tool known as imaging mass spectrometry (IMS). It makes it possible to see and map biomolecules inside of tissues and other complicated samples, giving important information about their location, distribution, and molecular makeup. In IMS, a mass spectrometer operating in raster scanning mode examines a sample, such as a tissue section or biological specimen. Many mass spectrometry platforms, such as secondary ion mass spectrometry (SIMS) and matrix-assisted laser desorption/ionization (MALDI), can be used to carry out the technique. The first step in the imaging process is sample preparation, which in the case of MALDI-IMS usually involves applying a matrix compound. When exposed to laser radiation, the matrix helps the molecules of interest desorb and ionize. The surface of the sample is ionized and sputter-deposited using a focused ion beam in SIMS-IMS. Following sample preparation, the mass spectrometer systematically scans the sample surface, collecting mass spectra at each location. The obtained mass spectra provide details about the molecular ions that are present at each site, which makes it possible to create an ion intensity map that shows the spatial distribution of different biomolecules in the sample.

Many different types of biomolecules, such as proteins, lipids, metabolites, and medications, can be seen thanks to IMS. Researchers can correlate the spatial distribution of different biomolecules and learn more about their interactions and roles within biological systems by comparing the ion intensity maps obtained from different molecules. IMS has a wide range of applications in various fields. IMS has been used in biological and medical research to analyze the spatial organization of cellular processes, find biomarkers, study disease pathology, and look into drug distribution within tissues. Among other fields, it has found use in pharmacology, microbiology, neurobiology, and oncology.

Applications for IMS can also be found in disciplines like materials science, forensics, and environmental science. It makes it possible to investigate material surfaces and interfaces, analyze trace evidence in forensic investigations, and characterize and map compounds within environmental samples. IMS does, however, have some drawbacks and difficulties. It needs careful sample preparation and optimization, which can take time. Because of the high dimensionality of the acquired datasets and the requirement for suitable data analysis techniques, interpreting IMS data can be challenging. Combining the molecular information from mass spectrometry with the spatial information from microscopy yields a potent technique known as imaging mass spectrometry. It provides important insights into the distribution and molecular makeup of biomolecules by enabling the mapping and visualization of these molecules within tissues and other complex samples. IMS continues to expand our knowledge of complex systems at the molecular level and has a wide range of applications in biological, medical, environmental, and forensic research (Watson and Sparkman, 2007)

10.2. ELECTROPHORETIC TECHNIQUES

In biochemistry and molecular biology, electrophoretic methods are frequently utilized to separate and examine biomolecules according to their size and charge. These methods allow for the separation and characterization of charged molecules by causing them to migrate through an electric field. Here are some commonly employed electrophoretic techniques:

- **Agarose Gel Electrophoresis:** The main applications of agarose gel electrophoresis are the separation and examination of nucleic acids, including DNA and RNA. The polymer agarose, which is obtained from seaweed, creates a gel matrix that allows the nucleic acids to move through. Size-based separation occurs as a result of smaller nucleic acid fragments moving through the gel more quickly and extensively. DNA-binding dyes can be used to visualize the separated fragments, or they can be transferred onto a membrane for additional analysis.
- **Polyacrylamide Gel Electrophoresis (PAGE):** Proteins and nucleic acids can be separated using a higher resolution method called polyacrylamide gel electrophoresis. To obtain the required resolution, polyacrylamide gels can be tailored with varying acrylamide concentrations. Before being separated according to size, proteins are usually denatured and treated with sodium dodecyl sulfate (SDS) to give them a uniform negative charge. Protein stains can be used to visualize the separated proteins, or they can be transferred to a membrane for Western blotting.
- **Capillary Electrophoresis (CE):** A high-resolution method called capillary electrophoresis uses capillaries with a narrow bore that are filled with an electrolyte solution. Under the influence of an electric field, analytes migrate through the capillary and are separated based on their charge-to-size ratio. CE provides low sample consumption, quick analysis times, and excellent separation efficiency. Numerous uses for it exist, such as small molecule analysis, protein analysis, and DNA sequencing.
- **Isoelectric Focusing (IEF):** Proteins are sorted by isoelectric focusing according to their isoelectric points (pI), or the pH at which a protein has no net charge. The method makes use of a pH gradient produced inside a capillary or gel. Proteins separate according to pI when they migrate until they get to the pH area of the gel where their net charge is zero. For thorough protein analysis, two-dimensional gel electrophoresis (2D-PAGE) frequently uses IEF as the first dimension.
- **SDS-PAGE:** SDS-PAGE, or sodium dodecyl sulfate-polyacrylamide gel electrophoresis, is a method for sorting proteins according to their molecular

weight. The detergent SDS gives the proteins a negative charge and denatures them. Smaller proteins travel farther and faster through the gel as a result, and proteins migrate through the gel in proportion to their molecular weight. In biochemical and molecular biology research, SDS-PAGE is frequently utilized for protein separation. For visualization and additional analysis, immunoblotting or protein staining is frequently performed after SDS-PAGE protein separation.

These are only a few applications of electrophoretic methods in molecular biology and biochemistry. Every method has unique benefits and can be applied to various kinds of biomolecules and scientific purposes. To separate, analyze, and characterize biomolecules and further our understanding of their structure and function, electrophoretic techniques remain essential.

10.2.1. Agarose Gel Electrophoresis

In biochemistry and molecular biology, agarose gel electrophoresis is a widely used technique for size-based nucleic acid separation and analysis, including DNA and RNA. When exposed to an electric field, the nucleic acids can migrate through a gel matrix made of agarose, a polysaccharide obtained from seaweed. Agarose gel electrophoresis is a multi-step procedure. To make a molten gel mixture, agarose powder is first dissolved in a buffer solution and heated. After filling a gel tray or horizontal gel apparatus with the gel mixture, a comb is placed to make wells where samples can be loaded.

The comb is taken out and the gel is put in an electrophoresis chamber with running buffer once it has solidified. The samples are combined with a loading dye that includes a tracking dye for electrophoresis visualization. The samples typically contain DNA or RNA fragments of varying sizes. Afterwards, a micropipette is used to carefully transfer the samples into the wells. The negatively charged nucleic acid molecules move through the gel in the direction of the positive electrode when an electric current is applied. While larger molecules encounter more resistance and migrate more slowly, smaller molecules pass through the gel matrix more readily and quickly.

Following the electrophoresis run, a DNA-specific dye—such as fluorescent dye or ethidium bromide—is applied to the gel to stain the nucleic acids and enable UV light visualization. With smaller fragments near the bottom and larger fragments near the top, the DNA bands show up on the gel as discrete bands or smears. DNA fragment analysis, DNA purification, and PCR product analysis are just a few of the uses for agarose gel electrophoresis. It is frequently used in genetic analysis, diagnostics, and molecular biology research. Through the use of this technique, scientists can evaluate the success of DNA amplification reactions, ascertain the size distribution of DNA or RNA fragments, and examine the outcomes of genetic engineering experiments.

All things considered, agarose gel electrophoresis is a useful and extensively applied method that offers important insights into the structure and size of nucleic acid molecules. It is an essential tool in the field of molecular biology and has made a substantial contribution to our knowledge of genetic material and the creation of different genetic engineering methods.

10.2.2. Polyacrylamide Gel Electrophoresis (PAGE)

In biochemistry and molecular biology, polyacrylamide gel electrophoresis (PAGE) is a commonly used technique for sorting and analyzing proteins and nucleic acids according to their size and charge. PAGE offers higher resolution and is appropriate for the separation of smaller molecules, such as proteins and short nucleic acids, in contrast to agarose gel electrophoresis, which is primarily used for larger nucleic acids. Although the gel matrix in PAGE is composed of polyacrylamide, a synthetic polymer, the methodology is similar to that of agarose gel electrophoresis. To obtain the appropriate resolution for a given application, polyacrylamide gels can be prepared in a variety of concentrations.

A free-radical initiator is used to polymerize a mixture of acrylamide, bisacrylamide (a cross-linking agent), and a buffer system to carry out PAGE. N,N,N,'N'-tetramethylethylenediamine (TEMED) and a catalyst, such as ammonium persulfate (APS), are added to start the polymerization process. As a result, a network of pores in the polyacrylamide gel forms, serving as a sieving matrix. To separate proteins, samples are typically denatured and then treated with a detergent (like sodium dodecyl sulfate, or SDS) to give the proteins a negative charge through binding. The most popular type of PAGE, SDS-PAGE, enables the separation of proteins according to their molecular weight. Once the gel's wells have been filled with the SDS-bound proteins, an electric current is applied. Larger proteins move more slowly and smaller proteins faster as the negatively charged proteins migrate through the gel matrix in the direction of the positively charged electrode. PAGE is frequently used for nucleic acid separation under native or denaturing conditions. The nucleic acids in native PAGE are separated according to size and charge while maintaining their original structure. Denaturing PAGE entails using chemical denaturants, like formamide or urea, to break nucleic acids' secondary structure and enable separation solely on the basis of size. Following electrophoresis, the separated molecules can be seen by staining them with particular dyes or by detecting them with labeled probes or antibodies. Analysis of the relative abundance, molecular weight, and other properties is made possible by the quantification of the bands or spots that represent the separated molecules.

PAGE finds extensive use in molecular biology and biochemistry. It is employed in research on protein-protein interactions as well as protein purification and characterization. PAGE is also frequently used in nucleic acid analysis, including RNA transcript analysis, DNA fragment analysis, and DNA sequencing. In conclusion, proteins and nucleic acids can be separated and analyzed according to their size and charge using the flexible and extensively used method known as polyacrylamide gel electrophoresis (PAGE). It works with a variety of biological molecules and offers better resolution than agarose gel electrophoresis. PAGE has greatly advanced our knowledge of biomolecules and made contributions to a number of scientific disciplines, including genetics, molecular biology, and protein biochemistry.

10.2.3. Capillary Electrophoresis (CE)

In biochemistry and molecular biology, capillary electrophoresis (CE) is a potent analytical method for the separation and examination of biomolecules, such as proteins, peptides, carbohydrates, nucleic acids, and small molecules. In comparison to conventional gel electrophoresis techniques, it provides high separation efficiency, quick analysis times, and minimal sample consumption. In CE, an electrolyte solution-filled narrow-bore capillary is where the separation takes place. Fused silica, which minimizes interactions with the analytes and offers high resolution, is used to make the capillary. The capillary usually has an inner diameter of 10–100 micrometers and a length of several tens of centimeters.

The differential migration of charged analytes in an electric field serves as the basis for separation in CE. Analytes migrate through the capillary according to their charge-to-size ratio when the electric field is applied. Disparities in size, charge, or both can be used to create the separation.

There are several modes of capillary electrophoresis, including:

- **Capillary Zone Electrophoresis (CZE):** CZE is the most common mode of CE and is used for the separation of charged molecules, such as ions, peptides, and small proteins. The separation is based on the differential migration of analytes according to their charge and size.
- **Capillary Gel Electrophoresis (CGE):** CGE involves the incorporation of a sieving matrix, typically a polymer solution, within the capillary. The sieving matrix helps to separate analytes based on their size. CGE is useful for the separation of large nucleic acids, proteins, and complex mixtures.
- **Capillary Electrochromatography (CEC):** CEC combines the principles of electrophoresis and liquid chromatography. It involves the use of a stationary phase within the capillary to enhance the separation based on both charge and size.

Compared to other separation methods, CE has a number of benefits. Because of the small capillary diameter and strong electric field, it enables faster separations, usually in a matter of minutes. CE is appropriate for the analysis of scarce or valuable samples because it also requires small sample volumes. Moreover, CE offers high sensitivity and resolution, making it possible to separate and identify analytes that are closely related. CE is frequently used in conjunction with a variety of detection techniques, such as mass spectrometry, UV absorbance, fluorescence, and electrochemical detection, to identify and quantify the separated analytes. By incorporating particular probes or labels or by using derivatization techniques, the detection sensitivity can be increased.

CE is widely used in biochemistry and molecular biology for tasks like drug analysis, environmental analysis, protein analysis, DNA sequencing, and fragment analysis. Its applications in research, clinical diagnostics, and industry quality control have greatly advanced our understanding of biological molecules. All things considered, capillary electrophoresis is a strong and adaptable method that provides high-resolution separation and examination of a variety of biomolecules. Its sensitivity, speed, and low sample consumption make it an invaluable tool for molecular biology and biochemistry research in a variety of fields.

10.2.4. Isoelectric Focusing (IEF)

Proteins can be divided and examined using isoelectric focusing (IEF), a potent method in molecular biology and biochemistry that is based on isoelectric points (pI). The protein's amino acid sequence and composition determine the pI, or the pH at which a protein has no net charge. In IEF, a mixture of buffering compounds with varying pKa values is usually used to create a pH gradient within a gel matrix. Polyacrylamide or other materials appropriate for electrophoresis can be used to create the gel. Applying an electric field across the gel with a positive electrode at one end and a negative electrode at the other establishes the pH gradient.

After being loaded onto the gel, proteins move along the pH gradient until they get to a place where the pH is equal to their pI. The proteins stop migrating and turn electrically neutral at this point. As a result, the gel is covered in a pattern of bands or spots, each of which represents a protein with a distinct pI value.

The electrophoresis principle, which states that charged molecules migrate in response to an electric field, forms the basis of the separation process in IEF. Nevertheless, in contrast to other electrophoretic methods, IEF makes use of the fact that proteins have distinct isoelectric points and will migrate to the gel's pH regions according to those points. IEF can be carried out in a variety of formats, such as capillaries, slab gels, or tube gels, based on the particular application and required resolution. To separate and analyze proteins according to both their pI and molecular weight, the technique is frequently used in conjunction with other techniques, such as sodium dodecyl sulfate-polyacrylamide gel electrophoresis (SDS-PAGE).

The proteins can be seen after IEF separation by employing immunoblotting methods or staining with particular dyes. Furthermore, by removing particular bands or spots from the gel, the separated proteins can be examined further to identify the proteins using mass spectrometry or other methods. In the fields of molecular biology and biochemistry, IEF is widely used. It is frequently used for post-translational modification identification, protein purification, protein characterization, and protein pI value determination. It is also helpful for examining complex protein mixtures and researching protein variations, such as isoforms or genetic variants.

All things considered, isoelectric focusing (IEF) is a strong and adaptable method for separating proteins according to their isoelectric points. Proteins with varying pI values can be separated and analyzed thanks to its high resolution and specificity. IEF is still a useful tool in many research fields, such as proteomics, biotechnology, and clinical diagnostics. It has made a substantial contribution to our understanding of protein structure, function, and diversity.

10.2.5. Protein Blotting Techniques (Western Blotting)

Western blotting, another name for protein blotting techniques, is a popular method in molecular biology and biochemistry for identifying and analyzing particular proteins in complex biological samples. Target proteins can be identified by Western blotting according to their size, quantity, and post-translational changes. There are several important steps in the Western blotting technique. Initially, gel electrophoresis—typically sodium dodecyl sulfate-polyacrylamide gel electrophoresis (SDS-PAGE)—is used to separate the proteins within a sample, such as cell lysates or tissue extracts. Next, the proteins are moved from the gel onto a solid membrane, which is typically made of PVDF or nitrocellulose.

To stop non-specific antibody binding after transfer, the membrane is blocked using a protein-based solution like non-fat dry milk or bovine serum albumin (BSA). After that, a primary antibody that is particular to the desired target protein is incubated on the membrane. An antigen-antibody complex is created when the target protein is recognized by the primary antibody and bound to it. The membrane is then cleaned to get rid of any primary antibody that isn't bound. A secondary antibody that is connected to a detection marker, like an enzyme or fluorescent dye, is then added. Target protein detection is made possible by the secondary antibody's binding to the primary antibody-antigen complex.

Lastly, depending on the detection marker, different techniques are used to visualize the presence of the target protein. Common techniques include fluorescence imaging for fluorescent detection and enzyme-linked chemiluminescence (ECL) for enzyme detection. X-ray film or specialized imaging systems are used to record the signal. Protein analysis benefits greatly from the use of western blotting. It makes it possible to identify particular proteins within a complicated mixture and gives details about the size, expression levels, and possible modifications of those proteins. Additionally, it makes it possible to evaluate how proteins interact, degrade, and alter in expression when subjected to various circumstances or therapies.

Numerous uses for western blotting exist in biotechnology, research, and diagnostics. It is frequently used to evaluate protein-protein interactions, look into disease markers, confirm antibody specificity, and study protein expression in cell and molecular biology. It is an important tool in immunology, drug development, and cancer research. In biochemistry and molecular biology, protein blotting techniques—particularly Western blotting—are crucial for the identification and examination of particular proteins. They offer important details about the size, expression, and modifications of proteins in intricate biological materials. Western blotting is still a vital tool in many fields of scientific research and clinical diagnostics, having contributed significantly to our understanding of protein biology.

10.2.6. DNA Gel Electrophoresis

A key method in molecular biology for sorting and analyzing DNA molecules according to size is DNA gel electrophoresis. It is a commonly used technique for a number of purposes, such as DNA purification, DNA sequencing, and DNA fragment analysis. There are several important steps in the DNA gel electrophoresis process. Firstly, a gel matrix is made, usually with polyacrylamide or agarose. Large DNA fragments are typically separated using agarose gels, whereas smaller DNA fragments can be separated with greater resolution using polyacrylamide gels. In a horizontal or vertical electrophoresis apparatus, the gel is cast, creating wells or slots that can hold DNA samples. The loading buffer, which includes a tracking dye to help track the electrophoresis's progress, is combined with the DNA samples.

After that, the DNA samples are put into the gel's wells, and the gel is subjected to an electric current. The negatively charged DNA molecules move in the direction of the electrode that is positively charged. Larger DNA fragments flow through the gel matrix more slowly, whereas smaller DNA fragments travel through it more quickly. The electrophoresis is halted and the DNA fragments are visible after an appropriate amount of time. This can be done by using fluorescent dyes to stain the DNA, like SYBR Green or ethidium bromide, which intercalate into the DNA and fluoresce when exposed to UV light. As an alternative, target DNA sequences can be found using particular antibodies or fluorescently labeled DNA probes. The size of the separated DNA fragments is inversely connected with their migration distance, and they show up on the gel as discrete bands or smears. The size of the unknown DNA fragments can be estimated by comparing their sizes to those of a DNA ladder, which is made up of DNA fragments of known sizes.

The technique of DNA gel electrophoresis is adaptable and has multiple uses. DNA fragments produced by restriction enzyme digestion, polymerase chain reaction (PCR), or DNA sequencing reactions are frequently analyzed in molecular biology research. Additionally, it is utilized in diagnostic labs for genetic testing and in forensic science for DNA profiling. Capillary electrophoresis-based systems, like capillary gel electrophoresis (CGE) and capillary array electrophoresis (CAE), have been developed in recent years as a result of improvements in DNA gel electrophoresis techniques. These methods provide enhanced throughput, improved resolution, and automated analytic capabilities.

10.2.7. RNA Gel Electrophoresis

In molecular biology, RNA molecules are frequently separated and analyzed according to their size and/or secondary structure using RNA gel electrophoresis. It is an essential technique that enables researchers to examine RNA integrity, identify RNA species, and study RNA expression. With a few adjustments, RNA gel electrophoresis is essentially the same as DNA gel electrophoresis. To separate RNA molecules according to size rather than secondary structure, an agarose or polyacrylamide gel is first made, usually under denaturing conditions. Because polyacrylamide gels have a higher resolution, they are frequently used for RNA electrophoresis.

To guarantee that the RNA molecules adopt a single-stranded conformation, denaturing agents, such as heat or chemical agents like formaldehyde or urea, are frequently used on the RNA samples before loading them onto the gel. The RNA samples may also be mixed with a denaturing loading buffer that contains intercalating dyes, like SYBR Green or ethidium bromide. After the RNA samples are put into the gel's wells, the gel is subjected to an electric current. Through the gel, the negatively charged RNA molecules move in the direction of the positively charged electrode. Larger RNA molecules flow through the gel matrix more slowly than smaller RNA molecules. Following electrophoresis, the RNA molecules are made visible on the gel by staining it with RNA-specific dyes like SYBR Green or ethidium bromide. When these dyes are exposed to UV light, they fluoresce, making it possible to see the RNA bands. RNA gel electrophoresis can be carried out in native conditions to preserve RNA secondary structures in addition to size separation. Studying RNA folding, RNA-protein interactions, or noncoding RNA molecules that depend on their secondary structure for function can benefit greatly from this. A flexible method with many applications in molecular biology is RNA gel electrophoresis. It is frequently used for RNA integrity analysis, the separation and purification of particular RNA species, and the analysis of total RNA extracted from cells or tissues. It is a crucial tool for research on gene expression, RNA sequencing, and noncoding RNA functional analysis.

Advancements in technology have also led to the development of capillary electrophoresis-based systems for RNA analysis, offering higher resolution and increased sensitivity. A key method in molecular biology for separating and analyzing RNA molecules according to their size or secondary structure is RNA gel electrophoresis. It is extensively used in gene expression studies, RNA characterization, and functional analysis of noncoding RNAs because it offers useful information about RNA expression, integrity, and interactions (Hames & Rickwood, 2005).

10.3. CHROMATOGRAPHIC TECHNIQUES

In biochemistry and molecular biology, chromatographic techniques are frequently employed to separate and examine intricate molecular mixtures according to their physical and chemical characteristics. These methods are predicated on the idea that components in a mixture migrate differently between a stationary phase and a mobile phase.

There are several types of chromatographic techniques commonly employed in biochemistry and molecular biology, including:

- **Liquid Chromatography (LC):** In this, a liquid solvent serves as the mobile phase, and the stationary phase is usually a solid or a liquid immobilized on a solid support. Depending on the type of stationary phase and the mechanism of separation, LC can be further divided into several modes, such as size exclusion chromatography (SEC) and reversed-phase chromatography (RP-HPLC).
- **Gas Chromatography (GC):** With gas chromatography, volatile substances are separated according to how they partition between a mobile phase (an inert gas) and a stationary phase (typically a liquid coated on a solid support). Small molecules and volatile organic compounds are frequently analyzed using gas chromatography (GC).
- **Ion Exchange Chromatography (IEC):** Molecules are separated according to their charges using ion exchange chromatography. An ion exchange resin that binds ions with opposing charges selectively makes up the stationary phase. IEC is frequently employed in the separation of charged molecules such as proteins and nucleic acids.
- **Affinity Chromatography:** The precise interaction between a target molecule and a ligand immobilized on the stationary phase is exploited in affinity chromatography. Proteins and other biomolecules can be highly selectively purified and isolated using this technique by determining how well they bind to a particular ligand or antibody.
- **Thin-Layer Chromatography (TLC):** TLC involves the separation of compounds on a thin layer of an adsorbent material, typically coated on a glass plate or a plastic sheet. The separation is achieved by the differential migration of compounds in a solvent system. TLC is often used for rapid qualitative analysis and can be a preliminary step for other chromatographic techniques.
- **High-Performance Liquid Chromatography (HPLC):** HPLC is a powerful variant of liquid chromatography that utilizes high-pressure pumps to improve separation efficiency and speed. It is widely used for the separation and quantification of

complex mixtures, such as protein or peptide mixtures, in research and analytical laboratories.

Chromatographic methods are adaptable and frequently used in many domains, such as metabolomics, environmental analysis, pharmaceutical analysis, and protein purification. They provide superior selectivity, sensitivity, and resolution for the analysis and separation of intricate molecular mixtures. Chromatographic techniques can be tailored for particular applications by varying the composition of the mobile phase and using different types of stationary phases. This can offer important insights into the makeup and characteristics of biological samples.

10.3.1. Reversed-Phase Chromatography

In biochemistry and molecular biology, reversed-phase chromatography (RPC) is a frequently employed chromatographic method for the separation and examination of biomolecules, especially hydrophobic substances like proteins, peptides, nucleic acids, and small organic molecules. It functions according to the principle of analyte differential partitioning between a polar mobile phase and a nonpolar stationary phase. The stationary phase in reversed-phase chromatography is usually a nonpolar hydrocarbon chain that has been bonded to a solid support, such as octadecylsilane (C18) or octyldecylsilane (C8). Hydrophobic interactions between the analytes and the nonpolar stationary phase are encouraged. Conversely, the mobile phase is typically an aqueous solvent with an organic modifier, like methanol or acetonitrile. The mobile phase's polarity is decreased by the addition of the organic modifier.

When the sample mixture is placed onto the column during the chromatographic procedure, the nonpolar analytes interact with the nonpolar stationary phase, which causes them to be retained. An analyte's duration in the stationary phase increases with its hydrophobicity. As a result of their weaker interactions with the nonpolar stationary phase, more polar components elute more quickly. High-performance liquid chromatography (HPLC) systems are frequently used for reversed-phase chromatography because they provide accurate temperature, flow rate, and solvent composition control. Typically, a UV-visible detector is used to monitor the separation by measuring the eluting compounds' absorbance at particular wavelengths. Reversed-phase chromatography has several benefits, such as high resolution, adaptability in mobile phase composition, and compatibility with a broad variety of analytes. It is especially helpful for the purification and separation of hydrophobic substances, including lipids, medications, and natural products. In the pharmaceutical and biotechnology sectors, reversed-phase chromatography is widely used for quality control, proteomics, metabolomics, and drug discovery.

In reversed-phase chromatography, variables like flow rate, column temperature, and mobile phase composition can all be changed to maximize separation. To improve resolution and separation efficiency, gradient elution—in which the mobile phase's composition is progressively altered during the separation process—is frequently used. In molecular biology and biochemistry, reversed-phase chromatography is an effective method for separating and analyzing hydrophobic biomolecules. It has a large sample capacity, great resolution, and compatibility with a range of detection techniques. In many fields of study, reversed-phase chromatography has proven to be an invaluable tool for characterizing, purifying, and quantifying a wide range of analytes in intricate mixtures.

10.3.2. Size Exclusion Chromatography

Size exclusion chromatography (SEC) is a popular chromatographic method for the separation and examination of biomolecules according to their size and molecular weight. It is also referred to as gel filtration chromatography or gel permeation chromatography. For the separation of macromolecules like proteins, nucleic acids, polysaccharides, and synthetic polymers, it is especially helpful. The stationary phase in size exclusion chromatography is a porous gel matrix. Spherical beads composed of a cross-linked polymer, like polyacrylamide or agarose, make up the gel matrix. Mobile phase and analyte molecules can pass through the different-sized pores in these beads. The differential exclusion or inclusion of analyte molecules in the pores of the gel matrix forms the basis of the separation principle in size exclusion chromatography. Smaller molecules enter the pores and stay in the gel matrix for a longer period of time than larger molecules, which cannot fit through the smaller pores and elute more slowly. Larger molecules elute first and smaller molecules elute later, indicating that the elution order is inversely related to the size of the analyte.

A sample comprising a mixture of molecules with varying sizes is injected into the chromatographic column to carry out size exclusion chromatography. After that, the mobile phase—typically a buffer or a suitable solvent—is continuously fed through the column. The sample components are subjected to differential partitioning based on size as they pass through the column and come into contact with the gel matrix. Smaller molecules enter the pores and are retained for a longer period of time, whereas larger molecules bypass the pores and move through the column more quickly. There are several ways to carry out detection in size exclusion chromatography, including UV absorption, fluorescence, refractive index detection, and multi-angle light scattering (MALS). The analyte concentration and elution profile are disclosed by these detection techniques. Size exclusion chromatography has several benefits, such as being easy to use, adaptable, and physiologically compatible. It can be used for both preparative and analytical purposes and does not require prior knowledge of the properties of the analyte. Size exclusion chromatography is widely used in molecular weight determination, protein purification, protein structure analysis, and sample purity and aggregation state assessment.

To attain the required separation and resolution, size exclusion chromatography optimization entails choosing the right gel matrix, column size, and mobile phase parameters. A variety of gel matrices with different pore sizes are offered, enabling the separation range to be customized. In molecular biology and biochemistry, size exclusion chromatography is a useful method for separating and analyzing biomolecules according to their molecular weight and size. To determine the molecular weight, purify proteins, and evaluate sample integrity, it gives information about the relative size distribution of analytes in a mixture. A flexible and dependable technique, size exclusion chromatography aids in the characterization and comprehension of biomolecular structures and interactions.

Size exclusion chromatography (SEC) is widely regarded as the most popular, effective, and rapid technique for obtaining information regarding the molecular weight distribution (MW) of polymers. Instead of using MW to separate the molecules in the sample, this method uses their hydrodynamic volumes. The SEC system involves injecting the mobile phase, which contains the sample solution, into the stationary phase, which is filled with a porous packing material. Because they can't fit through the pores, molecules larger than the pore volume will elute first. Conversely, because the smaller molecules can fit through the pores, they will elute more slowly (Figure 10 1). Nevertheless, since they will elute at the total permeation point, molecules that are able to freely diffuse through the pores (both in and out) cannot be separated.

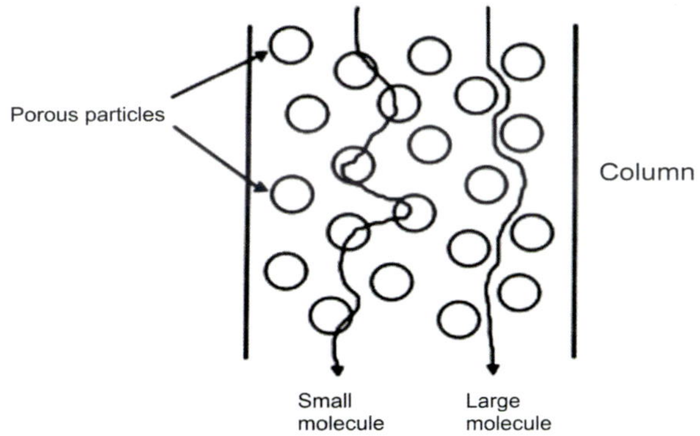

Figure 10.1. *Illustration of the separation process by size exclusion chromatography.*

Source: https://ars.els-cdn.com/content/image/3-s2.0-B9780128179093000066-f06-06-9780128179093.jpg.

Because it is challenging to achieve a pure size-exclusion separation for a variety of factors like concentration effects, osmotic effects, side processes, secondary retention mechanisms, secondary exclusion, SEC band broadening, parasitic processes, and preferential interactions, the main limitation of the SEC method is that it is poorly reproducible in terms of molar mass analysis.

10.3.3. Ion Exchange Chromatography

A popular chromatographic method for separating molecules according to their charge characteristics is ion exchange chromatography (IEC). It works especially well for separating and purifying charged molecules like ions, proteins, nucleic acids, and peptides. The stationary phase in ion exchange chromatography is made up of a solid support substance with charged functional groups, typically a resin or gel. Anion exchange chromatography uses negatively charged functional groups, whereas cation exchange chromatography uses positively charged groups. The target molecules' charge properties and the intended separation determine which ion exchange resin is best.

In ion exchange chromatography, the sample mixture is passed through the ion exchange column to initiate the separation process. The sample's charged molecules interact with the functional groups on the stationary phase that have opposite charges. In anion exchange chromatography, positively charged molecules bind to negatively charged functional groups; in cation exchange chromatography, negatively charged molecules bind to positively charged functional groups. The ionic strength and pH of the mobile phase determine how strongly the charged molecules interact with the stationary phase. Usually, the mobile phase's pH or ionic strength must change to elute the bound molecules. To break the ionic interactions between the sample molecules and the stationary phase, this can be accomplished by progressively raising the salt concentration or modifying the pH. The bound molecules can be separated and gathered according to their varying affinities for the stationary phase through the gradient or step-wise elution process.

Ion exchange chromatography offers a variety of detection techniques, including conductivity, fluorescence, and UV-visible absorption, which can be used based on the type of analyte and the particular requirements for detection. Numerous benefits come with ion exchange chromatography, such as high selectivity, adaptability, and a broad range of pH and salt conditions under which it can be used. It makes it possible to separate, purify, and analyze charged molecules in complicated mixtures for analytical as well as preparative uses. Additionally, for multidimensional separations, it is easily integrated with other chromatographic techniques. To achieve the desired separation and resolution, ion exchange chromatography must be optimized by choosing the right ion exchange resin, pH, salt concentration, and elution conditions. The efficiency and selectivity of the separation are also significantly influenced by the selection of column dimensions and mobile phase composition.

10.3.4. Hydrophobic Interaction Chromatography

A popular chromatographic method for the separation and purification of proteins, peptides, and other biomolecules based on their hydrophobic characteristics is called hydrophobic interaction chromatography, or HIC. It functions based on the idea that a hydrophobic stationary phase and hydrophobic analyte regions have different affinities. A hydrophobic resin or gel, like Sephadex or Sepharose, derivatized with alkyl chains, like butyl or phenyl groups, usually makes up the stationary phase in hydrophobic interaction chromatography. While the aqueous mobile phase solvates hydrophilic regions, the hydrophobic stationary phase encourages the binding of hydrophobic regions on the surface of the analyte molecules.

When the sample mixture is added to the hydrophobic stationary phase-containing column in HIC, the separation process takes place. Analyte molecules first attach themselves to the hydrophobic surface via hydrophobic interactions. By changing the mobile phase conditions, it is simple to reverse or modify the binding because the hydrophobic interaction is weaker than covalent or ionic interactions. In HIC, elution is usually accomplished by introducing a decreasing gradient of an organic solvent that is water soluble, like acetonitrile, isopropanol, or ethanol. The bound molecules are released and eluted as a result of the hydrophobic interactions between the analyte and the stationary phase being diminished by an increase in the concentration of organic solvent. Stronger hydrophobic interactions dictate a higher concentration of organic solvent to be used during elution, which in turn determines the order of elution.

In hydrophobic interaction chromatography, detection can be achieved through a variety of methods, such as fluorescence, UV-visible absorption, or refractive index detection. The analyte concentration and elution profile are disclosed by these detection techniques. There are numerous benefits to hydrophobic interaction chromatography in the fields of molecular biology and biochemistry. It makes it possible to separate biomolecules according to how hydrophobic they are, which is especially helpful for proteins and peptides that have comparable charges or sizes. Additionally, it is a mild separation method that keeps the analytes' natural structure and activity. To attain the intended separation and resolution, hydrophobic interaction chromatography optimization entails choosing the proper hydrophobic stationary phase, mobile phase composition, gradient conditions, and column dimensions. One way to modify the hydrophobicity of the stationary phase is to add specific functional groups or change the length of the alkyl chain.

10.3.5. Gas Chromatography

A popular analytical method for separating and analyzing volatile and semi-volatile substances in a sample mixture is gas chromatography (GC). The principles of analyte partitioning and differential migration between a stationary phase and a mobile phase serve as its foundation. The stationary phase in gas chromatography is usually a thin film coated on the inner surface of a capillary column, known as the capillary column, or a high-boiling liquid or solid immobilized on a solid support, known as the packed column. The characteristics of the analytes and the particular requirements for separation influence the stationary phase selection. In gas chromatography, the sample vapor is carried through the column by an inert gas, such as nitrogen or helium, known as the mobile phase. After the sample mixture is injected into the GC system, the various components interact with the stationary phase according to their respective affinities as it moves through the column. Stronger interactions with the stationary phase cause analytes to elute later in the column and spend more time there than weaker interactions, which cause them to elute earlier.

In gas chromatography, detection is usually accomplished by a detector at the end of the column. The thermal conductivity detector (TCD), electron capture detector (ECD), flame ionization detector (FID), and mass spectrometry (MS) detector are examples of common detectors. These detectors produce signals that are used to identify and quantify the separated components by measuring various analyte properties like ionization, heat conductivity, electron capture, or mass-to-charge ratio. Numerous industries, such as pharmaceuticals, forensics, food and beverage testing, environmental analysis, and petrochemical analysis, use gas chromatography extensively. It works especially well for analyzing volatile organic compounds (VOCs), which include pesticides, alcohols, esters, and hydrocarbons. GC enables quick analysis times, good sensitivity, and high-resolution separations. To achieve the best possible separation and resolution of the analytes, gas chromatography optimization entails choosing the right stationary phase, column temperature, flow rate, and injection technique. The particular requirements of the analysis, such as sensitivity, selectivity, and the kind of compounds to be detected, determine which detector is best.

10.3.5.1. How does Gas Chromatography Work?

As the name implies, GC uses a carrier gas in the separation, this plays the part of the mobile phase (Figure 10.2(1)). The carrier gas transports the sample molecules through the GC system, ideally without reacting with the sample or damaging the instrument components.

- The sample is first introduced into the gas chromatograph (GC), either with a syringe or transferred from an autosampler (Figure 10.2 (2)) that may also extract the chemical components from solid or liquid sample matrices. The sample is injected into the GC inlet (Figure 10.2 (3)) through a septum which enables the injection of the sample mixture without losing the mobile phase.

- Connected to the inlet is the analytical column (Figure 10.2 (4)), a long (10–150 m), narrow (0.1–0.53 mm internal diameter) fused silica or metal tube which contains the stationary phase coated on the inside walls.
- The analytical column is held in the column oven which is heated during the analysis to elute the less volatile components.
- The outlet of the column is inserted into the detector (Figure 10.2 (5)) which responds to the chemical components eluting from the column to produce a signal.
- The signal is recorded by the acquisition software on a computer to produce a chromatogram (Figure 10.2 (6)).

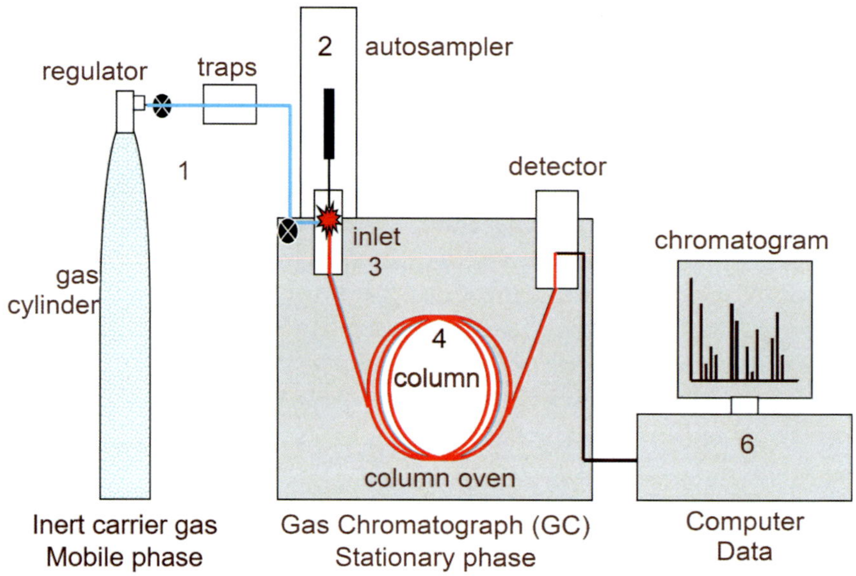

Figure 10.2. *A simplified diagram of a gas chromatograph showing: (1) carrier gas, (2) autosampler, (3) inlet, (4) analytical column, (5) detector and (6) PC. Credit: anthias consulting.*

Source: https://cdn.technologynetworks.com/tn/images/body/figure1gc1590068202170.png.

The chemical components of the sample mixture are first vaporized, if they aren't in the gas phase already, after injection into the GC inlet. In what is referred to as splitless mode, the carrier gas transfers the entire vapor cloud into the analytical column for low concentration samples. To prevent the analytical column from becoming overloaded, only a portion of high concentration samples is split mode transferred to the column and the remaining sample is flushed out of the system through the split line.

The sample components are separated once they are in the analytical column based on how they interact with the stationary phase differently. As a result, the analyte volatility and functional groups should be taken into account when choosing the type of column to use to match them to the stationary phase. The two primary types of liquid stationary phases are those based on polyethylene glycol (PEG) or polydimethylsiloxane (PDMS), the latter of which can contain variable amounts of dimethyl, diphenyl, or mid-polar functional groups like cyanopropylphenyl. Non-polar columns with dimethyl or a low percentage of diphenyl are useful for separating non-polar analytes because like separates like. Molecules with the ability to interact π-π can be separated on stationary phases that have phenyl groups in them. PEG columns work best for separating substances that can form hydrogen bonds, such as acids and alcohols, unless they have undergone derivatization to become less polar. Detecting the analyte molecules as they elute from the column is the last stage. There are many types of GC detectors, for example: those that respond to C-H bonds like the flame ionization detector (FID); those that respond to specific elements for example sulfur, nitrogen or phosphorus; and those that respond to specific properties of the molecule, like the ability to capture an electron, as is used with the electron capture detector (ECD).

10.3.6. High-Performance Liquid Chromatography (HPLC)

A potent analytical method for the separation, identification, and quantification of compounds in a sample mixture is high-performance liquid chromatography (HPLC). It is widely used in many different fields, such as biochemistry, forensics, environmental monitoring, food and beverage analysis, and pharmaceuticals. The interaction between the sample components and the stationary phase—typically, a packed column full of porous particles coated with a particular material—is what drives separation in high-performance liquid chromatography (HPLC). Depending on the type of sample and the intended separation mechanism, the stationary phase can be classified into several types, including reversed-phase, normal-phase, ion-exchange, size exclusion, and affinity chromatography.

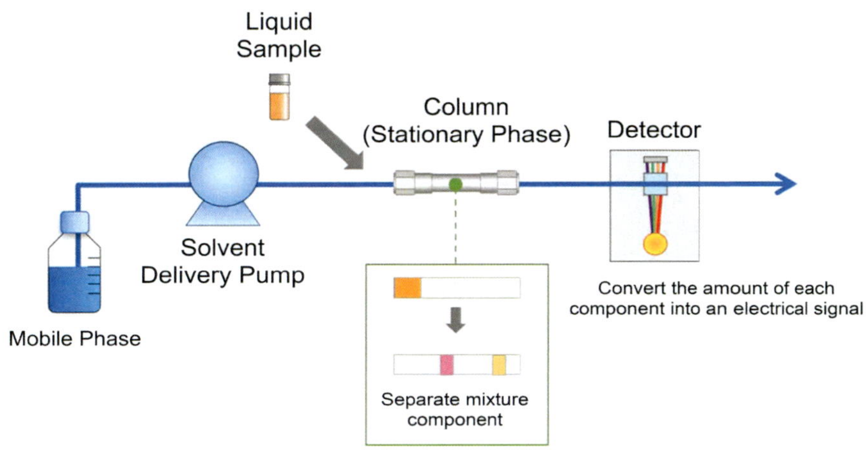

Figure 10.3. *Overview of HPLC.*

Source: https://www.shimadzu.com/an/service-support/technical-support/analysis-basics/basic/what_is_hplc.htm.

A liquid mobile phase, sometimes referred to as the eluent or solvent, is continuously pumped under high pressure through the column as part of the HPLC separation process. The components of the sample mixture interact with the stationary phase to varying degrees as it moves through the column after being injected into the mobile phase. While compounds with weaker interactions with the stationary phase elute more quickly, those with stronger interactions will be held in the column longer and elute more slowly. Ultraviolet-visible (UV-Vis), fluorescence, refractive index (RI), electrochemical (EC), and mass spectrometry (MS) detectors are among the many types of detectors commonly used in HPLC detection. These detectors produce signals that are utilized for identification and quantification by measuring various analyte properties (Figure 10.3).

High resolution, sensitivity, and versatility are just a few of the benefits that HPLC has over other chromatographic methods. It offers superior separation efficiency and can handle a broad variety of sample types, including big biomolecules and small molecules. Additionally, HPLC makes it possible to analyze complex mixtures and figure out the concentrations of individual compounds. To optimize HPLC and attain the required separation and resolution, the right stationary phase, mobile phase composition, flow rate, column temperature, and detection technique must be chosen. The stability, molecular weight, and polarity of the analytes all influence the choice of parameters.

10.4. SPECTROSCOPIC TECHNIQUES

Science uses spectroscopic techniques, which are very effective tools, to investigate how light interacts with matter. Scientists can better understand the characteristics and processes occurring within different materials by using these techniques, which offer insightful information about the makeup, structure, and behavior of molecules.

There are several types of spectroscopic techniques commonly employed in scientific research:

- **Ultraviolet-Visible Spectroscopy (UV-Vis):** The measurement of a sample's absorption or transmission of ultraviolet and visible light is known as UV-Vis spectroscopy. It gives scientists knowledge of the electronic transitions that occur within molecules, enabling them to identify specific compounds, measure their concentrations, and look into the structure of their chemical bonds.
- **Infrared Spectroscopy (IR):** IR quantifies how much infrared light molecules absorb or reflect. It gives scientists knowledge of the rotational and vibrational modes of molecules, which they can use to study chemical bonds, find functional groups, and ascertain the structure of compounds.
- **Nuclear Magnetic Resonance (NMR) Spectroscopy:** The magnetic characteristics of atomic nuclei, especially those of hydrogen and carbon, are exploited by NMR spectroscopy. Through the application of radiofrequency pulses and a magnetic field, NMR spectroscopy offers comprehensive insights into the molecular dynamics, interactions, and structure of a sample. It is extensively employed in the study of molecular processes in solution, structural elucidation, and compound identification.
- **Mass Spectrometry (MS):** With MS, the mass-to-charge ratio of ions produced by molecules is measured. It makes it possible to characterize chemical and isotopic compositions, identify compounds, analyze molecular structures, and determine molecular weights. Drug discovery, lipidomics, proteomics, and metabolomics all make extensive use of mass spectrometry.
- **Fluorescence Spectroscopy:** It quantifies the amount of light that molecules emit after being excited by a particular wavelength of light. It offers details on the electronic transitions and energy states that occur within molecules. Fluorescence spectroscopy is frequently used for molecular labeling, cellular imaging, and the study of biomolecules like proteins and nucleic acids.
- **Raman Spectroscopy:** The measurement of light's inelastic scattering by molecules is done using Raman spectroscopy. It offers details on rotational modes, structural characteristics, and molecular vibrations. Raman spectroscopy is employed in

the study of molecular dynamics, material analysis, crystal characterization, and chemical identification.

These spectroscopic methods provide important insights into the structural, chemical, and physical characteristics of materials, from intricate biological systems to tiny molecules. Through the application of various spectroscopic techniques, scientists can gain a thorough comprehension of the makeup, behavior, and interactions of molecules, contributing to the progress of disciplines like chemistry, biochemistry, materials science, and pharmaceutical research.

10.4.1. Ultraviolet-Visible Spectroscopy (UV-Vis)

In analytical chemistry and biochemistry, ultraviolet-visible spectroscopy, or UV-Vis spectroscopy, is a commonly used method for examining the absorption and transmission of light in the visible and ultraviolet portions of the electromagnetic spectrum. It gives researchers important details about the energy levels and electronic transitions of molecules, enabling them to examine the concentration, make-up, and chemical characteristics of a sample. The basis of UV-Vis spectroscopy is the idea that light is absorbed by molecules at particular wavelengths that are related to the energy differences between their electronic energy levels. A variety of wavelengths are applied to the sample, and the quantity of light absorbed or transmitted is measured. The resulting UV-Vis spectrum shows the intensity of absorption or transmission as a function of wavelength graphically.

Key features and applications of UV-Vis spectroscopy include:

- **Quantitative Analysis:** UV-Vis spectroscopy is commonly used for quantitative analysis, as the absorption intensity is directly proportional to the concentration of the absorbing species. By measuring the absorbance at a specific wavelength, researchers can determine the concentration of a compound in a sample using Beer-Lambert's Law.
- **Compound Identification:** UV-Vis spectra can be used for compound identification by comparing the characteristic absorption patterns of known compounds with those of an unknown sample. Each compound has a unique UV-Vis spectrum, enabling the identification and differentiation of similar compounds.
- **Determination of Purity:** UV-Vis spectroscopy is employed to assess the purity of a substance. Impurities or contaminants in a sample can alter the absorption profile, allowing researchers to quantify the extent of impurities or monitor the progress of purification processes.
- **Kinetic Studies:** UV-Vis spectroscopy is valuable for investigating chemical reactions and studying reaction rates. By monitoring the absorbance changes over time, researchers can determine reaction kinetics, including reaction rates, rate constants, and reaction mechanisms.
- **Protein and Nucleic Acid Analysis:** UV-Vis spectroscopy is widely used in biochemistry for the analysis of proteins and nucleic acids. The absorption characteristics of these biomolecules in the UV-Vis range provide insights into their concentration, structure, and interactions.

- **Environmental Analysis:** UV-Vis spectroscopy finds applications in environmental analysis, such as the detection and quantification of pollutants, monitoring water quality, and assessing the presence of contaminants in air and soil samples.

A flexible and easily available method, UV-Vis spectroscopy provides important information about the electrical characteristics and makeup of different materials. It is a vital tool in many scientific fields, from chemistry and biochemistry to pharmaceutical analysis and environmental monitoring, because of its ease of use, speed, and variety of applications.

10.4.2. Infrared Spectroscopy (IR)

A potent analytical method for examining how infrared radiation interacts with molecules is infrared spectroscopy or IR spectroscopy. It offers important details regarding the vibrational and rotational energy levels of molecules, making it possible to determine the molecular structure, identify functional groups, and characterize chemical bonds. With wavelengths longer than visible light, the infrared portion of the electromagnetic spectrum is where infrared spectroscopy operates. Its foundation is the idea that molecules absorb infrared light at particular frequencies that represent the energy differences between their rotational and vibrational states. A distinct fingerprint of the sample's molecular makeup and structure can be obtained from the resulting infrared spectrum.

Key features and applications of IR spectroscopy include:

- **Functional Group Analysis:** IR spectroscopy is widely used for the identification and characterization of functional groups in organic and inorganic compounds. Different functional groups exhibit characteristic absorption peaks in the IR spectrum, allowing for their rapid identification.
- **Chemical Bond Analysis:** IR spectroscopy provides information about the types of chemical bonds present in a molecule. It allows researchers to determine the presence of single or multiple bonds, as well as the presence of functional groups such as alcohols, amines, carbonyls, and carboxylic acids.
- **Structural Elucidation:** IR spectroscopy helps determine the structure and connectivity of molecules. By analyzing the positions and intensities of absorption peaks, researchers can deduce the presence and arrangement of atoms within a molecule.
- **Quantitative Analysis:** IR spectroscopy can be used for quantitative analysis, such as determining the concentration of a specific compound in a mixture. This is achieved by measuring the intensity of absorption at a specific wavelength and correlating it to the concentration using calibration curves.
- **Polymer Analysis:** IR spectroscopy is widely used in the analysis of polymers. It can provide information about polymer composition, structure, and degree of polymerization. It is also used to monitor polymerization reactions and assess the presence of impurities or additives in polymer samples.
- **Pharmaceutical Analysis:** IR spectroscopy plays a vital role in pharmaceutical analysis. It is used for the identification and characterization of drug substances

and formulations, assessment of drug purity, and detection of impurities or degradation products.

- **Environmental and Forensic Analysis:** IR spectroscopy finds applications in environmental and forensic analysis. It can be used to identify pollutants, analyze environmental samples, detect counterfeit drugs, and analyze forensic evidence.

IR spectroscopy is a versatile and widely utilized technique in various scientific fields, including chemistry, biochemistry, materials science, pharmaceuticals, and environmental analysis. Its ability to provide detailed structural and compositional information makes it an indispensable tool for both qualitative and quantitative analysis.

10.4.3. Nuclear Magnetic Resonance (NMR) Spectroscopy

NMR spectroscopy, a potent analytical method, is utilized to investigate how atomic nuclei interact with intense magnetic fields. It is an essential tool in the domains of chemistry, biochemistry, and materials science since it offers comprehensive information about the dynamics, structure, and chemical environment of molecules. Nuclear spin theory and the magnetic characteristics of atomic nuclei serve as the foundation for NMR spectroscopy. Atomic nuclei absorb energy and reemit it at different frequencies when they are exposed to radiofrequency pulses and placed in a strong magnetic field. Researchers can learn a great deal about the sample's physical and chemical characteristics by measuring these frequencies.

Key features and applications of NMR spectroscopy include:

- **Structural Elucidation:** NMR spectroscopy is widely used to determine the structure of organic and inorganic compounds. It provides information about the connectivity of atoms, stereochemistry, and molecular conformation. By analyzing the chemical shifts and splitting patterns in the NMR spectrum, researchers can deduce the arrangement of atoms in a molecule.
- **Compound Identification:** NMR spectroscopy is used for compound identification by comparing the NMR spectra of unknown compounds with those of known reference compounds. Each compound has a unique NMR fingerprint, allowing for accurate identification and differentiation of chemical species.
- **Quantitative Analysis:** NMR spectroscopy can be used for quantitative analysis, enabling researchers to determine the concentration of a specific compound in a mixture. By integrating the area under the NMR peaks, the relative concentrations of different components can be determined.
- **Determination of Molecular Dynamics:** NMR spectroscopy provides insights into the dynamics and motion of molecules. It can be used to study molecular rotations, vibrations, and conformational changes, providing valuable information about chemical reactions, protein folding, and molecular interactions.
- **Biomolecular Structure and Interactions:** NMR spectroscopy is extensively used in structural biology to determine the three-dimensional structures of proteins, nucleic acids, and other biomolecules. It can also study protein-ligand interactions, protein-protein interactions, and protein-nucleic acid interactions.

- **Metabolomics:** NMR spectroscopy is employed in metabolomics, the study of small molecule metabolites in biological systems. It allows researchers to identify and quantify metabolites, providing insights into metabolic pathways, disease biomarkers, and drug metabolism.
- **Solid-State NMR:** In addition to solution NMR, solid-state NMR spectroscopy is used to study the structure and dynamics of solids, including materials, catalysts, and complex biomolecular assemblies.

NMR spectroscopy is a versatile technique with applications in various scientific disciplines. Its ability to provide detailed structural and dynamic information about molecules and materials makes it an essential tool for research, drug discovery, materials characterization, and understanding biological processes.

10.4.4. Fluorescence Spectroscopy

Fluorescence spectroscopy is a quantitative method of measuring chemical products that relies on the fluorescence properties of the sample being studied. Based on a molecule's fluorescent characteristics, fluorescence spectroscopy examines its fluorescence. One kind of luminescence that results from photons stimulating a molecule and elevating it to an electronically excited state is fluorescence. To perform fluorescence spectroscopy, a light beam is used to excite the electrons in specific compounds' molecules, causing them to release light. This light is focused onto a filter and then onto a detector so that the molecule or any alterations to it can be measured and identified.

The term fluorescence refers to one type of luminescence. Luminescence, broadly defined, is light emission from a molecule. There are several types of luminescence.

- **Photoluminescence** is when light energy, or photons, stimulate the emission of a photon.
- **Chemiluminescence**, is defined as when chemical energy stimulates the emission of a photon, and this includes bioluminescence, as seen in fireflies and many forms of sea life.
- **Electroluminescence**, is when electrical energy or a strong electric field, stimulates the emission of a photon, such as in some lighting applications.
- **Fluorescence**, specifically, is a type of photoluminescence where light raises an electron to an excited state. The excited state undergoes rapid thermal energy loss to the environment through vibrations, and then a photon is emitted from the lowest-lying singlet excited state. This process of photon emission competes for other non-radiative processes including energy transfer and heat loss.

When the term "fluorescence" is used, the same methods of measurement can typically be applied to any of the above categories of luminescence.

10.4.5. Raman Spectroscopy

Raman spectroscopy is a powerful analytical technique used to study the vibrational and rotational energy levels of molecules. It provides information about the molecular structure, chemical composition, and bonding characteristics of a sample by measuring the scattering of light. Raman spectroscopy is widely used in various scientific fields, including chemistry, biochemistry, materials science, and pharmaceutical analysis. Raman spectroscopy is based on the Raman effect, discovered by C.V. Raman in 1928. When a sample is illuminated with a laser beam, a small fraction of the scattered light undergoes a change in energy due to interactions with the sample's molecular vibrations. This scattered light, known as the Raman scattering, contains information about the vibrational modes and energy levels of the sample.

Key features and applications of Raman spectroscopy include:

- **Molecular Identification:** Raman spectroscopy is used for the identification and characterization of chemical compounds. Each compound exhibits a unique Raman spectrum, allowing researchers to identify unknown substances and differentiate between similar compounds.
- **Structural Analysis:** Raman spectroscopy provides insights into the molecular structure and bonding of a sample. It can detect and analyze functional groups, identify the presence of specific chemical bonds, and determine the conformation of molecules.
- **Chemical Imaging:** Raman spectroscopy can be coupled with microscopy to generate chemical images of samples. By raster scanning the laser beam across a sample, researchers can obtain spatially resolved Raman spectra, allowing for the visualization of chemical distributions and mapping of molecular components.
- **Biological and Medical Applications:** Raman spectroscopy finds applications in biological and medical research. It can be used to analyze biomolecules, study cellular structures, and detect diseases at the molecular level. Raman spectroscopy has been employed in cancer diagnosis, tissue imaging, and monitoring drug delivery systems.
- **Materials Science:** Raman spectroscopy is widely used in materials science for the characterization of materials. It can provide information about crystallinity, phase transitions, and defects in materials. Raman spectroscopy is particularly useful for the analysis of carbon-based materials, polymers, catalysts, and semiconductors.
- **Quality Control and Forensic Analysis:** Raman spectroscopy is employed in quality control processes to verify the composition and quality of products. It is also used in forensic analysis to identify trace evidence, analyze counterfeit substances, and investigate crime scenes.
- **Non-destructive Analysis:** Raman spectroscopy is a non-destructive technique, allowing for the analysis of samples without altering their chemical or physical properties. This makes it suitable for the analysis of valuable or sensitive samples.

Raman spectroscopy offers several advantages, including high specificity, non-destructiveness, and the ability to analyze samples in various states (solid, liquid, or gas). It is a versatile technique that provides valuable insights into molecular structures, chemical compositions, and material properties, making it an important tool in scientific research and analysis (Snyder et al., 2010).

CASE STUDY

UNDERSTANDING CANCER BIOLOGY THROUGH BIOCHEMISTRY AND MOLECULAR BIOLOGY TECHNIQUES

Background: Cancer Research Institute (CRI) is dedicated to investigating the molecular mechanisms underlying cancer development and progression. They are focusing on a specific type of cancer, pancreatic adenocarcinoma (PDAC), known for its aggressive nature and poor prognosis.

Problem Statement: PDAC is notoriously difficult to treat due to its late diagnosis and resistance to conventional therapies. CRI aims to unravel the molecular pathways driving PDAC growth and metastasis using advanced biochemistry and molecular biology techniques.

Solution: To address these challenges, CRI employs a multidisciplinary approach integrating various biochemistry and molecular biology techniques. This approach allows them to elucidate the complex interactions between cancer cells, their microenvironment, and therapeutic targets.

Implementation

- Cell Culture and Molecular Profiling:
  - **Cell Lines:** CRI uses established PDAC cell lines and primary patient-derived cell cultures to mimic tumor behavior *in vitro*.
  - **Molecular Profiling:** Cells are characterized using techniques such as Western blotting, immunofluorescence microscopy, and flow cytometry to assess protein expression, signaling pathways activation, and cellular phenotypes.
- Genetic and Epigenetic Analysis:
  - **DNA and RNA Sequencing:** Whole-genome and transcriptome sequencing are performed to identify genetic mutations, gene expression profiles, and epigenetic modifications associated with PDAC.
  - **Methylation Analysis:** Bisulfite sequencing or methylation-specific PCR (MSP) is used to investigate DNA methylation patterns that regulate gene expression in PDAC cells.
- Protein-Protein Interactions and Signaling Pathways:
  - **Protein Interaction Studies:** Techniques like co-immunoprecipitation (Co-IP) and yeast two-hybrid (Y2H) assays are employed to identify protein-protein interactions critical for PDAC progression.

  - **Pathway Analysis:** Bioinformatics tools are used to analyze data from proteomics and genomics studies, mapping signaling pathways altered in PDAC and identifying potential therapeutic targets.
- Drug Screening and Functional Assays:
  - **High-Throughput Screening (HTS):** CRI conducts HTS using libraries of small molecules to identify compounds that selectively target PDAC cells or modulate specific pathways identified through molecular profiling.
  - **Functional Assays:** Cell viability assays, migration assays, and 3D culture models are used to validate the efficacy of candidate drugs and understand their mechanisms of action in PDAC cells.
- Clinical Translation and Therapeutic Development:
  - **Preclinical Models:** Findings from biochemical and molecular studies are validated in preclinical animal models of PDAC to assess therapeutic efficacy and safety.
  - **Clinical Trials:** Promising drug candidates are advanced to clinical trials, where their efficacy is evaluated in patients with PDAC, guided by biomarkers identified through molecular profiling.

Results: Through the application of biochemistry and molecular biology techniques, CRI identifies key molecular drivers and therapeutic vulnerabilities in PDAC. They uncover novel biomarkers, signaling pathways, and potential drug targets that pave the way for targeted therapies and personalized treatment approaches.

Conclusion: This case study highlights how biochemistry and molecular biology techniques are indispensable in cancer research, particularly in deciphering the complex biology of PDAC. By integrating these advanced methodologies, CRI advances our understanding of cancer mechanisms and accelerates the development of innovative therapies to improve patient outcomes.

CLASS ACTIVITY

Students will engage in hands-on demonstrations of essential biochemistry and molecular biology techniques. They'll perform tasks like DNA extraction, PCR (Polymerase Chain Reaction), gel electrophoresis, and protein analysis. Through these activities, students will gain practical skills in handling biological samples, interpreting results, and understanding the significance of these techniques in research and diagnostics. The activity aims to deepen their understanding of fundamental concepts while fostering critical thinking and problem-solving abilities in the field of life sciences.

SUMMARY

- Mass spectrometry is a powerful analytical technique widely used in biochemistry and molecular biology for the identification, characterization, and quantification of biomolecules. It provides information on the mass-to-charge ratio (m/z) of ions, enabling the determination of molecular weight, structural elucidation, and analysis of molecular interactions.
- High-Resolution Mass Spectrometry (HRMS): HRMS is characterized by its ability to accurately measure the mass-to-charge ratio of ions with high precision and resolution. It allows for the discrimination of closely related ions and provides enhanced accuracy in determining molecular weights.
- Electrospray Ionization (ESI) is a widely used ionization technique in mass spectrometry that enables the analysis of large biomolecules such as proteins, peptides, nucleic acids, carbohydrates, and lipids. ESI is particularly valuable in the field of biochemistry and molecular biology because it allows the transfer of labile biomolecules into the gas phase while preserving their native structure and noncovalent interactions.
- Matrix-Assisted Laser Desorption/Ionization (MALDI) is a powerful ionization technique used in mass spectrometry for the analysis of biomolecules, particularly large proteins and peptides.
- Tandem mass spectrometry (MS/MS) is a powerful technique used in analytical chemistry and proteomics to obtain detailed structural information about biomolecules. It involves the sequential analysis of ions, where a selected precursor ion is isolated and fragmented, and the resulting fragments are then analyzed to provide insights into the molecule's structure, composition, and interactions.
- Quantitative mass spectrometry is a technique used to determine the abundance or concentration of specific molecules in a sample. It provides a quantitative assessment of analytes based on their mass-to-charge ratio (m/z) and allows for the measurement of the relative or absolute amounts of target molecules in complex mixtures.
- Imaging mass spectrometry (IMS) is a powerful analytical technique that combines the spatial information of microscopy with the molecular information provided by mass spectrometry. It allows for the visualization and mapping of biomolecules directly within tissues or other complex samples, providing valuable insights into their distribution, localization, and molecular composition.
- Electrophoretic techniques are widely used in biochemistry and molecular biology to separate and analyze biomolecules based on their charge and size.
- Agarose gel electrophoresis is a commonly used technique in biochemistry and molecular biology for separating and analyzing nucleic acids, such as DNA and RNA, based on their size. Agarose, a polysaccharide derived from seaweed, is used to create a gel matrix through which the nucleic acids can migrate when subjected to an electric field.

- Polyacrylamide gel electrophoresis (PAGE) is a widely used technique in biochemistry and molecular biology for separating and analyzing proteins and nucleic acids based on their size and charge.
- Capillary electrophoresis (CE) is a powerful analytical technique used in biochemistry and molecular biology for the separation and analysis of biomolecules, including nucleic acids, proteins, peptides, carbohydrates, and small molecules. It offers high separation efficiency, short analysis times, and low sample consumption compared to traditional gel electrophoresis methods.
- Isoelectric focusing (IEF) is a powerful technique used in biochemistry and molecular biology to separate and analyze proteins based on their isoelectric points (pI). The pI is the pH at which a protein has no net charge, and it is determined by the amino acid composition and sequence of the protein.
- Protein blotting techniques, also known as Western blotting, are widely used in biochemistry and molecular biology to detect and analyze specific proteins within complex biological samples. Western blotting allows for the identification of target proteins based on their size, abundance, and post-translational modifications.
- DNA gel electrophoresis is a fundamental technique used in molecular biology to separate and analyze DNA molecules based on their size. It is a widely employed method for various applications, including DNA fragment analysis, DNA sequencing, and DNA purification.
- RNA gel electrophoresis is a commonly used technique in molecular biology for the separation and analysis of RNA molecules based on their size and/or secondary structure. It is a fundamental method that allows researchers to study RNA expression, identify RNA species, and analyze RNA integrity.
- Chromatographic techniques are widely used in biochemistry and molecular biology to separate and analyze complex mixtures of molecules based on their physical and chemical properties. These techniques are based on the principle of differential migration of components in a mixture between a stationary phase and a mobile phase.
- Size exclusion chromatography (SEC), also known as gel filtration chromatography or gel permeation chromatography, is a widely used chromatographic technique for the separation and analysis of biomolecules based on their size and molecular weight. It is particularly useful for the separation of macromolecules such as proteins, nucleic acids, polysaccharides, and synthetic polymers.
- Gas chromatography (GC) is a widely used analytical technique that separates and analyzes volatile and semi-volatile compounds in a sample mixture. It is based on the principles of partitioning and differential migration of analytes between a stationary phase and a mobile phase.
- Ultraviolet-visible spectroscopy (UV-Vis spectroscopy) is a widely used technique in analytical chemistry and biochemistry for studying the absorption and transmission of light in the ultraviolet and visible regions of the electromagnetic spectrum. It provides valuable information about the electronic transitions and energy levels of molecules, allowing researchers to analyze the composition, concentration, and chemical properties of a sample.

REVIEW QUESTIONS

1. What is Electrospray Ionization (ESI), and how does it work?
2. Describe the principle behind Tandem Mass Spectrometry (MS/MS) and its applications in structural elucidation.
3. What are the advantages of High-Resolution Mass Spectrometry (HRMS) over conventional mass spectrometry? Give examples of applications where HRMS is particularly useful.
4. Explain the concept of Quantitative Mass Spectrometry. Compare and contrast different quantitative MS techniques.
5. How does Imaging Mass Spectrometry (IMS) differ from traditional MS techniques? What are its main applications in biological and clinical research?
6. What is Polyacrylamide Gel Electrophoresis (PAGE), and how is it used in protein separation? Discuss its limitations and advantages compared to other protein separation techniques.
7. Describe the principle of Isoelectric Focusing (IEF) in protein separation. What factors influence the resolution and effectiveness of IEF?
8. Outline the steps involved in Protein Blotting Techniques (Western Blotting). How does Western Blotting complement other protein analysis methods such as MS?
9. What is the role of RNA Gel Electrophoresis in molecular biology? How does it differ from DNA gel electrophoresis in terms of methodology and applications?
10. Explain the principles of Hydrophobic Interaction Chromatography (HIC) in protein purification. Compare HIC with other chromatographic techniques such as Ion Exchange Chromatography (IEX) and Size Exclusion Chromatography (SEC).

REFERENCES

1. Aebersold, R., & Mann, M. (2016). Mass spectrometry-based proteomics. *Nature, 537*(7620), 347–355. https://doi.org/10.1038/nature19949.
2. Ausubel, F. M., Brent, R., Kingston, R. E., Moore, D. D., Seidman, J. G., Smith, J. A., & Struhl, K. (2005). *Current Protocols in Molecular Biology*. John Wiley & Sons.
3. Halket, J. M., & Zaikin, V. G. (2017). *Derivatization in Mass Spectrometry: Principles and Applications*. John Wiley & Sons.
4. Hames, B. D., & Rickwood, D. (2005). *Gel Electrophoresis of Proteins: A Practical Approach*. Oxford University Press.
5. Marriott, P. J. (2012). *Gas Chromatography: Principles, Techniques, and Applications*. CRC Press.
6. Snyder, L. R., Kirkland, J. J., & Glajch, J. L. (2010). *Practical HPLC Method Development*. John Wiley & Sons.

7. Watson, J. T., & Sparkman, O. D. (2007). *Introduction to Mass Spectrometry: Instrumentation, Applications, and Strategies for Data Interpretation*. John Wiley & Sons.

8. Westermeier, R., & Naven, T. (2006). *Proteomics in Practice: A Laboratory Manual of Proteome Analysis*. Wiley-VCH.

9. Wu, J., & Liang, G. (2008). *Chiral Separations by Liquid Chromatography and Related Technologies*. CRC Press.

Index

A

Abiotic stresses 13, 14
Acrosome 82
Adoptive cell transfer (ACT) 12
Aerobic cell 70
Aerobic metabolism 70
Aerobic prokaryotes 70
Agricultural ecosystems 14
Agricultural landscapes 14
Agricultural residues 15, 18
Agricultural soils 14
Agroecosystems 14
Amino acid 257
Animal behavior 9
Animal cells 71, 87, 308
Annealing 175, 194
Antibiotic production 6
Anticodon 38, 39, 40, 44, 52, 53, 54, 55, 61, 259, 260, 261, 262, 267
Antisense oligonucleotides (ASOs) 11
Assisted reproduction technologies (ARTs) 16
Autophagosome 82

B

Bacteria 3, 6, 10, 13, 15, 16, 18, 31, 37, 38, 50, 53, 56, 66, 68, 70, 83, 84, 97, 105, 123, 124, 150, 158, 159, 160, 161, 199, 202, 203, 212, 229, 230, 243, 245, 247, 254, 264, 265, 279, 281, 288, 294, 300, 304, 308, 312, 317, 322, 329, 333
Bacterial artificial chromosomes (BACs) 168
Bacterial genetics 1, 6
Bacteriophages 162
Biocatalytic cascades 19
Biochemistry 1, 49, 63, 148, 149, 150, 151, 152, 153, 154, 155, 156, 157, 158, 159, 160, 161, 162, 163, 164, 165, 166, 335, 337, 340, 369
Biodegradability 18, 104
Biodiesel 15, 18
Biodiversity conservation 14, 15, 17
Bioenergy 15, 17
Biofortified crops 13
Biofuel production 15, 17, 18
Biological catalysts 68, 69
Biological control agents 13, 14
Biological phenomena 110, 130
Biological sciences 189
Biological synthesis 6
Biological systems 2, 15, 17, 92, 128, 130, 258, 338, 342, 345, 364, 367
Biomass production 14
Bioremediation 15, 17, 18, 204
Biotinylation 109
Bromo-deoxyuridine (BrdU) 101

C

Cancer biomarkers 1

Cancer diagnosis 10, 12, 368
Cancer predisposition 10
Carbon atom 92
Carbon fixation 14, 18
Carbon sequestration 14, 17
Catalytic molecules 74
Cell biology 9, 25, 34, 74, 75, 88, 101, 108, 310
Cell chemistry 73
Cell cycle 65, 97
Cell division 28, 30, 36, 72, 75, 82, 95, 96, 97, 98, 99, 111, 124, 141, 158, 167, 223, 224, 314
Cell nucleus 8, 30, 44, 45
Cell reproduction 72
Cell science 74
Cell signaling 65, 70, 88
Cell theory 70, 71, 74, 75
Cellular architecture 88
Cellular organism 8
Cell wall 65, 80
Centriole 65, 82
Chaperones 276, 277, 279, 303, 305
Chemical engineering 3, 4, 291
Chimeric antigen receptor (CAR) 12
Chloroplasts 31, 70, 84, 107, 264, 281
Chromosome 27, 36, 65, 168
Circulating tumor cells (CTCs) 11
Collision-induced dissociation (CID) 339, 342
Computational biology 12
Conventional biotechnologies 3, 310
Crop breeding 14
Crop diversity 14
Crop genomes 12
Crossbreeding 1
Cyanobacterial systems 18
Cytoblastema 72
Cytochemistry 75
Cytokinesis 84, 95, 101
Cytoplasm 34, 38, 39, 43, 45, 54, 61, 67, 76, 77, 78, 80, 84, 88, 96, 99, 105, 258, 294, 303, 305
Cytoplasmic division 84, 95
Cytosol 65

D

Data analytics 13
Daughter cells 73, 84, 95, 99, 105
Deoxyribonucleic acid (DNA) 1
Deoxyribose 31, 32, 37, 41, 42, 90, 178, 179
Digital PCR (dPCR) 196, 217
Disease management 16
Disease modeling 12
Disease resistance 12
Distributive justice 21
DNA extraction 60, 203, 337, 370
DNA molecule 23, 30, 37, 49, 105, 147, 149, 151, 157, 176, 186, 189, 190, 196, 197, 200, 202, 206, 208, 211, 214, 215, 327
DNA polymerase 31, 109, 113, 117, 145, 151, 153, 156, 175, 187, 189, 190, 191, 192, 193, 194, 197, 200, 206, 209, 214, 243, 245, 324
DNA replication 29, 30, 36, 37, 46, 61, 62, 78, 84, 96, 97, 158, 176, 314, 316
DNA sequencing 10, 14, 17, 170, 183, 211, 212, 213, 215, 216, 242, 253, 338, 346, 348, 349, 351, 352, 372
DNA technology 1, 6, 7, 11, 17, 25, 145, 146, 170, 171
DNA template 28, 39, 61, 175, 194, 197, 198, 245, 289
Drought resilience 14
Drought tolerance 12, 13, 170, 171
Drug discovery 11, 12, 128, 173, 174, 340, 355, 367

E

Ecological restoration 17
Ecosystem health 16, 18
Ecosystem resilience 14, 17
Electricity generation 18
Electron capture dissociation (ECD) 339, 342
Electron microscopy 73, 106, 107, 108, 240, 295
Electron transfer chain reactions (ETC) 80
Electron transfer dissociation (ETD) 342
Electropherogram 175
Electrophoresis 38, 122, 125, 131, 133, 134, 137, 142, 143, 145, 153, 159, 164, 175, 182,

195, 197, 234, 236, 337, 338, 346, 347, 348, 349, 350, 351, 352, 353, 370, 371, 372, 373
Electrospray ionization (ESI) 137, 337
Elongation 39, 54, 195, 257, 265, 268
Endoplasmic reticulum 66, 78, 81, 84, 105, 278, 279, 280, 281, 282, 283, 284, 287, 303, 305, 306, 317, 318, 319, 320, 334
Energy metabolism 70
Entomopathogenic fungi 13
Environmental monitoring 16, 361, 365
Environmental performance 16
Environmental pollution 14, 18, 19
Environmental quality 16
Environmental restoration 18
Environmental science 10, 15, 202, 345
Environmental stressors 15, 16
Environmental sustainability 2
Enzymatic hydrolysis 18, 19
Enzyme chemistry 73
Erythropoietin 8, 318, 320
Ethanolamine 89
Eukaryote 65
Eukaryotic cells 6, 31, 53, 66, 70, 77, 78, 79, 80, 84, 90, 97, 105, 133, 230, 262, 263, 264, 266, 267, 270, 278, 279, 280, 281, 283, 285, 305, 308, 319
Eukaryotic chromosomes 34
Eukaryotic expression system 303
Eukaryotic organism 8, 117
Expression vector 303
Extension 175, 194

F

Flagellum 65
Flow cytometry 337
Forensic research 345
Fourier transform ion cyclotron resonance (FT-ICR) 339, 343
Fungicides 14

G

Gas chromatography (GC) 337, 343, 354, 359, 372
Gene cloning 145, 146
Gene editing 14, 20
Gene expression 10, 27, 36, 45, 51, 62, 84, 110, 111, 112, 114, 115, 116, 117, 123, 125, 155, 196, 199, 201, 292, 294, 328, 331, 337, 338, 353, 369
Gene library 145, 172
Gene silencing 12, 45
Gene therapies 17, 22
Genetically modified organisms (GMOs) 12
Genetic analysis 145
Genetic code 7, 27, 28, 30, 36, 43, 47, 48, 49, 52, 62, 74, 258, 259, 261, 288, 292
Genetic disorders 10, 11, 199, 201, 218, 296
Genetic diversity 14, 15, 16, 30, 202, 204
Genetic engineering 6, 7, 12, 13, 14, 15, 18, 31, 103, 347
Genetic resources conservation 14
Genome editing 10, 12, 16
Genomics 9, 107, 109, 110, 111, 121, 124, 130, 139, 140, 141, 142, 143, 200, 201, 203, 213, 214, 291, 370
Genotyping 217, 238, 242, 246
Glycerol phospholipids 89
Glycosome 82
Glycosylation 283, 284, 303, 306, 335
Glyoxysome 82
Golden Rice 13, 26
Gowth hormone 6, 26, 300
Greenhouse gas emissions 14, 16, 18

H

Habitat restoration 14, 16
Haemoglobin 50
Half hydrophobic 68
Herbicides 14
Herbicide tolerance 12, 13
Higher-energy collisional dissociation (HCD) 342
High-throughput screening (HTS) 11
High-throughput sequencing 175, 201
Homogeneous cells 101
Homology 49, 127

Host cell 158, 303
Human chromosomes 8, 24, 62, 314
Human gene microarrays 8
Human genes 50
Human genome 8
Human Immunodeficiency Virus (HIV) 29
Hydrocarbon chains 88, 89
Hydrogen 15, 18, 29, 30, 31, 32, 33, 44, 82, 83, 85, 86, 88, 91, 92, 93, 148, 194, 272, 273, 274, 275, 298, 316, 361, 363
Hydrogenosome 82
Hydrophobic 68, 85, 88, 89, 92, 93, 133, 273, 274, 281, 285, 298, 355, 358
Hydrophobic portions 68

I

Imaging mass spectrometry (IMS) 344
Immunoprecipitation 337
Induced pluripotent stem cells (iPSCs) 12
Industrial production 4, 17, 19
Infectious disease 3, 252, 253, 254
Infectious pathogens 10, 243
Inflammatory disorders 11
Initiation complex 257
Inorganic ions 85
Insect cells 303, 327
Insect resistance 12, 13
Integrated pest management (IPM) 14
Intellectual property rights (IPRs) 21
Interaction networks 126
In vitro fertilization (IVF) 16

L

Licensing agreements 21
Liquid chromatography (LC) 137, 337, 343
Loop-mediated isothermal amplification (LAMP) 13
Lysosomes 84, 105, 281

M

Macromolecules 6, 28, 73, 85, 91, 99, 263, 291, 356, 372
Mammalian artificial chromosomes (MACs) 168
Mammalian cell 17, 95, 100, 290, 331, 332
Marker-assisted selection (MAS) 13
Mass spectrometry 242, 247, 337, 371, 373
Medicine 2, 10, 11, 12, 22, 25, 74, 103, 110, 127, 130, 139, 140, 197, 198, 199, 201, 202, 218, 296
messenger RNA (mRNA) 11, 28, 31, 38, 61, 68, 90, 258, 288, 298, 305, 333
Metabolic engineering 15
Metabolic functions 1
Metabolomics 109, 367
Metagenomic analysis 16
Methylation 37, 369
Microarray analysis 1, 141, 145
Microbial fermentation 18, 19
Microbiology 3, 4, 25, 42, 73, 74, 202, 204, 213, 240, 245, 246, 260, 345
Micronutrients 13
Microorganisms 1, 3, 73
Microtubule 65, 82
Mitochondria 29, 31, 66, 70, 79, 80, 83, 84, 105, 264, 277, 278, 279, 281
Mitosis 73, 84, 95, 97, 101, 105
Modern biology 23, 140
Molecular biology 1, 2, 5, 6, 24, 25, 26, 28, 31, 46, 58, 60, 61, 74, 121, 131, 146, 170, 173, 189, 197, 214, 217, 254, 288, 338, 339, 340, 341, 342, 343, 346, 347, 348, 349, 350, 351, 352, 353, 354, 355, 356, 358, 369, 370, 371, 372, 373
Molecular diagnostic 10, 13
Molecular evolution 9
Molecular level 1
Molecular markers 13, 15
Molecular mechanisms 12, 101, 110, 128, 130, 139, 140, 369
Molecular oxygen 70
Molecular probe 217
Molecular tools 14, 16, 17
Monitor biodiversity 16
Monoclonal antibodies (mAbs) 11
M phases 95, 102
Multiple cloning site (MCS) 161, 166, 309, 325

Mutations 30, 54, 109, 295
Mycoplasma 67

N

Natural bioreactors 6
Near-infrared (NIR) 341
Neurodegenerative disorders 12, 128, 130
Next-generation sequencing (NGS) 10, 139, 175, 198, 201, 212, 218, 251
Nuclear division 84
Nuclear magnetic resonance (NMR) 337
Nucleic acid enzymology 6
Nucleic acids 23, 25, 28, 31, 33, 46, 47, 48, 50, 52, 53, 54, 55, 62, 63, 67, 68, 75, 85, 90, 91, 104, 105, 141, 223, 242, 246, 338, 339, 340, 346, 347, 348, 349, 354, 355, 356, 357, 363, 364, 366, 371, 372
Nucleolus 65, 83
Nucleosome 34, 35, 276
Nucleotide 48, 175, 197
Nucleus 28, 29, 30, 34, 38, 43, 50, 51, 54, 66, 71, 72, 75, 76, 78, 79, 80, 83, 84, 90, 96, 105, 107, 258, 294, 305, 323
Nutrient cycling 14, 17, 202, 204
Nutritional quality 12, 13

O

Oligosaccharide 87, 284, 287, 319, 320
Oligosaccharides 88, 284, 285, 287, 306, 307
Organelle 65, 80, 82, 83
Organic molecules 70, 77, 85, 342, 355
Organic pollutants 15, 18
Organic wastes 15, 18, 19

P

Parasitoid wasps 13
Parenthesome 83
Patent policies 21
Pathogen detection 217
Patient-derived xenografts (PDX) 12
Peptidoglycan 77, 84, 105
Peroxisome 65, 83
Pest control 13
Pharmaceuticals 19, 310, 359, 361, 366
Pharmacogenomics 11
Phenylalanine 48, 52, 92
Phospholipid bilayer 65
Photosynthesis 6, 29, 31, 70, 81
Phycoerythrin 109
Phylogenetic 109
Phytoremediation 15, 18
Plant cells 31, 71, 84, 101
Plant growth 17
Plant species 15, 170
Plant tissues 71
Plasma membrane 65, 70, 77, 78, 80, 84, 281, 283, 285, 286, 287
Plasmids 31, 158, 160
Point-of-care testing 217
Policy interventions 22
Pollution remediation 15
Polyadenylation 109
Polymerase chain reaction (PCR) 10, 111, 146, 176, 189, 190, 197, 218, 240, 243, 251, 255, 326, 338, 352
Polymorphisms 109
Polynucleotide synthesis 91
Polysaccharide 77, 87, 347, 371
Post-translational modification (PTM) 129
Power generation 18
Precision agriculture 2, 13
Preproinsulin 303
Pricing strategies 22
Probiotics 3
Prokaryotic chromosomes 34
Proteasome 83
Protein electrophoresis 109, 131
Protein expression 131, 132, 337
Protein folding 278, 298, 303, 334
Protein purification 337
Protein structure 57, 59, 92, 94, 128, 131, 257, 271, 274, 298, 350, 356
Protein synthesis 29, 39, 43, 44, 45, 49, 50, 52, 53, 54, 56, 61, 62, 66, 68, 69, 74, 77, 90, 145, 156, 257, 258, 259, 262, 263, 264, 267,

279, 288, 289, 290, 291, 292, 293, 295, 296, 297, 298, 299, 300, 305, 333
Proteolysis 109, 123
Proteomics 109, 110, 121, 122, 123, 124, 125, 126, 127, 128, 129, 130, 131, 132, 134, 137, 139, 140, 141, 142, 143, 292, 340, 341, 342, 343, 350, 355, 363, 370, 371, 373
Protoplasm 71, 73
Pyrophosphate (PP) 53

Q

qPCR (quantitative PCR) 217
Qualitative expression 109
Quantitative mass spectrometry 337, 371

R

Remote sensing 13
Renewable biogas 18
Renewable biomass 15, 18
Renewable energy production 15
Resistant genotypes 13
Restriction enzyme 27, 175
Ribosomes 28, 38, 39, 43, 45, 53, 56, 74, 78, 79, 81, 84, 90, 155, 259, 262, 263, 264, 270, 277, 288, 294, 295, 297, 298, 305, 318
Ribozymes 44
RNA catalysts 68, 69
RNA extraction 109, 114
RNAi technology 13
RNA molecules 13, 41, 45, 49, 68, 69, 70, 117, 198, 259, 262, 263, 314, 352, 353, 372
Robotic platforms 11
RT-PCR (reverse transcription PCR) 217

S

Secretion signal 303
Selected reaction monitoring (SRM) 340
Serial analysis of gene expression (SAGE) 111
Single nucleotide polymorphisms (SNPs) 16, 139, 199
Small nuclear ribonucleoproteins (snRNPs) 45
Social justice 21
Soil conservation 14
Soil fertility 14, 17
Soil health and productivity 14
Soil microbes 3
Soil microbiomes 14
Stress tolerance mechanisms 14
Sugar polymers 86
Sustainable chemicals 19
Sustainable farming 14
Synthetic biology 15, 18, 49

T

Tandem mass spectrometry 337, 342, 371
Taq polymerase 191, 194, 239, 255
Time-of-flight (TOF 340, 341, 342
Traditional biotechnology 3, 5
Transcriptomics 110, 111, 130, 146, 200
Transfection 303, 327, 331
Transfer RNA (tRNA) 28, 38, 61, 68
Transgenic plants 15, 18
Triacylglycerols 88, 89
tRNA biosynthesis 44
Trypsinogen 305
Tryptophan 92, 134, 312, 315

U

UV radiation 4

V

Vaccines 11, 17, 22, 310
Vacuole 65, 82
Vanishing White Matter Disease (VWM) 54
Vesicle 65, 82, 83
Viral infections 11

X

X-ray crystallography 53, 259, 338

Y

Yeast artificial chromosomes (YACs) 168

FSC
www.fsc.org
MIX
Papier | Fördert
gute Waldnutzung
FSC® C083411

Kolibri 360 GmbH
Ferdinand-Jühlke-Straße 7
99095 Erfurt, Deutschland
produktsicherheit@kolibri360.de